I0032710

Royal Commission on Forest Reserv. and Nat. Park

Papers and Reports upon Forestry, Forest Schools

Forest Administration and Management, in Europe and America

Royal Commission on Forest Reserv. and Nat. Park

Papers and Reports upon Forestry, Forest Schools
Forest Administration and Management, in Europe and America

ISBN/EAN: 9783337156145

Printed in Europe, USA, Canada, Australia, Japan

Cover: Foto ©berggeist007 / pixelio.de

More available books at **www.hansebooks.com**

PAPERS AND REPORTS

UPON

FORESTRY, FOREST SCHOOLS,

FOREST ADMINISTRATION AND MANAGEMENT

IN

EUROPE, AMERICA, AND THE BRITISH POSSESSIONS,

AND UPON

FORESTS AS PUBLIC PARKS AND SANITARY RESORTS

COLLECTED BY

MR. A. KIRKWOOD,

(Chief Officer of the Lands Branch of the Department of Crown Lands) under the direction of the Commissioner of Crown Lands,

TO ACCOMPANY THE REPORT OF THE ROYAL COMMISSION ON FOREST RESERVATION AND NATIONAL PARK.

PRINTED BY ORDER OF THE LEGISLATIVE ASSEMBLY.

TORONTO:
PRINTED BY WARWICK & SONS, 68 & 70 FRONT STREET WEST
1893.

TABLE OF CONTENTS.

INTRODUCTION.

— —— -

" A great State was a desert, and the land
Lay bare and lifeless under sun and storm,
Treeless and shelterless.   Spring came and went,
And came, but brought no joy ; but, in its stead,
The desolation of the ravening floods,
That leaped like wolves or wild cats from the hills,
And spread destruction over fruitful farms ;
Devouring as they went the works of man,
And sweeping seaward Nature's kindly soils,
To choke the water-courses worse than waste.

" The forest trees, that in the olden time—
The people's glory, and the poet's pride—
Tempered the air and guarded well the earth,
And, under spreading boughs, for ages kept
Great reservoirs to hold the snow and rain,
From which the moisture thro' the teeming year,
Flowed equably but freely—all were gone.
Their precious bales exchanged for petty cash,
The cash that melted and had left no sign ;
The logger and the lumberman were dead ;
The axe had rusted out for lack of use ;
But all the endless evil they had done
Was manifested in the desert waste.

" Dead springs no longer sparkled in the sun ;
Lost and forgotten brooks no longer laughed,
Deserted mills mourned all their moveless wheels
The snow no longer covered, as with wool,
Mountain and plain, but buried starving flocks
In Arctic drifts ; in rivers and canals
The vessels rotted idly in the mud,
Until the spring flood buried all their bones ;
Great cities that had thriven marv'lously,
Before their source of thrift was swept away,
Faded and perished, as a plant will die
With water banished from its roots and leaves ;
And men sat starving in their treeless waste,
Beside their treeless farms and empty marts,
And wondered at the ways of Providence ! "

THE UTILITY OF FORESTS.[*]

———   — · ——

The effects of forests may be looked at from the point of view of the owner or from that of the State. The owner considers, in the first place, the benefits which he personally derives from his forests ; the State appreciates the effects which they have upon the country and the nation as a whole.

The important direct effects of forests are due to the produce which they yield, the capital which they represent and the work which they provide.

Wood is used as timber in construction, ship-building, machinery, agriculture, for tools, furniture, etc., and as fuel for domestic and industrial firing. The quantity of wood required in a country depends on various considerations. In modern times iron and other materials have, to a considerable extent, replaced timber, while coal, lignite and peat compete with firewood ; nevertheless, wood is still indispensable and likely to remain so. The more general introduction of substitutes for firewood has, however, drawn increased attention to the production of timber in preference to firewood. For instance, of the total produce of the Saxon State Forests, only 35 per cent. were classified for timber in 1850, but the proportion has risen to 75 per cent. in 1880. At the same time, new demands for the consumption of wood have sprung up, such as the preparation of wood-pulp for the manufacture of paper. It is estimated that the annual consumption of wood in this industry in Germany alone amounts to upwards of 40,000,000 cubic feet.

FORESTS AS OBJECTS OF INDUSTRY.

The capital employed in forestry consists principally of the soil and the growing stock of wood. When the working is of an intermittent nature, the amount of capital fluctuates from time to time ; when the working is so arranged that an equal annual return is secured, the capital remains of the same amount, and consists of the soil plus the permanently present growing stock.

The soil is called the fixed, the growing stock the movable or shifting capital of forestry. The proportion of the one to the other depends chiefly on the method of treatment. In forests treated as coppice woods the fixed may be greater than the movable capital ; but in high forests where the object is to produce timber of some size, the shifting capital is generally of considerably greater value than the soil. An example will illustrate this :—Assuming that an area of one hundred acres is treated as a Scotch pine timber forest, under a rotation of one hundred years, with the object of obtaining an annually equal return ; in that case one acre must be stocked with one year old seedlings, another with two year's old seedlings, another with three years' old trees, and so on to the last acre which would be stocked with trees one hundred years old. Every year the oldest wood one hundred years old is cut over and the area at once restocked. Immediately after the cutting ninety-nine acres remain stocked with trees ranging in age from

* A Manual of Forestry : By William Schlich, Ph.D., Principal Professor of Forestry at the Royal Indian Engineering College, Cooper's Hill, England. Dr. Schlich spent upwards of twenty years in the Forest Department of the Government of India, and succeeded Sir Dietrich Brandis as Inspector-General. On his return from India he was appointed Professor of Forestry at Cooper's Hill College.

one year to ninety-nine years old, and this is called the normal growing stock. Without the presence in the forest of this series of age gradations it would be impossible to obtain a regular annual yield of trees one hundred years old.

The subjoined table gives the capital invested in a forest worked upon the principle of a sustained annual yield. The data for the growing stock are taken from the Yield Tables for the Scotch Pine, by M. Weise, converted into English measures. In calculating the value of the growing stock it has been assumed that fagots would not yield any money return, and that timber, including all pieces of three inches diameter and upwards at the thin end, would yield two pence per cubic foot under a rotation of thirty years, gradually rising to five pence per cubic foot under a rotation of one hundred and twenty years. Soil adapted for the growth of Scotch pines is generally light, and the value of such land of the I. or best quality cannot, on an average, be estimated at more than £25 per acre, while land of the III. or middling quality may be estimated at £12 per acre, and land of the V. or lowest quality at £4, though land of the latter quality is not worth more than a few shillings per acre.

CAPITAL INVESTED IN FORESTRY—POUNDS STERLING PER ACRE.

Length of rotation in years.	I. Quality best.			III. Quality middling.			V. Quality lowest.		
	Land.	Growing stock.	Total.	Land.	Growing stock.	Total.	Land.	Growing stock.	Total.
	£	£	£	£	£	£	£	£	£
30	25	8	23	12	2	14	4	1	5
40	25	17	42	12	7	19	4	3	7
50	25	26	51	12	12	24	4	5	9
60	25	35	60	12	17	29	4	8	12
70	25	45	70	12	22	34	4	11	15
80	25	56	81	12	28	40	4	15	19
90	25	68	93	12	35	47	4	20	24
100	25	80	105	12	42	54			
110	25	93	118	12	50	62			
120	25	106	131	12	58	70			

1). This table shows that the capital increases with the length of the rotation.

(2). That the value of the growing stock is at first smaller than the value of the land, equal to it under a rotation of from forty to fifty years, and greater after that period.

(3). That the capital invested in timber forests is considerably greater than that of the land only.

PRECIPITATIONS OR RAINFALL.

The question whether, and in how far, forests affect the rainfall is one which has been actively discussed for many years past, but so far no final decision has been possible. That forests can affect precipitations follows from the facts that forest air is relatively moister than air in the open and that the trees mechanically affect the movement of the air; but, on the other hand, the rainfall depends chiefly on other much more powerful agencies, in comparison with which the effect of forests is small. Numerous comparative observations have been made, but only a certain portion has so far been published and unfortunately those which seem to indicate a decided effect of forests on the rainfall are not always very reliable. The great difficulty in comparing the results of observations at forest stations (that is to say stations situated inside a forest) with those of the ordinary meteorological stations consists in the fact that elevation above the sea affects the rainfall most powerfully because air cools on rising and precipitations become more frequent with elevation.

Although further observations are necessary before a final conclusion can be arrived at, the following data may prove interesting :—

In the Prussian system the forest stations have shown the subjoined increase of rainfall over the average rainfall of the open country as taken from the ordinary meteorological stations :

Excess of rainfall in forest station over that of open
country, in per cent. of the latter rainfall—

Between sea level and 328	feet elevation....	1.25 per cent.
"	328	and	556	"	"	....	14.2	"
"	1,969	"	2,297	"	"	....	19.	"
"	2,297	"	2,625	"	"	....	43.	"

Although these figures may not represent the absolute facts of the case, they seem to indicate that in the plains forests have very little effect upon the rainfall, if any at all, but that their influence becomes considerable with increasing elevation in mountainous countries.

The results of seven years' observations made at two stations near Nancy show a decided increase of rainfall in the forest. The stations are situated 1,247 feet above the sea, one in the middle of an extensive forest five miles to the west of Nancy, the other in an almost woodless country six miles to the north-east of Nancy.

The results were as follows :

Increase of rainfall in forest over that in the open in per
cent. of the latter—

February to April	7 per cent.
May to July	13 "
August to October	23 "
November to January	21 "
Mean of year	16 "

EVAPORATION.

Owing to the lower temperature, the greater humidity of the air, and the quieter state of the atmosphere, evaporation must be considerably smaller in forests than in the open. This has been conclusively proved by direct observations. Those made in Bavaria and Prussia show the following results :—

Stations.	Quantity of water evaporated from a free surface of water, height in inches.			Less in forest expressed in per cent. of the total quantity evaporated in the open.
	In the open.	In forest.	Less in forest.	
Bavarian.	23.53	8.61	−14.92	−63
Prussian	13.16	5.98	−7.18	−55
Mean	18.34	7.29	−11.05	−60

These data show that evaporation in the forest was only two-fifths of that in the open country.

The effect of this action is that of the water which falls on the ground in a forest a considerably larger proportion is secured to the soil than in the open. That water is available to be taken up by the roots, while any balance goes to the ground water and helps to feed springs. Of considerable importance in this respect is the covering of forest soil. Dr. Ebenmayer's observations on this point, extending over five years, show the following results :—

Water evaporated from soil in the open. . . . . 100 parts.
Evaporation from forest soil, without leaf mould. . 47 "
"   "   " with full layer of leaf
mould . . . . 22 "

In other words, forest soil without leaf mould evaporated less than half the water in the open, while forest soil covered with a good layer of humus evaporated even less than one-fourth of that evaporated in the open.

The result of these peculiarities is that, at any rate up to a certain elevation, the forest soil retains, after allowing for evaporation, more water than open soil, although some 23 per cent. of the rainfall are intercepted by the crowns of the trees In order to illustrate this the following table, taken from Dr. Weber's calculations, is inserted, as it shows the balance of rainfall over evaporation according to elevation ; it is based upon Prussian observations :

Altitude of stations in feet.	Excess of rainfall over evaporation in inches.		Percentage of rainfall which evaporated.	
	In the open.	In forest.	In the open.	In forest.
0 328	12.02	12.32	55	37
328— 656	12.69	13.84	53	30
984—1,312	12.20	17.65	58	25
1,969—2,297	36.96	30.79	22	13
2,540	47.10	43.08	15	9
3,050	56.77	46.34	19	11

This table shows that the balance of water retained by the soil increases rapidly with altitude, and that the evaporation in mountain forests may be reduced to about 10 per cent of the rainfall. If it be remembered that the moisture is most effectually preserved in forests, it will easily be understood why the mountain forests have from time immemorial been looked upon as the preservers of moisture and feeders of springs. No doubt, a certain portion of the water is again taken out of the ground by the roots of the trees and evaporated through the leaves. The quantity thus consumed is not known at present, but it cannot be more than 12 inches, the total quantity available in plain forests, and probably it becomes less with elevation, so that a considerable balance remains available in hill forests for the feeding of springs.

MECHANICAL EFFECT OF FORESTS.

The mechanical effect of forests makes itself chiefly felt in regard to the distribution of the rain-water, the preservation of the soil on sloping ground, the binding of moving sand, the prevention of avalanches, and the moderation of air currents.

(a) *Feeding of springs and rivers.*

Most of the rain-water falling on a bare slope rushes down into the nearest water course in a comparatively short time, thus causing a rapid rise in the level of the stream. Only a comparatively small portion sinks into the ground, so as to become available for the feeding of springs. Of the rain falling over a forest, close on one-fourth is intercepted by the crowns of the trees, and the other three-fourths fall upon a layer of humus, which possesses a great capacity to absorb water and to retain it for a time. It has been shown, for instance, that mosses of the species *Hypnum*, which grow under the shade of conifers, can absorb up to five times their own weight of water, and peat mosses of the genus *Sphagnum* up to seven times, while the leaf-mould to be found in a middle-aged well-preserved beech wood can absorb and retain for a time a rainfall of five inches. Part of the water thus absorbed penetrates into the ground and becomes available for the feeding of springs, while the rest gradually finds its way into the nearest stream. In this manner well-preserved forests must have a decided effect upon the sustained feeding of springs, and the moderation of sudden floods in rivers. When, however, the humus has been saturated with water and rain continues, the effect of forests as regards inundations must cease, because the additional water follows the laws of gravity, and finds its way into the valleys.

(b) *Protection of the soil.*

Water rushing down a bare slope possesses a great mechanical power, by means of which it loosens the soil, and carries it down hill. In this way land-slips are often caused, ravines are formed, and fertile land situated at the foot of the ravine may be covered with silt and rendered valueless. Frequently the *debris* collects in rivers and forms obstructions, which are followed by a diversion of the bed and erosion of fertile lands. The rate at which this process proceeds depends on the geological origin and the formation of the surface; the less binding the soil and the looser the formation the greater will be the damage. If, on the other hand, such a slope is covered with a well-preserved forest the roots of the trees and the layers of humus keep together, and protect the soil against the action of water, besides the crown intercepts and retains, at any rate for a time a considerable portion of the water. On the whole a series of obstacles are, opposed to the movement of the water, which reduce its velocity and force, or at any rate divide it into numerous small channels. The beneficial effect of tree vegetation in this respect can be observed in most mountain ranges, and especially in the Alps from France to Austria. Wherever, in these parts, extensive deforestations have taken place, the consequence has been the gradual formation of a series of torrents, in all places where the surface did not consist of hard rock ; the *debris* brought down has covered more and more fertile land at the base of the torrents, and this evil has grown to such an extent, that not only in France but also in the other Alpine countries great efforts are now made to re-afforest the denuded areas at a great outlay. When once the evil has been created immediate afforestation is not possible ; it must be preceded by the construction of dams, dikes, walls, etc., to steady the soil until the young forest growth has had time to establish itself, and once more to lay hold of the surface soil. Forests protect the soil not only in the hills, but also in low lands, wherever it consists of so-called moving or shifting sand along the sea coast, as well as in the interior of countries. The action in this case is due partly to their moderating the force of the air currents and partly by keeping the soil together through their roots, by the formation of humus and the retention of moisture. In this way the Landes of France have from a dreary waste been converted into extensive forests intersected by cultivated fields.

(c) *Protection against air currents.*

Forests break or moderate the force of air currents, and in this way afford protection to lands lying beyond them against cold or dry winds. Woodlands afford also shelter to game and useful birds. Their importance in this respect should not be overlooked ; the presence of birds, which are the great enemies of injurious insects, depends often on that of woodlands.

HYGIENIC EFFECTS OF FORESTS.

Forests, in forming a substantial part of the vegetation of the earth, are an important agency for the production of oxygen obtained by the decomposition of carbon dioxide. Direct observations have also shown that forest air (like sea air) is much richer in ozone than the air of open countries, and especially of towns.

If forests, then, produce oxygen and ozone and protect human habitations against injurious air currents, they may exercise a beneficial effect upon the healthiness of adjoining lands. Instances are not wanting where forests are said to have given protection against the germs of malaria, but there are others where they are said to have had the opposite effect. As far as India is concerned, in some cases the medical authorities of military cantonments ordered forests to be planted, and in others to be cut down. Whether certain species, such as eucalyptus, really possesses the quality of drying up soil, and thus remove swampiness, has yet to be proved.

No general rule can be laid down showing whether forests are required in a country or what percentage of the area should be so used. The forest question must be determined on the special circumstances of each country. By way of illustration the areas at present under forest in a number of countries are shown in the following table:

Countries.	Area under forest, in acres.	Percentage of total area of country under forest.	Forest area per head of population, in acres.	Distribution of forest area, according to ownership, in percentage of total forest area.		
				State and crown forests.	Forests of corporations, endowments, etc.	Private forests.
Servia	5,166,000	48	3.3	..	..	
Russia-in-Europe	527,427,000	42	6.1	60	40	
Sweden	42,366,000	42	9.1	20	80	
Austria proper	24,161,000	33	1.1	6	94	..
Hungary	22,603,000	29	1.4	16	52	32
Germany	34,350,000	26	.8	33	19	48
Norway	18,920,000	25	9.9	12	3	85
Turkey, including Bulgaria, Bosnia and Herzegovina....	20,512,000	22	3.5			
Roumania	4,883,000	22	.9	..	..	
Italy	14,235,000	22	.5	4	43	53
Switzerland	1,930,000	19	.7	4	67	29
Spain	21,345,000	17	1.3	82	..	18
France	20,750,000	16	.6	11	23	66.
Greece	2,026,000	16	1.2	80		20
Belgium..........	1,073,000	15	.2			..
Holland.................	554,000	7	.1			
Denmark.........	477,000	6	.2			
Portugal	1,166,000	5	.2			
Great Britain and Ireland.....	2,790,000	4	.1	..	..	..
Total for Europe	766,824,000	31	2.5	..	..	..
United States of N. America..	380,000,000	17	7.6	..	..	..
East India, British...........	140,000,000	25	.6	50	50	

The percentage of forest area varies from 48 to 4, and the area per head of population from 9.9 to .1 acres. This shows that the general conditions in the various countries must make different demands in respect of afforestation. Servia, Russia, Sweden and Norway may as yet have more forests than they require for their own population. On the other hand Great Britain and Ireland, Portugal, Denmark, Holland, Belgium, and even France and Italy have a smaller forest area than is necessary to supply them with a sufficient quantity of forest produce At the same time, they are all sea-bound countries, and consequently subject to conditions, which differ altogether from those found in continental countries; most of them are under the influence of moist sea winds, and all are favorably situated in respect of importation by sea.

Intimately connected with the area under forest in a country is the state of ownerships. Forest owners in Europe may be grouped into the following three great classes:

(a) The State or the Crown.

(b) Corporations, endowments, etc.

(c) Private persons.

Where forests are not required on account of their indirect effects, and where importation from other countries is easy and assured, the government of a country need not, as a rule, trouble itself to maintain or acquire forests, but where the opposite conditions exist, that is to say, where forests are necessary to produce climatic and mechanical effects, and where the cost of transport over long distances becomes prohibitive, a wise administration will take measures to assure the maintenance of a certain proportion of the country under forest. This can be done either by maintaining or constituting a certain area of State forests, or by exercising a certain amount of control over private forests. In most of those countries where corporation forests exist they are subject to the control of the State, though the degree to which such control is exercised may differ. Private forests are free from control in some European States, and subject to it in others. In all such cases the State is only justified in interfering when the welfare of the general community requires it. The extent to which interference may be carried depends on the special conditions of each country, and on the proportions of the forest area belonging to the State. Thus, of the Swiss forests only 4 per cent. belong to the State, while 67 per cent. belong to corporations, and 29 per cent. to private owners; at the same time a large proportion of them are so-called protection forests, and in consequence the Government exercises an extensive control over both corporation and private forests. Of the German forests, 33 per cent belong to the State, 19 per cent. to corporations, and 48 per cent. to private persons; the corporation forests are under State control, making with the State forests 52 per cent. This being more than one-half of the area, the control over private forests has of late years been considerably reduced, and in some parts abolished altogether. It is worthy of notice that only 20 per cent. of the Swedish and 12 per cent. of the Norwegian forests belong to the State, while the bulk are private forests, over which little or no control is exercised by the State. Large quantities of timber are exported annually from these countries to Great Britain and other countries, and it may safely be expected that these supplies will considerably decrease in the course of time.

DEFORESTATION IN RUSSIA.

The following article appeared in a recent number of the *Literary Digest*. It was translated from *Preussiche Jahrbucher* for July.

When treating of the Russian famine of 1891-92 in the April number of this magazine, we remarked that this was not to be regarded as a passing incident, but rather as the inauguration of a chronic condition of affairs traceable to

unsystematic farming, to the general withdrawal of capital from the land for investment in manufacturing enterprise, under the aegis of a protective tariff, and to
the general deforestation of the country, in great part to provide fuel for railroads
and protected enterprises. The fatal consequences of this general deforestation
are now generally appreciated, the shrunken state of the once noble rivers of
the country, and growing aridity of the climate, affording evidence that can
neither be overlooked or gainsaid.

The regions of the mighty rivers, the Don, Volga, and the Dneiper, the great
arteries of Russia, were formerly fringed with wide-spreading forests, along their
whole upper and middle courses, which sheltered their sources and tributaries
from evaporation throughout the year. These forests have now for the most part
disappeared. Mile after mile the traveler sees nothing but low scrubs and
melancholy stumps in unbroken succession; the "Mother Volga" grows yearly
shallower; the steamers find scarcely seven or eight feet of water in mid-stream,
and the ferries pursue their snake-like course from bank to bank in search of the
ever-shifting channel. The Don, with its tributaries is choked; the sources of
the Dneiper creep downward and its chief tributary, the once noble Worskla,
with a flow of some 220 English miles, is now dry from source to mouth.

The city of Poltawa lies on its banks, and it was at its mouth that the
Swedish Army surrendered to Peter the Great. This stream, which fertilized a
broad region, supporting a numerous population, exists no more—not temporarily
run dry, but with all its springs exhausted, so that in future it may be stricken
from the map. Of the Bitjug, another river in the Don region, the upper course
has wholly disappeared—valley and bed are filled to the bank with sand and earth.
As if by magic, wide, fertile lands are buried under the sands, and whole villages
desolated. "There has been," says Wiestnik Jewropy, "an unparalleled revolution of natural conditions, which threatens a great part of the country with
the heat and aridity of the Central Asian steppes. The present condition of
our black earth region is so serious, and its future so dangerous, that it cannot
possibly escape the serious attention of the Government, the scientist and the
husbandman, to whom the further development of the situation is perhaps a
question of life and death."

There is perfect unanimity in attributing the threatening catastrophe to the
denudation of the forests. Innumerable factories sprang into existence, and, in
the absence of any systematic provision for coal-supply, they were erected in the
heart of the forest, and, after having consumed all the available fuel within easy
distance, their plant was actually sometimes transferred to fresh fields. Thus
originated the system of wholesale destruction, which was liberally furthered by
the network of railways built to maintain their communication with the great
marts of commerce and provide generally for the transport of produce. For the
past forty years thousands of locomotives and factories have been run almost
wholly with wood without a thought being given to any provision for reproduction. The extension of the railways afforded an opportunity for extracting colossal
fortunes from the "worthless" forests. These were the manufacturers' views
also; so the fate of the Russian forests was sealed. "The machines have
devoured the woods."

The recently passed law for the protection of the forests has come fifteen,
twenty, or twenty-five years too late to avert the destruction of the agricultural
region.

And the Government and people of Russia had already been warned. Forty-
two years ago—that is, shortly after the famine of 1847-49—we find the following
in a letter from the Charkowski Government to the Imperial Society of Econo-

mies : " There are now living people who remember when the present limitless expanse of sand-waste along the banks of the Donez was covered with almost impenetrable forest, interspersed with lakes, which have since dried up, or are fast drying up. Our region is flat, deforested, and exposed to all winds. The fatal east wind finds no impediment, and brings ruin in its train. This wind will perhaps at no distant date prove fatal. The Grecian colonies went under probably from the same cause. Protect the forest ; so plant forests ; protect them with rigorous laws. The Volga and Don and all the rivers of southern Russia will be silted up and disappear unless the forests be protected."

More fatal even than the drying up of the streams is the cessation of the spring and summer rains. This is the immediate cause of last year's harvest failure, and on it even depends the current year's harvest. There have been local rains, but not nearly enough.

This reversal of old conditions has been coming on gradually with the denudation of the forests ; and emphatic warnings, as we have seen, have been uttered. The only result has been the appointment of commissions which have done nothing. Remedial measures on a large scale are now contemplated. Are they too late ?

A PLEA FOR PLANTING.

The sixth Earl of Haddington, in a work in the form of letters to his grandson, published in 1773, says : " When I came to live here (1770) there were not above fourteen acres set with trees. I believe that it was a received motion that no trees would grow here on account of the sea air, and the north-east wind ; so that the first of our family, who had lived here, either believed the common opinion, or did not delight in planting." He continues : " I had no pleasure in planting, but delighted in horses, and dogs, and the sports of the field ; but my wife did what she could to engage me to it, but in vain. At last she asked leave to go about it herself, which she did, and I was much pleased with some little things, which were well laid out and executed. These attracted my notice, and the Earl of Mar, the Marquis of Tweedale, and others admired the beauty of the work, and the enterprise of the lady." After his lady had planted several ornamental clumps in the shape of wildernesses, she proposed to plant a field of about three hundred Scotch acres, called the Muir of Tynningham, a waste common of very little value. From this all her ladyship's friends, as well as her lord, tried to dissuade her, but in vain ; she planted this likewise. In 1707 she began Benningwood ; the prejudice of the country being still against her, they continued to deride her, telling her it could be of no use. Success, however, always gave her encouragement. The next was a large tract of ground, mostly dead sand with very little grass, and very near the sea. Here her ladyship participated in the common prejudices, and thought it would be of no use, but as a gentleman from Hamburgh, being there on a visit, told her he had seen timber growing on such land, she immediately formed a resolution of putting it to a test ; planted sixty-seven acres of it ; and the trees grew to the astonishment of all who saw them. Thus her ladyship, to the honor of her sex and benefit of her lord and her country, overcame the prejudices of the sea and the barren moor being pernicious, and of horses and dogs being the best amusement for a nobleman ; converting a dashing son of Nimrod into an industrious planter, a thoughtless spendthrift into a frugal patriot. His lordship goes on to say the next was a field, which he had often let to tenants, who could do nothing with it ; and further, that he had a great deal more waste land, and intended to plant it all. These woods were of all the usual sorts of timber, fir, beech, chestnut, larch, etc. But oaks were the favorite and succeeded extremely well in every sort of soil.

> " Thus can good wives, when wise, in every station,
> " On man work miracles of reformation,
> " And were such wives more common, their husbands would endure it,
> " However great the malady, a loving wife can cure it.
> " And much their aid is wanted, we hope they'll use it fairish,
> " While barren ground, where wood should be, appears in every parish."

TREES FOR SHELTER.

Fuller in his "Practical Forestry" very truthfully says that pioneers in heavily wooded regions are usually anxious to make a clearing, and as every tree felled not only increases the area which he is to cultivate, but extends his view, the axe is often kept in use long after there is any necessity for the purpose of obtaining land for cultivation. In a few years the settler who was at first so anxious to open up the country, finds he has gone a little too far in this direction, for his own comfort and that of his animals, for on taking down the screen, he has not only admitted the cold winds of winter, but those of summer sweep over his fields, driving away needed moisture—whip the fruit from his trees before it is ripe, and otherwise cause loss that might have been prevented.

It is then that he begins to feel the need of protection, and to wish that his house and outbuildings were located by the side of some friendly forest or grove.

The hygroscopicity of humus or vegetable earth is much greater than that of any mineral soil, and consequently forest ground, where humus abounds, absorbs the moisture of the atmosphere more rapidly and in larger proportion than common earth. The condensation of vapour by absorption develops heat, and consequently elevates the temperature of the soil which absorbs it, together with that of air in contact with the surface. Von Babo found the temperature of sandy ground thus raised from 68° to 80° F., that of soil rich in humus from 68° to 88° F.

The question of the influence of the woods on temperature does not, in the present state of our knowledge, admit of precise solution, and, unhappily, the primitive forests are disappearing so rapidly before the axe of the woodman that we shall never be able to estimate with accuracy the climatological action of the natural wood, though all the physical functions of artificial plantations will, doubtless, one day be approximately known.

But the value of trees as a mechanical screen to the soil they cover, and often to ground far to the leeward of them, is most abundantly established, and this agency alone is important enough to justify extensive plantation in all countries which do not enjoy this indispensable protection.

FORESTRY BYE-PRODUCTS.*

It has been said that there is in the British Isles an immense area of land that either never has yielded, or at the present time does not yield, any agricultural rent, but which might become of value were capital invested in planting it with timber trees.

Though, in common with other trades, the British production of timber has been rendered far less remunerative than formerly by keen foreign competition, it can be shown that timber will yet, in many cases, yield a very fair return for capital.

The forest produce of Great Britain is mainly applied in the following ways : —

1. Ship and boat-building, piers, bridges, etc., requiring much large and sound timber.

2. Building, scaffolding, etc.

3. Railway sleepers.

4. Pit props.

5. Fencing.

6. Furniture ; mainly chairs of beech, yew, etc.

7. Hop poles and agricultural implements.

8. Bobbin wood.

9. Fagots and firewood.

10. Charcoal for gunpowder, pitch, etc.

11. Bark for tanning.

In the first four of these branches, the produce of the forests of Scandinavia, for the present apparently inexhaustible, shipped at the very smallest modicum of profit to the producers, has almost driven British timber out of the market. The rent charges on land, the costliness of labor and of overland transit in Great Britain, may be contributing causes to this result ; but it is also apparently the fact that the sending of crooked or heavily-shaped timber into the market by British foresters is another reason for the success of Scandinavian trade. The best means of meeting foreign competition is by looking to increased economy of production, coupled with excellence of quality ; and as in other trades, it is probable that the utilization of waste substances and bye-products may prove the chief key to economical production. Hop poles, agricultural implements, and bobbin-wood are locally among the most remunerative outlets for coppice produce, the main question with reference to them being economy of production and utilization of waste.

Though hops form a very uncertain and costly crop, the profits on hop-growing are so considerable that the extension of its culture should be considered by our farmers. Chemical substitutes for hops cannot be successfully used to the exclusion of the natural bitter ; whilst the "bine" could be sold in the manufacture of textile products, or of paper. The Spanish chestnut, ash, and larch are largely grown for hop-poles in the south of England.

Ash is also in constant demand for the handles of ploughs, spades, axes, and other implements, and is, like sycamore, the wood of which is considerably used by wheelwrights, a rapidly growing tree.

---

* By G. S. Boulger, F.L.S., F.G.S., *London*, in **Forestry and Forest Products** (Edinburgh), 1894.

The cultivation of hornbeam for such purposes might also be extended more especially on poor gravelly land.

There are undoubtedly many trades not thought of by the timber merchants in which considerable quantities of small woods are consumed; thus it is alleged by United States statisticians that besides 300,000 new telegraph poles, and 3,000,000 cords of wood used in brick-burning, the making of shoe-pegs alone uses 100,000 cords of soft maple annually, that of lucifer matches 390,000 cubic feet of pine, and that of boot-lasts and tool-handles 1,000,000 cords of birch.

Such facts suggest the possibility of a remunerative production of larger quantities of coppice woods.

It being man's highest intellectual function to utilize to the full all the latent powers of nature, we may well direct our attention to such bye-products as bark, charcoal, wood-spirit, turpentine, tar, sawdust, leaf-manure, and wood-ashes.

BARK.

Bark is used for tanning, i.e., for the conversion of hides or skins into a strong, supple, impenetrable, and durable material known as *leather*, by the union of albumen, gelatine, or collagen of their connective tissue, with a substance in the bark known as *tannin*, so as to form insoluble tannates. Tannin is widely distributed throughout the vegetable kingdom, especially in barks, fruits, and galls. It is characterised by a slightly acid reaction, and astringent taste, a blueish or greenish black coloration (ink), with ferric salts, the precipitation of gelatine and albumen from their solution, and its union with them as above mentioned. Sumach leaves form a valuable material for white morocco leathers; divi-divi, the seed-pods of cœsalpinia: "hemlock extract," a decoction of the bark of the hemlock-spruce (*Abies canadensis*); and "mimosa bark" from the Australian "wattles" (*Acacia*) are largely used, but the chief British tanning material is oak bark. In France the young bark of the cork oak (*Quercus suber*) is largely used, and for fine leather that of the evergreen oak (*Q. ilex*); in the eastern United States the white oak (*Q. alba*), the quercitron (*Q. tinctoria*), and the red oak bark are employed, while California and the western territories depend on the chestnut oak for tanning purposes.

When coppice was largely grown for bark a rotation of twenty-four years was common, the stools being eight feet apart, the trees are more productive in proportion at twelve than at double that number years' growth. Branches down to an inch in diameter should be carefully peeled, since their bark contains a higher proportion of tannin than that of the trunk.

Since the low prices for bark have made many foresters doubt the expediency of felling their oak in May when the timber is almost at its worst, it may be well to bear in mind that French and Prussian experiments have shown that bark of good quality may be obtained at any season by steaming the wood for from one-and-a-half to two-and-a half hours according to the season, after which the bark peels easily.

Willow bark is largely and successfully used in Russia for the best leather, and the bark of young alder shoots, not a third of an inch in diameter, yields sixteen per cent. of tannin. The bark of pine and larch is only used for roughly tanning sheepskins.

There is a possibility of chrome-tanning superseding the use of bark of any kind.

CHARCOAL.

The value of charred wood-fibre, deprived of its liquid and volatile portion by destructive distillation, for smelting and heating purposes, has long been recognized. Though now in England, Belgium and other coal producing countries, cheapness of production outweighs considerations of quality, the value of Swedish iron is, probably with justice, attributed to its being smelted with charcoal; and many of the furnaces of the United States are as dependent upon the wood supply, as were those of Sussex in the sixteenth and seventeenth centuries. For charcoal making, the hard woods are mainly used; beech charcoal being preferred in the mineralogical laboratory.

In Great Britain the chief use of charcoal is in the manufacture of gunpowder, for which purpose a highly inflammable quality, which is obtained from light spongy woods of various broad-leaved species is generally required; it requires to be as free from earthy or mineral matter as possible, though no charcoal is absolutely pure carbon, generally retaining as it does, some hydrogen and oxygen, as well as mineral ash. For this reason, though still largely prepared by the primitive method of pits or heaps covered with turf, charcoal is preferably manufactured in iron cylinders or retorts—a method which is far more economical and yields a more uniform result. The inflammable gases distilled from the wood are conveyed by pipes into the furnaces below the retorts, so that an immense saving in fuel is effected, while the tar, pyroligneous acid, etc., are condensed and collected. The temperature at which the wood is charred exercises a great effect upon the properties of the charcoal. The higher the temperature the more completely are the hydrogen and oxygen of the wood driven off, and the denser and blacker is the resulting charcoal, while its temperature of ignition is also higher in proportion. Slack-burnt charcoal retains more volatile matter, is softer, reddish, more readily inflammable and more hygroscopic.

It has been found by experiment that, with sixty grains of saltpetre, twelve grains of each of the following kinds of charcoal give the number of cubic inches of gas (CO_2) in the table :—

Dogwood (*Rhamnus frangula*) ....82 cubic inches.
Willow (*Salix alba*) ....77 "   "
Alder.......................................74 "   "
Filbert.....................................72 "   "
Fir, chestnut, hazel........................66 "   "

The three first named species are accordingly preferred for the purpose, though *Cornus sanguinea, Euonymus europæus, Rhamnus catharticus,* and perhaps other species are not uncommonly substituted for the alder-buckthorn, berry-bearing alder, or true "dogwood" of gunpowder manufacturers. *Rhamnus frangula*; this is a slow-growing shrub, being cut, when about an inch in diameter and under ten years of age, in lengths of not more than six feet. It is grown in Prussia, Belgium, and Sussex. It forms a very explosive powder, used for military small arms, and for sporting purposes. Willow and alder are of quicker growth, especially the former, and are cut when about four inches in diameter. With reference to the use of the two last named species an important fact is that they can be cut in the spring, when their bark is in the best condition for tanning purposes. Charcoal is also of great value as a filtering and deodorising agent.

Volatile Products.

Besides a certain amount of tar, and the inflammable gases which, as has been stated, are utilized as fuel in the charcoal manufacture, even the smoke has proved of considerable value. It contains methylic alcohol, which also distils over from the retort in a liquid form, accompanied by acetic acid. This crude distillate is known as "wood vinegar," and is redistilled and rectified over quick-lime, yielding "wood-spirit" (crude methylic alcohol). The acid portion is then saturated with slaked lime, so as to form a solution of calcium acetate, which is evaporated, the salt being used in the manufacture of acetic acid and metallic acetates, especially that of lead, as as a step towards the formation of white lead.

Dr. Hough describes a charcoal iron-smelting factory in Michigan, where a cord of wood yields forty-two bushels of charcoal, worth 7 cents per bushel, or $2.99 cents per cord, besides 2,800 cubic feet of "smoke," valued at over 30 cents, and the inflammable gas which supplies three-fourths of the fuel consumed. The "smoke" yields two gallons of wood-spirit, worth 85 cents (3s. 4½d.) per gallon in the Chicago market, and 200 lb. of acetate of lime, worth 2½ cents (1¼d.) per lb. in Philadelphia. Thus the "smoke" for which 30 cents (1s. 3d.) are paid, yields $6.70 (£1 6s. 9d.) the total yield of a cord of wood being $9.69 (£1 18s. 8d.) These products are best obtained from dry hard woods, especially beech.

Tar and Pitch.

The mixture of heavy non-volatile hydrocarbons known as tar, though obtained in small proportions in the destructive distillation of all kinds of wood, is yielded mainly by the roots, boles, branches, and waste timber of pines, especially *Pinus palustris*, *P. sylvestris*, and *P. pinaster*. It is mainly imported from the southern United States, from Archangel, and from Riga and other Baltic ports.

The pitch pine (*P. palustris*) covers extensive tracts, and springs up spontaneously in the disused cotton fields of the Southern States, while the Scots pine (*P. sylvestris*) forms enormous forests in the North of Europe and in Siberia. The preparation of tar is still virtually the same as that described by Theophrastus. A hole is dug in the side of the bank in which billets of wood are heaped up and covered closely with turf or earth, a fire is then kindled from below, and the slow combustion causes the tar to exude from the wood and flow out from the heap into barrels placed below to receive it. On distillation, tar yields wood vinegar, creosote, and oil of tar, leaving a residue of pitch. The black, brittle, glossy solid which we know as pitch, and which is mainly home manufactured, is usually obtained by simply boiling the tar, so as drive off the volatile oils.

Turpentine, Resin, etc.

In addition to being our chief sources of tar and pitch, the firs are the exclusive commercial source of the oleo-resin known as turpentine. This is a solution of a resin in a volatile oil which exudes from incisions made in the stems of these trees. On distillation it yields from 14 to 16 per cent. of colourless essential oil, known as oil or spirits of turpentine ($C_{10} H_{16}$), the residue being resin or colophony (the formula of which is probably $C_{44} H_{62} O_{04}$). The greatest quantity of turpentine imported is the produce of the pitch pines (*Pinus palustris*, or *P. australis*), the swamp pine, and *Pinus tœda*, the frankincense pine of Virginia, Carolina and other Southern States, but a considerable quantity imported from Russia and Sweden, is the produce of the Scots pine (*P. sylvestris*), and from the south of France, under the name of "Bordeaux turpentine," where it is obtained from the cluster pine (*P. pinaster*), and other species. Strasburg turpentine is obtained from the silver fir (*Abies pectinata*), and "Venice turpentine" from the larch (*Larix europœa*). Canada balsam is a similar product from *Abies balsamea* and *A. canadensis*. The cultivation of the cluster pine (*P. pinaster*) on the sand dunes of the Landes of Bordeaux is a good example of the conversion of an originally merely protective measure into a source of profit from soil formerly worse than useless. Whether the felling of the forests of Southern France, by producing floods and droughts, was or was not originally the cause of the arid barren sands, certain it is that by damming up the natural drainage and shifting inland, these dunes produced swamps and wastes, the advance of which was only stopped by the binding roots of these pines. Originally planted with this protective object, their yield of timber, bark, turpentine and tar, has rendered them a source of profit, which should remind us that we have in our own country considerable stretches of sand wastes. A useful illuminating oil containing from 80 to 92 per cent. of carbon has been obtained by M. Guillemara from the resin from the Bordeaux area.

MINOR PRODUCTS.

Besides such substances as bark, charcoal, wood-spirit, acetic acid, tar, pitch, turpentine, and resin, which are important articles of commerce, chemical discovery has demonstrated in the past, and may be expected to show still more frequently in the future, the presence of substances in trees which might well form sources of profit. For example : from the sap of the Scots pine and of the larch felled in summer, barked and scraped, a substance known as "coniferin" is obtained, which yields "vanillin" the essential constituent of "vanilla." Though an expensive substance to prepare, this is considerably cheaper than common vanilla, the sole source of which is the inner pulp of the pods of one or two species of orchid.

Another similar product, not as yet much developed, is the "rubber" obtained by distillation from the bark of the common birch, a black, gummy "latex," which resists the action of air and acids, and which will considerably increase the durability of india-rubber or gutta-percha, even if mixed with them in only a small proportion. As in the salt mines of Stassfurt it has been found that the formerly wasted bye-products, the salts of potash, are as valuable as the rock salt, if not more so, and as in Michigan the "smoke" has proved more valuable than the charcoal, so the development of new chemical industries may render such products as this vanillin and gutta-percha more remunerative than the timber itself.

LEAVES, SAWDUST, ETC.

In Styria young pine and fir needles, from loppings or thinnings made in spring, are dried in ovens or kilns, ground, mixed with one-twenty-fifth of salt, and used with advantage as a food for cattle. Similarly, in the north of Italy the dried leaves of the poplar have long been used as cattle food, and chemical analysis bears out their value for this purpose. When not used for fodder, however, dried leaves make an excellent litter, and analysis proves their value as manure.

The following table, prepared in Bavaria, exhibits the composition of the leaves produced annually by an acre of forest under beech, pine, or spruce, as compared with that of a ton of wheat straw :—

	Dry matter.	Ash.	Potash.	Lime.	Magnesia.	Phosphoric acid.	Sulphuric acid.
	lb.	lb.	lb.	lb.	lb.	lb.	lb.
Beech, per acre of land..................	2,972	165.5	1.80	17.18	10.90	9.32	3.28
Pine, per acre of land.....	2,842	41.5	4.32	16.84	4.28	3.28	1.51
Spruce, per acre of land.;.............. .	2,683	121.3	4.30	57.37	6.20	5.72	1.87
Wheat straw per ton.....	2,240	85.2	9.80	5.20	2.20	4.60	2.41

Saw dust also forms a good litter for cows and horses ; and though destitute itself of manurial value, being very absorbent of liquid manure, can, when thus soaked, be used as a valuable top dressing.

Finally, if the forester has any waste that he cannot utilize in the tan pit or the charcoal retort, as paper pulp, or as firewood, it is probably best to burn it so as to avoid harboring insects and fungus life. By so doing he will lose little of the manurial value of the refuse, and leached ashes, being rich in potash, is a valuable dressing for old grass land, orchards, market gardens, onions, rye, and other crops. The potash might even be recovered by lixiviation of the ashes and used for many purposes.

THE PRODUCTION OF WOOD PULP.[*]

The wood-pulp industry may be said to have commenced in the year 1846. But its development during the first thirty years was decidedly slow. Since 1876, however, the production of this material has increased rapidly. Its pre-industrial period was known only to the chemist. Cellulose was made in the laboratory in 1840, but it was not manufactured commercially till 1852. Ground wood was first used for paper-making about the year 1846, when it was manufactured by Keller under a patent taken out in Saxony in the previous year. Since that date many improvements have been made in the machinery and methods used in grinding, the main object being to produce a longer and finer fibre. The fibres of the wood are torn away by mechanical pressure against a revolving grindstone in contact with water. No chemical treatment of the wood is necessary, the only requirements of this industry being cheap wood, abundant water-power and suitable machinery.

Processes, such as Sinclair's, have long been in use for pulping very finely cut coniferous wood, and in the Paris Exhibition of 1880 one of the most prominent objects exhibited in the Norwegian Section was a *pate de bois* or *papier maché*, made in this way from pine wood, and worked into cardboard and various moulded pannellings, etc. It has been found, moreover, that in this way the whole of a pine tree trunk—branches, needles, and all—can be converted into paper without waste. Saplings, which it would not pay to cut for firewood, are now profitably worked up in this way into pasteboard.

[*] By G. F. Green, C. F. Cross and E. J. Bevan, in Forestry and Forest Products (Edinburgh), 1884.

By the chemical processes for manufacturing wood-pulp, a good class of pulp is made from the quick-growing poplar and from spruce. The wood of the slower growing linden or basswood makes an equally valuable white paper pulp.

Oak can also be used, though yielding an inferior product that requires bleaching. One great advantage in the method is that the tannin in the oak is obtained as a bye-product, and the chemicals with it in the lye being rather an aid than a hindrance to the tanning process, it is found that hides can be perfectly tanned in it in ten days. This seems to offer to the cultivator of oak coppice, or the enterprising planter of poplars, a most important source of income ; whilst in coniferous plantations there need be absolutely no waste.

The chemical preparation of fibre has given rise to two distinct processes— the soda process and the acid process.

Chemical pulp (cellulose) is used as an adjunct, with esparto, rags, or mechanical pulp, in the manufacture of news, printings, colors, and some kinds of wrapping-paper. It forms (according to Mr. Routledge) an excellent *succedane,* or filler up, and bleaches to a high color. Fine prints are also manufactured exclusively from acid pulp.

Mechanical pulp is chiefly used as an adjunct in the manufacture of news, cheap printings, and wall-papers ; but there are several distinct classes of paper made from it without any other ingredient, viz., wood-pulp middles from white pine pulp, and various self-colored wrappings and tinted wall-papers from brown, sometimes styled patent pulp.

Another important use is for wood-pulp boards and so-called " patent " or brown boards, the latter being produced from brown pine-pulp and the former from white pine-pulp.

The consumption of wood-pulp boards is increasing rapidly, chiefly for making paper boxes, for which they possess certain advantages over straw boards.

Although almost any wood can be converted into pulp, experience has hitherto decided in favor of conifers of a certain age.

For chemical pulp, trees on an average of twenty years' growth, and a thickness of six to eight inches at the base of the stem are said to be the best. Younger wood is more tractable by chemical means, but produces a fibre of inferior quality. Older wood requires stronger chemicals to remove the incrusting matter, and possesses no compensating advantages.

In Canada many species of wood have been utilized, amongst which may be mentioned pine, poplar, spruce, willow, basswood, cedar, hemlock, maple, and birch.

Poplar pulp remains white, birch becomes pink, maple turns of a purple tint, and basswood reddish after grinding.

The practical operations concerned in the manufacture of pulp from wood by the caustic soda process may be divided into the following :—Barking, sawing, chopping, crushing, boiling or digesting, washing and bleaching, treatment for sale as half-stuff, and soda recovery.

HEMLOCK EXTRACT.

In the final Report of the Select Committee of the House of Commons(Canada) in 1868 on the best means of protecting Hemlock Timber from destruction, the conclusion was come to, after a most careful consideration of the question, that unless some steps were speedily taken to check the wasteful and extravagant rate of consumption then going on, really for the benefit of other countries at the

expense of Canada, many years would not elapse before our own tanneries would
bo seriously crippled, and we no longer able to compete successfully with other
countries in the manufacture of leather.

Answers from Galt, Guelph, etc.:—Hemlock getting scarce ; not more than
ten years' supply ; timber, sawn into lumber, used for rough work, such as roof-
ing, etc.

Large quantities in Drummond and Arthabaska ; five bark factories in the
counties ; timber used for scantling and rough boarding ; in settled districts
timber worth four to eight dollars per M ; in remote districts not of sufficient
value to bear transportation ; about one cord of bark is procured from 1,000 feet
of lumber; bark in some districts used for domestic purposes only, in others, as
Eastern Townships, manufactured into extract; five factories produced twenty
thousand barrels of 400 pounds each, worth 2½ cents per lb. at the factory and 5
cents per lb. in Boston; 1½ to 3 cords of bark will produce 1 bbl. of extract,
worth in Boston $20 a barrel.

On non-resident lands the timber is allowed to rot on the ground, and squat-
ters, as a rule, either burn or allow it to rot if not close to a market.

In some districts the best is sawn into lumber or cut into lathwood for
exportation to England, the balance for railroad ties or cordwood.

Wherere a farmer makes bark on his own land, he cuts the peeled trees into
sawlogs and clears the land. Where trespassers peel bark they leave the trees to
rot.

After the destruction of a hemlock forest it is generally succeeded by a
mixed growth of maple, poplar, cherry, and balsam, when not cleared for farming
purposes ; even if replanted, it would take one hundred years' growth to render
it available for tanning or extract purposes.

About 500 lbs. or more of green raw hide can be converted into leather by
one cord of bark. Hemlock extract is said to convert raw hide and leather in
one-sixth of the time, and at one-half of the cost.

The effect of the manufacture and export of hemlock extract will exhaust
the hemlock forests at all accessible points and compel manufacturers to remove
their tanneries into the rural districts to obtain a supply of bark, raise the price
of bark, and consequently of leather, and diminish the quantity of leather produced.

For the protection of our hemlock forests it was suggested that the extract
should be manufactured under a license upon Government land, an export duty
charged on bark, and a considerable excise duty on extract manufactured for
exportation.

An acre of good hemlock land would produce from ten to twelve cords of
bark, worth from $30 to $36 delivered at the factory, at a cost, say, of $1 per
cord for felling and $1 per cord for carting.

The cord of hemlock lathwood (128 cubic feet) is worth $8 at Montreal or
Quebec.

The conclusion arrived at from this evidence was that our hemlock forests are
being and have been rapidly depleted and destroyed, and that without proper
care and forethought the language put into the mouth of the Indian many years
ago (referring to stripping the soil of its trees) may in a degree become true—

"The realms our tribes are crushed to get,
May be a barren desert yet."

And the remedy seems to be that active measures be resorted to for select
cutting, natural reforestation and planting, and systematic forestry management,

THE GEOGRAPHICAL DISTRIBUTION OF THE FOREST TREES OF CANADA.

In the report of the Geological Survey of Canada for 1880, there is a paper by Dr. Bell, the Assistant Director, accompanied by a map on which the general northern limits of the principal forest trees of Canada east of the Rocky mountains are represented. Dr. Bell says :—

The Continent of North America posesses a great variety of forest trees. About 340 different species occur within the United States. All the kinds which we have in Canada, amounting to about ninety, including those of the Pacific slope, are also met with in that country. Some species are not only very widely diffused, but are also persistent over great areas, being found almost everywhere within the limits of their distribution, while others, although having an extensive range, are nowhere very common, and are sometimes absent for a considerable interval. Others again are confined to comparatively small tracts. As a general rule, the more northern species occupy the greatest extent of country, while the southern ones are progressively more and more restricted, even in a more rapid ratio than would be implied by the narrowing of the continent from north to south ; this is owing to the great differences experienced in climatic conditions in going from east to west in the more southern latitudes. Along the northern borders of the forests of the continent, the elevation of the land above the sea is comparatively slight, and regular, and the other physical conditions are tolerably uniform. As a consequence, we find the most northern group of trees extending from Newfoundland into Alaska, a distance of about 4,000 miles.

An inspection of the map accompanying Dr. Bell's report show some interesting features as to the general distribution of our forest trees, as well as regarding almost every individual species of timber. For example, it will be observed that there is no material change in the woods throughout the great triangular area, embracing about 600,000 square miles, of which the national boundary line between the Rocky Mountains and Lake Superior forms the base, and the Rocky Mountains and Laurentian Hills respectively the west and east sides, the apex being at the mouth of the Mackenzie River. In the southern part of of this area, a number of species are added to the kinds which everywhere throughout it make up the bulk of the forest ; and again, few trees of any kind are found to the south of the North Saskatchewan ; still, making allowance for local peculiarities of condition, there is a remarkable uniformity in the timber of this enormous area. It includes, however, only a few species, of which the aspen, balsam poplar, and willows are more abundant towards the western and the spruces, larch, balsam, fir and Banksian pine toward the eastern side of the area

It will be observed that the lines marking the northern limits of about a dozen species turn southward and become their western limits on reaching the eastern side of the valley of Lake Winnipeg and the Red River ; while the boundaries of the species occurring next to the south of these also manifest a tendency to turn southward in approaching the prairies of the west. The species above referred to are the white cedar, black ash, white pine, red pine, sugar maple, yellow birch, red oak, white ash, hemlock, beech, ironwood, red cedar (arborescent variety) and white oak. They are to a great extent replaced by other species before the region of open plains is reached. Had the great forests originally extended further west, and been destroyed by fire or other causes, in comparatively recent times, we should have found the northern limits of these species continuing their general course through the prarie regions, and ending abruptly there, instead of which, they all curve gradually round, in a more or less concentric fashion, and other trees occupy the intervening ground. These well-marked features of forest distribution show that the present divisions of prairie and woodland are of very ancient date. The evidence of the smaller plants, and also of certain superficial geological conditions, all point to the same conclusion.

The State of Minnesota is situated in a very interesting region in regard to forest distribution. Here we find the northern limit of the group to which the most southern trees of Ontario belong, such as the black walnut, shell-bark hickory, hackberry, and Kentucky coffee tree; the north-western limit of the commoner trees of the northern states and of Quebec and Ontario, such as white oak, red cedar (arborescent variety), ironwood, beech, hemlock, white ash, rock elm, red oak, yellow and black birch, sugar maple, red maple, wild plum, etc. The western boundaries of some of the trees whose northern limits pass through Northern Ontario, such as the white cedar, black ash, white pine and red pine; the southern limits of the most northern group, including the white spruce, the large Banksian pine, balsam fir, balsam poplar and canoe birch; and the general eastern limits of some of the western species, such as the ash-leaved maple, green ash, burr oak, and cottonwood.

It will be observed that in Labrador peninsula the tree-lines tend northward midway between the eastern and western shores. This is due partly to the unfavorable influence of the sea on either side, and partly to the beneficial effect of the central depressions in which the rivers run northward into Ungava Bay.

From Mingan to Lake Superior, the height of land, north of the St. Lawrence, is rudely parallel to the general course of the lines marking the northern boundary of the trees. And it may have had some effect in limiting the northward range of a number of species. A southward curve in the watershed about the longitude of Ottawa is marked by a corresponding curve in the tree lines. Again, where a great depression occurs in this dividing plateau, some of the trees which in such cases may be approaching their northern boundaries, are found to extend, in the lower levels, beyond their general outline on either side. As examples of this the Lake Temiscaming and Abittibi District, and the valley of the Kenogami, or principal south branch of the Albany, may be mentioned.

On the Missinaibi, or west branch of the Moose River, the white elm reappears, 130 miles north of its general boundary on descending to a sufficiently low elevation above the sea. The Saguenay, for about 100 miles from the St. Lawrence is really a narrow arm of the sea, and the country in the vicinity of Lake St. John at the head of the river, is only slightly elevated above its level and has a fertile soil, although surrounded by a mountainous region. Here we find an isolated colony of basswood, sugar maple and other trees, considerably removed from the rest of their species. On the north side of Lake Huron, and to the north of the City of Quebec, the land rises somewhat rapidly, and in both instances the tree lines near these latitudes are more closely crowded together than elsewhere.

Some kinds of trees, in approaching their northern limits, show a tendency to diminish gradually in size, and to become more and more scattered, rendering it difficult to draw any definite boundary of the species, while others vanish abruptly. The latter habit is more characteristic of southern than northern species so far as the Dominion is concerned. The various species appear to die out more gradually as they range northward in the western than in the eastern regions.

Forest trees east of the Rocky Mountains may be divided into four groups, as regards their geographical distribution within the Dominion.

1. A northern group, including the white and black spruces, larch, Banksian pine, balsam fir, aspen, balsam poplar, canoe birch, willows and alder. These cover the vast territory down to about the line of the white pine.

2. A central group of about forty species occupying a belt of country from the white pine line to that of the buttonwood.

3. A southern group, embracing the buttonwood, black walnut, the hickories, chestnut, tulip tree, prickly ash, sour-gum, sassafras, and flowering dogwood, which are found only in a small area in the southern part of Ontario.

4. A western group, consisting of the ash-leaved maple, burr oak, cottonwood, and green ash, which are scattered sparingly over the prairie and wooded regions west of Red River, and Lake Winnipeg.

In the western peninsula of Ontario the forests present a remarkable richness in the number of species to be found growing together. In some localities as many as fifty different kinds may be counted on a single farm lot. A more varied mixture is probably not to be met with in any other part of the continent, or perhaps in the world.

LEVELS OF THE OTTAWA.

The following tables from the reports of the Geological Survey (Canada) show the levels at various points on the St. Lawrence, Ottawa and Mattawa rivers, from Three Rivers to Lake Nipissing.

Levels of the Ottawa above the waters of the St Lawrence at Three Rivers, which is about the highest point affected by the action of the tides :--

	Distance. miles.	Rise. ft. & in.	Total rise. ft. & in.	
Rise from Three Rivers to Montreal harbour. as stated in a report of the Hon. H. H. Killaly, President of the Board of Works in 1845	90	12.9	12.9	Montreal.
Rise from Montreal harbour to lake St. Louis at Lachine, from the same report :				
1 lock 13.3				
2 " 13.3				
3 " 8.6				
4 " 9.0				
5 " 0.9	10	44.9	57.6	Lachine .
Rise in lake St. Louis from Lachine to Ste. Anne ..	13	0.6	58.0	
Rise in the lock at Ste. Anne......................		3.0	61.0	
Rise in the lake of Two Mountains from Ste. Anne to Carillon	23	0.8	61.8	Carillon.
Rise from Carillon to Blondeau :				
1 lock, up..............................10.0				
2 " 11.0				
———— 21				
3 lock down 13	4½	8.0	69.8	
Rise in Chute à Blondeau		4.0	73.8	
Rise in the Grenville canal from the head of Blondeau to the head of Grenville canal				
1 lock 3.0				
2 " 3				
3 " 8				
4 " 8				
5 " 7				
6 " 6	6½	35.0	108.0	Grenville.
Rise in the navigable part of the Ottawa, between Grenville and the entrance to the Rideau canal...	58½	9.4	118.0	Ottawa.
Rise from the entrance of the Rideau canal to the Chaudiére lake, viz :—Rise in the Rideau canal to Dow's Swamp :				
1 lock........................... ....11.0				
2 " 10				
3 " 10				
4 " 10				
5 " .... .10				
6 " 10				
7 " 10				
8 " 10				
———— 81				
Fall from Dow's Swamp to Chaudiére lake..... .18	6	63.0	181.0	Chaudiére.
Rise in Chaudiére lake from the foot to Fitzroy harbour at the head, supposed to be one inch per mile.	25	2.1	183.1	
Rise from Fitzroy harbour to Chats lake, as ascertained by levels taken up the Mississippi channel by the Board of Works in 1845, 49.96, say	3	50.0	233.1	Chats.
Rise in Chats lake from the head of the Rapides des Chats to the foot of the Chenaux, supposed to be one inch per mile..............................	15	1.3	234.4	

Levels of the Ottawa above the waters of the St. Lawrence at Three
Rivers.—*Continued.*

—	Distance. miles.	Rise. ft. & in.	Total rise. ft. & in.	—
Rise from the foot of the Chenaux to Portage du Fort, a strong current prevailing all the way, supposed to be 12 inches per mile.	5	5.0	239.4	Portage du Fort.
Rise in the rapid at Portage du Fort		17.0	256.4	
Rise between the head of Portage du Fort rapid and the foot of the Sable, a strong current prevailing all the way, say one foot per mile.	5	5.0	261.4	
Rise in the Sable rapid and two small ripples above.	0½	6.2	267.6	Sable.
Rise between the Sable and the Mountain Chute.	1¼	1.0	268.6	
Rise from the boom at the foot to dead water at the head of the Mountain Chute, according to Mr. Gerrard Nagle.		15.0	283.6	Mountain.
Rise from the head of the Mountain Chute to the foot of D'Argis rapid, say 8 inches per mile	1	0.8	284.2	
Rise in the D'Argis rapid		5.0	289.2	D'Argis.
Rise from the head of D'Argis to the foot of the Calumet Falls, say 8 inches per mile	1¼	0.10	290.0	
Rise in the Calumet Falls, according to Mr. Gerrard Nagle :—from dead water at the foot of the falls to the foot of the middle slide ... 26.3				
From the foot of the middle slide to dead water at the head ... 39.7	1	65 10	355.10	Calumet.
Rise from the head of the Calumet Falls to the head of the Calumet island, a considerable current prevailing the whole distance, say 6 inches per mile.	13	6.6	362.4	
Rise from the head of the Calumet island to Fort Coulonges, including about 1 foot at La Passe rapid.	5	2.8	365.0	Fort Coulonges.
Rise in Fort Coulonges lake from Fort Coulonges to the mouth of the Black river, quiet water all the way, say 2 inches per mile.	8	1.4	366.4	Black river.
Rise from the mouth of the Black river to the Chapeau rapid, swift water, say 6 inches to the mile ...	6	3.0	369.4	
Rise in the Chapeau rapid.		2.0	371.4	Chapeau.
Rise from the Chapeau to the Culbute, swift water all the way, say 6 inches per mile.	5	2.6	373.10	
Rise in the Chute Culbute from the foot of the current to dead water at the head, according to the Board of Works		19.7	393.5	{ Culbute and lake Allumettes.
Rise from the head of Culbute rapid by Upper Allumettes lake and the Deep river to the foot of the Joachim Falls. The current in the deep river is so moderate that with a very gentle wind rafts are sometimes carried up stream without sails. The rise is supposed to be 2 inches per mile	32	5.4	398.9	Joachim.
Rise in the Joachim Falls from the Deep river to dead water at the head, according to Mr. Gerrard Nagle.	1	23.3	422.0	
Rise from the head of the Joachim Falls to the mouth of Bennett's brook, say 3 inches per mile.	4	1.0	423.0	Bennett's brook.
Rise from Bennett's brook to the mouth of the Riviere du Moine, a strong current prevailing most of the way, say 6 inches per mile	3½	1.9	424.9	Moine river.
Rise from the Riviere du Moine to the foot of Islet rapid, a strong current prevailing at Riley's clearing and at McSwirley's clearing, say 5 inches per mile	8	3.4	428.1	
Rise from the foot of Islet rapid to the Roche Capitaine rapids, or that part of them called the Maribou, allowing one foot for the Islet.	1	1.5	429.6	Islet.
Rise from the head of Roche Capitaine	2	42.10	472.4	Roche Capitaine.
Rise from the head of Roche Capitaine to the foot of the Deux Rivieres, quiet water nearly the whole way, say 3 inches per mile	11	2.9	475.1	

Levels of the Ottawa above the waters of the St. Lawrence at Three Rivers.—*Continued.*

	Distance. miles.	Rise. ft. & in.	Total rise. ft. & in.	
Rise from the foot of the Deux Rivieres rapid to the head of the Levier rapid, viz :— Difference of level between smooth water at the foot and smooth water at the head of the Deux Riviere Portage....13.38 Difference of level between the heads of the Deux Rivieres Portage and the mouth of the Maganisipi ....8.55 Difference of level between the mouth of the Maganisipi and the head of the Levier rapid .8.09	3	30.4	505.5	Levier.
Rise from the head of the Levier to the foot of the Mattawa rapids, being swift water nearly the whole distance, supposed to be 6 inches per mile........	18	9.0	514.5	
Rise from the foot of the Mattawa rapids to the mouth of the Mattawa river..............................	1½	5.0	519.5	Mattawa.
Rise from the mouth of the Mattawa to the foot of the Cave rapid, a considerable current about midway up, say 4 inches per mile	2½	0.10	520.3	Cave.
Rise from the foot of the Cave to the head of the Chaudron rapid viz :-- Rise in the cave ....5.75 Rise in the Chaudron.....................6.00	¾	11.9	532.0	Chaudron.
Rise from the head of the Chaudron to the foot of the Erables rapid, say 3½ inches per mile.........	3½	1.0	533.0	
Rise from the foot to the head of the Erables rapid.	½	13.0	546.0	Erables.
Rise from the head of the Erables to the foot of the Mountain rapid, say 3½ inches per mile..........	3½	1.0	547.0	
Rise from the foot to the head of the Mountain rapid.	¼	5.5	552.5	Mountain.
Rise in the Seven League lake from the head of the Mountain rapid to the foot of the Long Sault rapids, say 2½ inches per mile	17	3.6	555.11	
Rise from the foot to the head of the Long Sault rapids :— 1st or lower leap............................6.92 Intermediate 1⅛ mile ....2.50 2nd leap ....6.16 Intermediate 1 2-18 mile ....2.20 3rd Crooked rapid.......................6.38 Intermediate 1 2-12 mile ....0.23 4th leap.......... ....15.82 5th Upper rapid 8.34	6	48.5	604.4	Long Sault.
Rise from the head of Long Sault rapids to the mouth of the Opinika river above the Galere current ; there is a perceptible current only in two places, say 3 inches per mile...........................	12	3.0	607.4	Galere.
Rise from the mouth of the Opinika river to the head of lake Temiscaming, say 1 inch per mile....	55	4.8	612.0	Temiscaming.
Miles	492½			

Levels of the Mattawa from its junction with the Ottawa, 519 feet 5 inches above the surface of the St. Lawrence at Three Rivers, to Trout or Turtle Lake.

	Miles.	Feet and inches.	Total.	
Height above Three Rivers......................			519.5	
Rise from the mouth of the Mattawa to the foot of Plain-Chant rapids, including a rise of 1 foot 8 inches in 2 small rapids, allowing 4 inches per mile	2½	2.6	521.11	
Rise from foot to head of Plain-Chant rapids :				
1 rise......................................15.98				
2 "....................................1.60	½	17.7	539.6	Plain Chant.
Rise in Long lake from the head of Plain Chant rapids to the foot of Portage a la Rose, say 3 inches per mile	5½	1.4	540 10	
Rise from the foot of Portage a la Rose to the head of Portage du Rocher above Amable du Fond river :				
1 Portage a la Rose rise 5.90				
Intermediate 20				
2 Portage de la Compagnie................5.80				
Intermediate 0.80				
3 Portage du Rocher 5.05	2	17.9	558.7	Du Rocher.
Rise from the head of Portage du Rocher to the foot of Portage des Parresseux, say 3 inches per mile in addition to a small fall of 4 inches............	3¼	1.2	559.9	
Rise from the foot of Portage des Parresseux to the foot of the Talon or Hang falls :				
1 Portage des Parresseux rise.... ....33.9				
Intermediate 0.25				
2 Portage de la Prairie rise 8.55				
Intermediate 0.95				
3 Portage rise...........................6.30				
Intermediate 0.10				
4 No Portage rise........................3.31				
Intermediate 0.33	2¾	53.9	613.6	Foot of Talon.
Rise from the foot of Talon or Hang falls to the foot of Talon lake :				
1 Portage de Talon rise 42.23				
Intermediate 0.25				
2 No Portage rise 0.85	¾	43.4	656.10	
Rise from the foot to the head of lake Talon by the old canoe route, say 1 inch per mile.............	7	0.7	657.5	Lake Talon.
Rise from the head of lake Talon to the foot of Lower Trout lake, the difference of level ascertained by the new canoe route, the distance by the old route, viz :				
Rise from lake Talon to Lac des Pins.....42.19				
Fall from Lac des Pins to Lower Trout lake.10.89	4¼	31.3	688.8	Lower Trout lake.
Rise from Lower to Upper Trout lake, say 1 inch per mile in addition to a rise of 1.1 at the outlet of the Upper lake..............................	3½	1.4	690.0	
Rise from foot to head of Upper Trout lake	8	0.0	690.0	Upper Trout lake.
	39¾			
Levels from the surface of Upper Trout lake, 690 feet above the waters of the St. Lawrence at Three Rivers, to the surface of lake Nipissing				
Height of Upper Trout lake.			690.0	
Rise from Trout lake to the height of land between it and the Vase river on the canoe Portage	½	24.5	714.5	Height of land.
Fall from the height of land to the Riviere a la Vase, at the end of the Portage	¼	22.11	691.6	Vase.
Fall from Trout lake Portage, on the Vase, to lake Nipissing :				
Fall at 1st Portage................ ....3.14				
Intermediate 1.00				
Fall at 2nd Portage.....................20.88				
Intermediate 1.50	4¼	26.6	665.0	Nipissing.

Levels from the surface of Lake Nipissing, 665 feet above the waters of the St.
Lawrence at Three Rivers, to that of Lake Huron, at the mouth of the
French River.

	Miles.	Feet.	Total rise.	
Height of lake Nipissing agreeably to the estimate of Mr. Wm. Hawkins in his report to the Commissioners of the lake Huron and Ottawa Survey in 1838, the falls on the French river are :—			665.0	
1 Chaudiere falls (upper)........ ....10.0				
2 Chaudiere " (lower)..................15.0				
3 Rapids......... .... 3.0				
4 " 3.6				
5 " 3.0				
6 " 8.0				
7 " 2.0				
8 " 3.0				
9 " 6.0				
10 " 3.0		56.6		
Allowance for the supposed general slope of intermediate parts of the river, say 6 inches per mile..	55	27.6	84.0	
To the level of lake Huron			581.0	
The ascertained height of the surface of lake Huron above the sea, according to the Michigan Surveyors is.....................................			578.0	
Making a difference of			3.0	

FOREST PROTECTION AND TREE CULTURE ON WATER FRONT-
AGES WITH THE VIEW OF PROVIDING A CONSTANT AND
STEADY SUPPLY OF WATER, FOOD, SHADE, AND SHELTER FOR
FRESH-WATER FISH.

At the conference of the International Fisheries Exhibition, London, on Wed-
nesday, July 18th, 1883, a paper on the above subject was read by D. Howitz,
Esq., Forest Conservator, and Commissioner for Denmark. It is in substance as
follows :—

Professional foresters take a great interest in this question, as it is of much
importance to the success of pisciculture, and to all fresh-water fishing. Its
value may not at first appear so great as it really is, but it is sincerely hoped it
may become a question of interest to all, and a special subject for future legislation. It is the question of the protection, proper management, and cultivation of
forest and forest trees in localities where are found the sources of creeks, rivers,
and a supply of water to lakes and other fresh waters. The greatest part of the
forest land in Canada with which this question has to deal is in the possession of
the State, but there are no laws in existence giving a guarantee for the preservation and proper management of these forests.

That the forests regulate the flow of the water in water-courses, and insure a
steady supply during dry seasons, while they prevent sudden and disastrous
floods, is a fact so often discussed and proved, that it need only be referred to
here. There is still a great deal of uncertainty as to the extent of the effect of a
forest on the rainfall, and it is only by very minute observations of forests, consisting of the same species of trees in various altitudes, that series of trustworthy
results can be obtained. Still, there is no longer any doubt as to the effect of the
forest in conserving the water that falls, or that the humidity of the air above a
forest is considerably larger than that of the air of the open country. Experiments in the south of France showed that the rainfall in a forest, as compared
with that in the open country, was in the proportion of 100 to 92.5, while the
evaporation in the forest was only one-third of the evaporation in the open. The
result of this is that the actual water received and retained from the atmosphere
is nearly 50 per cent. greater in a forest than that received and retained by the
plains. Numerous observations have also established the fact that the forests, as
ready conductors of electricity, influence the current of vapours, and that their
action is felt far above the actual height of the trees. Also that they condense
the clouds into rain by lowering the temperature, and act as bulwarks against the
severity of storms; all this we know by daily experience and observation. That
want of forest protection may have most fearful results has been so often and
sadly proved, and I need only remind you of the disasters caused by great floods
and long droughts in Spain, South of France, Sicily, Chili, Peru, Mauritius, and
many other places, and you will grant the importance of the question. In the
Murcia Valley the river was reduced to a succession of stagnant pools, which dur-
the summer heat developed malaria, fever, and miasmatic exhalations, detrimental
to life and health, and furnishing but scant and bad accommodation for the few
remaining fish.

But as soon as the winter rains came, the river, in fact nearly all the valley,
became a raging torrent, destroying life and property, and all because the forests
on the ranges and mountains had been devastated, no legal restrictions protecting
them. As a question of national economy ; as a question of protection to life and
property ; and as a question of prosperity, forest protection has the greatest

claim to the attention of the Legislature. The forest, with its number-less roots and decaying vegetation, retains the rain water, and prevents it from rushing to the rivers and the sea, while it gives it off to these slowly and steadily. It acts like a great sieve and retains the fine particles of the soil, which the influence of the air and sun, the frost and rain, and the action of the numberless roots have decomposed, thereby fertilizing the land and forming a layer of mould or humus, in which insects, worms, larvæ, and other animalcules live and breed.

In his most interesting paper on fish diseases, Prof. Huxley said that drought or flood did not seem to affect the *saprolegnea,* but that a steady flow was beneficial to the fish.

Mr. Wilmot, Superintendent of Fish Culture, Canada, in the discussion which followed, pointed out that the disease nearly always appeared where the regularity of the supply of water had been disturbed by the destruction of the forests.

I presume, therefore, that both these learned and practical gentlemen will agree with me in the importance of forest protection as a means of preserving the health of the fishes.

The branchlets, leaves, decaying and decayed vegetation, produce a vast amount of nourishment for the fish, and one most agreeable to them. Each breeze drops into the water numberless grubs, caterpillars, beetles, flies, and other insects, the food most relished by the fishes, while from the banks and roots worms and grubs are constantly supplying them with delicacies.

The shade of the overhanging trees is also agreeable to the fish, and one needs only to place a board in a stream and see the fish gather underneath it to be convinced of this.

We all know that a shady deep pool is a good place in which to seek for fish, and have often observed the predilection fish have for the shady side of a stream. But not only as regards fresh-water fishing can this be said. In Denmark it is a well-known fact, that the best fishing is where a forest is close to the shore, and in particular where the trees, as is often the case in that country, overhang the very sea. The shadowing trees have another, and, perhaps, the far more important effect of preventing a large evaporation, and at the same time, keeping the water clear and cool in summer, while on the same account the winter frosts do not deal so severely with them. In all forest country the changes of temperature are not so severely felt as in a treeless country or on the open plains, and the effect upon the water is even greater. It is a popular saying in Denmark of the forest streams, that they are cool in the summer and warm in the winter, this, of course, meaning that they present that feeling in comparison to the atmosphere. The forests not only regulate the flow of the water, but they purify the water. This is an experience often demonstrated in Australia in cases where streams have been polluted by wool-washing establishments. After having passed a few miles through a shady and dense forest, the water will appear as clear and pure as it was above the woolwash.

I need not here enter upon more reasons for the conservation of existing forests to insure a steady supply, or to draw your attention to the danger in not protecting them by legislation. But I will draw your attention to the advisability of cultivating forests on places suitable for the supply of water, and especially along water courses and lakes, as means of purifying these, preventing too great

---

*A fungus or mould.

evaporation, supplying food for fish, and providing these with shade against the rays of the summer sun, and shelter from the pelting rains, the hail and the tempests.

Salmon fishing and all fresh-water fishing depend upon proper attention to this matter, and I feel certain that if the true causes were properly investigated where fish were said to disappear from a stream, in half the cases it would be found that the shade and shelter of the forests or protecting border trees had been taken away. It was said at the reading of Sir James Gibson Maitland's excellent paper on the "Salmonidæ" that it was not enough to place spawn and fry in a water; they must be provided with proper food, and the best means to do this is to preserve the border trees and insure a steady supply of water and food by preserving the forests from whence the supply of water is derived. But, as before remarked, it is not enough to preserve the present forest. New forest must be cultivated on the barren ranges, and many a stream; now nearly empty during dry seasons, will be refilled and soon teem with fish and food for the many. So far for the principle of the conservation of the forest.

I will now briefly mention the most suitable trees and their culture. But, before entering upon this, I must draw your attention to the important condition to be observed in the management of such forest areas as are preserved for the sake of conservation of water. This condition is density. In the dense shade of a well closed forest are developed all these atmospheric conditions on which depend the greatest effects of the forest in regard to climate and water conservation. The so-called periodical thinning out in these areas should be carried on with the greatest care, and might with advantage be nearly dispensed with, if the economy of the management would permit it. The result would be, besides the effect on the water conservation, that tall straight trees would be reared, yielding timber most valuable for all practical purposes. Nature itself would do the thinning out, and do it in a better way than we could hope to do, while the ground would be kept moist and in a state favorable to the decomposition of vegetable matter. It is desirable, therefore, to frame regulations regarding such forests, deciding the minimum to be preserved of the number of trees per acre, due regard being, of course, paid to age, species, altitude and locality. For these reasons it is highly important that all such forests, whether private property, commons, or belonging to the State, should be placed under the control of the State.

The different trees have naturally a different effect as regards conservation of water and production of food and shelter for fish, as I will here briefly point out. To simplify matters, we may divide all forest trees into two large groups, the deciduous and the evergreen trees. The deciduous trees, of which, so far as Great Britain is concerned, the oak, elm, beech, plane, larch, willow, and poplar, are the most prominent, have a decided advantage over the evergreens. I need not here enlarge upon the fact that the full shady foliage during summer is far more effective in preventing a large evaporation, and that the branches of the trees of this group are more spreading than those of the other. The energy of life seems to be far greater in these trees towards effecting our objects, and, for direct border trees to a water, they are undoubtedly the best suited. The great amount of foliage and branchlets yearly thrown by these trees forms a prominent factor in the economy of nature, and their decaying vegetation is full of teeming life and food for fish.

That this group is eminently suited for water conservation, was illustrated in a forest in Denmark, where an area of firs and pines was cultivated with beech and oak. After a lapse of about fifteen years, a mill stream, which, during the time of the evergreen trees, had dwindled down considerably, assumed such proportions that the irrigation of a considerable area was affected by it, besides supplying the mill with an abundance of water. As regards the evergreen trees, the first cultivation of barren ranges on high plateaus might advantageously be undertaken with these, on account of their ability to resist the severity of the climate in those exposed localities, and to grow on stony and poor soil. But, even on rocky ground and in high altitudes, the larches, birches, and other deciduous trees, will often do well and serve better for the end which we have in view, the water storage and the pisciculture.

In such localities, where only the most hardy trees can be reared, it would be practicable to cultivate along the watercourses, in the valleys and ravines, or any lower ground, a few rows of deciduous trees as soon as the other trees had attained sufficient height to protect them from the storms and the frosts. Several objects may be gained by doing so. First, the shade, shelter, and other beneficial effects for the fishes ; secondly, that more valuable timber could be reared, as these trees have, as a rule, a greater preference for damp and moist localities than the evergreens; and thirdly, because the deciduous trees permit more freely a luxuriant undergrowth of shrubs and annuals.

All fresh-water fishermen will agree with me in the advantage of having a good growth of annuals as watercress, nettles, etc., near the bank, and have observed that during feeding time the fish always seek such places. There is a vast variety of shrubs and annuals that might easily and with great advantage be introduced and grown on the river banks, but it would be outside the bounds of this paper to enter fully on the theme. However, I may only mention that many fodder plants and grasses from other countries might be a source of wealth to the population, and greatly benefit the fish as well as the owners of the land, if cultivated on the banks.

The Prickly Comfrey, e. g. (Symphytum asperrimum) which yields such a splendid forage by its abundant foliage, and many others, are easily reared, both from seed and cuttings, and should do well in the low lands, while on the sandy beaches, near the outlet of rivers and creeks, the cabbage radish (Pringlæ antiscorbutica) would cover these barren and desolate places with vegetation, and furnish an object of merchandise by packing them for the use of fishermen and sailors in the Arctic regions. The plant, when cooked, is a good substitute for cabbage, and has a most wholesome effect on persons suffering from scorbutica.

By a judicious forest management, the land can be kept covered constantly and always in a state favorable to the purpose of storing the water, but it is important that both sides of the stream should be planted instead of cultivating twice the distance on one side. A great many American trees might well be introduced, as, for example, the Swamp Cypress (Taxodium distichum), a great tree yielding a fine-grained timber, hard and durable, and the Leverwood tree, Hop-hornbeam, Ironwood (Ostrya virginica), which, besides excellent timber, furnishes a relished forage from its rich foliage: these, and a great many more might have a good effect on the river fishing, besides other advantages But it is particularly the willows to which our attention should be drawn. The preference which these trees have for water, and particularly for running water, is well known, and points directly to the practicability of placing them in those localities so well suited for them. The fish like willows, and I have oftentimes in Australia seen the best fishing places close to where some weeping willows (Salix babylonica) had taken the place of the indigenous and even more shady wattles (Acacius).

The yearly consumption of osiers in England is far greater than the national supply, and as the basket industry is constantly on the increase, it would also on this account be advisable to further the cultivation of the osier willows. For light, sandy banks, the best willow should be *Salix purpurea*, and as it is so easily propagated, it will well repay the cost of cultivation, besides binding the banks, making them firm and adding to the health of the locality as well as that of the water. For more clayey soil, *S. viminalis* and the more celebrated *S. caprea*, so much sought for powder factories, should be the best. The cuttings must be taken from the one to two-year-old shoots, and be put 1 to 1½ foot apart, in double or treble rows 2 to 3 feet apart, care being taken to leave only half an inch or less above ground.

There are many localities where comparatively valueless land, close to the mouths of rivers and canals, might be made highly profitable, at the same time as the cultivation of it with the before mentioned trees and plants would improve the state of the fishing, and, before placing spawn and fish in any water, I consider it important to pay great attention to this question. Where few or no trees exist it will be necessary to cultivate them, and I feel certain that such proceeding will enhance the chances of the success of pisciculture. I will not here enter further upon the practical details of the question. These are bound to vary with the locality, and the local foresters will know how to deal with them.

In drawing the attention of the conference to this question, it is with the sincere hope that it may enlist your sympathy, and that the public opinion may be won for it. That it is important for all fresh-water fishing is evident. That is one more reason added to the many why we should regard the forest as a precious heirloom to be deeply revered, properly used, and, through careful maintenance, descend improved and enriched to posterity.

SYSTEMATIC MANAGEMENT OF FORESTS.*

THE MODEL FOREST.

Imagine a uniformly productive tract, divided into any number (*n*) of divisions, or compartments of equal area; the first stocked with trees one year old, the second with trees two years old, and so on in an ascending series up to the *nth* compartment stocked with trees *n* years old. And let the revolution or age at which the trees of any compartment are to be cut, be *n* years. The land will then be parcelled out into a number of compartments equal to the number of years in the revolution and each one will be stocked with trees one year older than those of a compartment immediately proceeding it in age, so that there will be a complete series of groups of all ages from one to *n* years old If, now, all trees *n* years old, that is those in the *nth* compartment, be cut, and the land immediately restocked with young growth, it is evident that, at the end of twelve months, the group of trees next in order of age, or *n* minus one year at the time of the first cutting, will have advanced to maturity, while the plants on the first coupe will have taken the place of the youngest group in the series, and the plants of all intermediate compartments have advanced one year in age. At the expiration of twelve months from the time of the first cutting, we may therefore again cut a group *n* years old, and so on forever, cutting a group *n* years old once a year without demolishing the standing stock.

The yearly produce thus obtained is, in fact, the annual growth, or interest, of the material standing on *n* compartments, and is called the sustained yield, and a forest so organized is called a *model*, or ideal forest, because it represents a state of things which is theoretically perfect, if never quite attainable in practice.

If, in the case just considered, we were to cut more than the sustained yield in any year, we would be trenching on the capital stock and unable to maintain an unvarying yield. If on the other hand, we were to cut less, we would not be working up to the full capability of the forest and would have a certain amount of capital, in the form of trees, lying idle, and for the time being unremunerative.

A forest may, therefore, be regarded in the light of a capital producing by its yearly growth a certain interest in wood, just as a sum of money which is lent out produces interest; and, in estimating the growth of a forest viewed as a productive money capital the rate is calculated in precisely the same way as in ordinary money transactions.

Trees of about the same age and height, growing together in a mass, or trees growing in a sub-compartment are called a *group*. A compartment may contain one or more groups; if more than one, the area occupied by each group is called a sub-compartment. The group is the smallest unit of mass, and the sub-compartment is the smallest of area, in regular forests.

THE REVOLUTION.

The term *revolution* is used to donate the period of years which is being *fixed* to elapse from the time of the production of a tree, or group, to the time of its being cut down. It does not necessarily correspond to the age at which a tress is harvested, because trees sometimes have to be cut, or fall from natural causes, before the revolution fixed upon is completed.

---

* Macgregor ; Organization and Valuation of Forests.

The length of the revolution may depend on many things; such as the kind of tree, and the method of regeneration to be followed—subjects which are fully examined in books on sylviculture—and the special objects of the proprietor.

The principal objects of the latter may be classed as follows:—

To obtain from the land the largest possible average annual return, (1) of material, (2) of money, (3) of interest on his capital invested; or, to adopt the revolution best suited to (4) natural regeneration, or some (5) special, technical purpose. Revolutions fixed with a view to make such special requirements are called, respectively:—The revolution of the largest mean yearly yield, (1) in wood and (2) money, (3) the financial revolution, (4) the physical, and (5) the technical.

CHOICE OF A REVOLUTION.

For private owners there can be no doubt as to the most favorable revolution—the financial. But when it is a question of forests belonging to the State, it is frequently urged that cost what it may, it is the duty of a government to provide for all possible requirements of the community, and to prevent a diminution of the supply of any kind of material. No doubt a good deal may be said in favor of this view. In the first place, it is undeniable that forests that can be cut down any day may take years or even centuries to replace, and that it would never do to rely on private enterprise for the supply for the largest timber, more particularly as it seldom pays to grow it. Again, experience teaches that private individuals cannot be relied upon to provide even small timber, or fire wood, which *does* pay; the temptation to exceed the capability of the forest, or to convert all the standing stock into gold, whenever money is required by the proprietor, is irresistible, and not to be restrained by other people's ideas of moral obligations to themselves and posterity.

Now, without denying that circumstances (as in the case of protective forests) are conceivable which would render it advisable for a State to keep a forest standing after it had reached financial maturity, advocates of the financial revolution may reply as follows:—As a general rule, it is the business of a government to make the most of the property entrusted to its charge, rather than to anticipate and provide for highly improbable contingencies which, if they ever did threaten to arise, would certainly not in these days take everybody by surprise.

The government timber forests of all civilized countries are of vast extent, Spain perhaps alone excepted. They are are all systematically managed, or in a fair way to be so, and could not therefore, be swept away as if by magic, nor the standing stock suddenly reduced to a great extent, because that would involve the sale of largely increased quantities of wood, which could not be quickly disposed of without greatly depreciating its value. In a well-regulated forest, therefore, the financial revolution would act as a self-adjusting measure of the requirements of the people, and act as a regulator of the supply in sympathy with their most pressing wants.

NATURAL REGENERATION OF WOODS.*

In forests naturally regenerated by seed, the mother trees are only gradually removed, and several cuttings go on at once. In every rational method of working a forest, reproduction ought to be the result of the cuttings themselves.

---

*Bagneris: Elements of Sylviculture.

This is one of the essential objects of the science and art of sylviculture. Thus in the different kinds of high forest, reproduction is obtained from seed shed by the trees under conditions favorable to germination, while in coppices it is obtained just as naturally, by means of the shoots principally and secondarily by means of the seeds furnished by the standards.

But whatever the precautions taken, in both descriptions of forest there are often spots where seedlings do not come up, or where stools die and leave blanks. At other times it may happen that the reserve does not contain a sufficient proportion of a given species, a mixture of which is necessary, or that this species has disappeared owing to indiscreet operations or the total absence of all operations. In each of these different cases recourse must be had to artificial means in order to restore the good condition of the forest or a satisfactory composition of the crops. But such means ought to be the exception not the rule. It cannot become general and take the place of natural methods. To abandon natural reproduction is only to retrograde, to return to the infancy of the art; it is tantamount to claiming to supersede the forces of nature; above all it is simply wasting money under the false idea of economy, only to arrive in the end at results which are at the best doubtful.

Nevertheless, artificial restocking cannot be totally proscribed. It forms the necessary complement of natural regeneration, but it must remain only its complement. Hence it is necessary for the forester to know how to do it well. Besides this, it is the only method of stocking extensive treeless wastes.

REGENERATION BY SEED.[*]

Regeneration by seed is applicable to all species; that by shoots and suckers applies only to broad-leaved species; since the power of reproduction of conifers by shoots is either absent altogether, or so feeble that it is useless for sylvicultural purposes.

Under natural regeneration by seed is understood the formation of a new wood by the natural fall of seed, which germinates and develops into a crop of seedlings. The trees which yield the seed are called the *mother trees;* they may either stand on the area which is to be restocked, or on adjoining ground. A distinction is made between—

(1) Natural regeneration under shelter-woods;

(2) Natural regeneration from adjoining woods.

In natural regeneration under shelter-woods the area is stocked with seed-bearing trees, and the new generation springs up under their shelter; for some time at any rate, the area bears the new crop and part of the old one.

The system is that which occurs in primeval forests. When a tree falls from old age, or other cause and an opening is thus formed in the cover overhead, the seed falling from the adjoining trees germinates and develops into seedlings; these grow up under the shelter of the older trees, until they in their turn become mother and shelter trees. In this manner primeval forest, if undisturbed, goes on on regenerating itself for generations. The process is a slow one, as the young crop will only develop when sufficient light is admitted by the fall or death of the old trees. In sylviculture it is accelerated by the artificial removal of a

* Schlich : A Manual of Forestry.

portion of the old trees, when they have become fit for economic purposes. By degrees, modifications have been introduced which lead to a number of distinct methods :—

(1) *The selection system.*

(2) *The group system.*

(3) *The compartment system.*

(4) *The strip system.*

In each of these there are certain general conditions of success which hold good for all.

Under the selection system, regeneration goes on in all parts of the forest by the removal of the oldest, largest, diseased or defective trees, wherever they are found. No part of the forest is ever at rest; advantage is taken of all seed years for the restocking of small holes cut into the cover here and there by the removal of one or a few trees. Of the large quantities of seed which fall annually or periodically to the ground, only a small portion finds conditions favorable for the development of young trees; the latter are found chiefly in those parts where old trees are standing, or where the cover has been interrupted. Here little groups of seedlings spring up, which must be assisted by cuttings either final or intermediate, to afford them the necessary light.

Choice Between Direct Sowing and Planting.

Formerly the artificial formation of woods was chiefly effected by direct sowing, planting being restricted to special cases where the other method was not likely to succeed. The reasons for this were that sowing was considered to be more certain, cheaper, and that it was generally the custom to use too large transplants. In the course of time the raising of plants was elaborated, smaller plants were used, and the expense considerably reduced, so that now far more planting than direct sowing is done.

Yet it is not always a foregone conclusion that planting is better or more suitable than direct sowing, since many different conditions and factors affect the ultimate results. The effect of some of these factors is as yet somewhat obscure, but in many respects experience has taught the forester which of the two methods is preferable under a given set of conditions.

Sowing and planting are costly. The outlay on the latter can, however, be considerably reduced by planting small plants according to a simple and cheap method.

Where artificial regeneration follows clear cutting, the young plants are exposed to damage by frosts, drought, insects and weeds in a far higher degree than if the regeneration is conducted under a shelter-wood. In fact, tender species must be raised in the latter way, so that for them clear-cutting is excluded. Insects frequently become formidable to coniferous woods raised in clear-cuttings, while experience has shown them to be less dangerous to natural seedlings, especially when these are raised under a shelter-wood.

In the case of clear-cuttings, the laying bare of the ground for a series of years may seriously affect the fertility of the soil, so much so that the method is hardly admissible on inferior soils.

Natural regeneration involves less expenditure than sowing or planting. In some cases the outlay may be absolutely *nil*, but in most cases some artificial help has to be given either by working (wounding) the soil, or by sowing and planting. Still the outlay is considerably smaller. It must, however, not be overlooked that in the majority of cases natural regeneration requires much time ; as long as the shelter trees increase sufficiently in size and quality so as to make up for any loss on this account no harm is done, but where this is not the case artificial regeneration may be actually more profitable.

Damage by frost, drought, and weed growth is avoided, or at any rate considerably reduced. The same may be said as regards damage by insects, though perhaps not to an equal extent.

SUMMING UP.

Neither the artificial nor the natural method of regeneration is the best at all times and under any circumstances; only a consideration of the local conditions can lead to a sound decision as to which is preferable in a given case. In forming such a decision the forester must chiefly take the following points into consideration :—

(*a*.) General objects of management.

(*b*.) Species to be grown.

(*c*.) Condition of locality.

(*d*.) Available funds.

(*e*.) Skill and capacity of the staff.

LABOR REQUIRED IN FORESTRY.

Forests require labor in a great variety of ways, which may be brought under the following three headings :—

(1) General administration, creation, tending, harvesting, etc., or work done in the forest.

(2) Transport of produce.

(3) Industries which depend on forests for their prime material.

(1) *General Administration.* The quantity of labor required in the forests differs considerably according to circumstances, the value of the produce, and the consequent degree of the minuteness of the system of management. Great difficulty is experienced in obtaining accurate statistics on this point, but five days' work annually for every acre of land under forest may be accepted as an approximate estimate all round. From the available data it has been calculated that in the forests of Germany about $39,000,000 are paid annually for administration, creation, preservation, road making, cutting of wood, and collection of minor forest produce, on which about 200,000 families exist, or about 1,000,000 people. This estimate refers to forests which are already in existence, and in which fencing is done only in very rare instances. When new forests are created, additional labor is required at the outset. Nevertheless it is beyond doubt that forests require considerably less labour than land under field crops.

(2) *Transport of produce.* Owing to the bulky nature of forest produce its transport forms a business of considerable magnitude. Timber and firewood are carried by water wherever practicable, but also extensively overland. Under this head the sum of at least $19,480,000 is paid annually in Germany.

(3) *Forest industries.* The labour which is required to work up the raw material yielded by forests is of a much greater extent than that employed in managing the forests and in transport. There are the workmen employed in

saw-mills, building, ship-building, carpentry, coach-building, engineering, turning, carving, paper pulp manufacture, match-making, the manufacture of cases, and boxes, round and square, from the largest packing case to the smallest toy box, frames of sieves, drums and cask hoops, wooden-ware for table covers, blinds, pencils, wooden nails, instruments, tools, plates, shovels, spoons, shoes, lasts, saddle-trees, brushes, harrows, and gunstocks, toys of thousands of patterns, and endless other branches of industry, some of which can only exist in and around forests.

The wages earned under this head amount in Germany to something like $146,100,000 a year, maintaining 600,000 families or 3,000,000 people.

Taking now the three heads of labour together, it has been estimated that something like 12 per cent. of the population of Germany is employed in forest work, transport of forest produce, and the working up of the raw material yielded by the forests. An important feature of the work connected with the forests and their produce is, that a greater part of it can be made to fit in with the requirements of agriculture; that is to say, that it can be done when field crops do not require attention. Hence forest work offers an excellent opportunity to the rural labourer or small farmer of earning some money when he has nothing else to do, and when he would probably sit idle, if no forest work were obtainable.

Organization of the Personnel.[*]

This will depend in a great measure on the extent of the forest concerned. It is evident that the degree of division of labour which is possible in the management of forests comprising a million acres could not be applied with advantage to an estate of a thousand acres, and that private individuals will seldom be in a position to adopt the elaborate systems followed in the State Forests of European countries.

The following plan is that usually adopted for the management of forests of large extent, such as those of most European countries.

The establishment consists of an inferior and a superior branch. The former consists of (1) guards and (2) rangers.

(1) *Guards or Under-Foresters.*—The duty of these is, as the name implies in the first place, protective. But, besides this, they are employed in the executive work of their beats, as, for instance, in supervising work of regeneration and felling.

(2) *Rangers*, or, range-foresters, who have immediate charge of the executive work of a *range*, and are responsible for its proper conduct to the assistant conservator.

The superior branch consists of (1) Assistant-Conservators, (2) Deputy-Conservators, (3) Conservators, and, in certain cases, of (4) an Inspector-General.

(1) *Assistant-Conservators.*—An assistant-conservator has charge of several ranges, called collectively, a *sub-division*. Besides the general management of the work of the sub-division, the accounts of each range are audited, and have to be passed by him before payment is made.

(2) *Deputy-Conservators.*—A Deputy-Conservator has charge of several sub-divisions, called collectively, a division. His duty is purely to control, and he does not, as a rule, interfere with the executive work of the Assistant-Conserva-

---

[*] Macgregor ; Organization and Valuation of Forests.

tors; but it is his business to see that the general provisions of the sanctioned working schemes and yearly budget of his division are properly carried out, and to audit and pass the accounts of the sub-divisional officers.

(3) *Conservators.*—A Conservator has general control of several divisions, collectively called a *circle*, comprising all the forests of the State, or, if they are very extensive, of a Province only. He is the immediate adviser of government in all forest matters concerning his circle ; holds in fact, in this respect, much the same position as an under-secretary of State, and usually has his headquarters at the seat of Government.

(4) *Inspectors-General.*—An Inspector-General stands in the same relation to a supreme government as a conservator to its local government, and exercises a general supervision over the whole system of a country.

It will be observed that by this system the administration is divided into an executive and a controlling branch, the former consisting of Assistant-Conservators and their subordinates, and the latter of Deputy-Conservators and officers of superior rank.

Members of the inferior establishment do not, as a rule, rise higher in the service. A much lower standard of technical an l general education is demanded from them than from the members of the superior branch, and they are, therefore, generally unfitted for the higher appointments.

The size of ranges, sub-divisions, divisions, or circles, depends on local circumstances, such as the degree of intensiveness of the working, compactness of the forest area, mode of treatment and means of communication. It is, for instance, evident that, other things being equal, a Deputy-Conservator could manage a larger division where there was railway communication than where there was none. It is equally obvious that a ranger could manage a much larger forest worked by the method of equal areas, and solely with the view to producing firewood coppice, than a seedling forest worked by the combined method with a view to the production of large timber and naturally regenerated.

CHOICE OF AN ORGANIZER.

Should the sub-divisional officer who has been in immediate charge of the forest, perhaps for many years, be intrusted with a preparation of a plan, or should a special branch of the executive be employed, whose sole business is to prepare plans of management ?

In regard to this question, opinions are divided. Of course it is one which can only arise in regard to large tracts of forest belonging to one proprietor— the State for example. A small proprietor would not be able to keep a special staff fully employed.

It has been urged in favor of the local officials conducting the organization and revision of a forest, that he must know the special conditions far better than other people, and that he would take much more interest in the carrying out of his own programme than that of another.

On the other hand, it has been maintained that the special practical knowledge and skill necessary to organize a forest successfully cannot be acquired in the ordinary routine of an executive officer, who would probably not be called upon to carry out a work of this kind more than a few times during his whole career ; that by constant practice a special branch would attain the necessary proficiency ; that if the work is done by a small body of men, it is more likely to be uniformly carried out than by a number of different persons ; that the officer in charge is not the proper person to revise his own work ; that he will be always there to assist and advise the organizer.

A large majority of countries, including India, have adopted the system of having works of organization carried out by a separate branch of the service ; and some have gone still further and constituted a distinct survey branch as well as an assessment branch. As a rule the separation of these two departments is not desirable. Perhaps it conduces towards efficiency, if a part of the staff is exclusively employed in surveying and the other in assessment, but the work of the two is so intimately connected that it is expedient they should both be under one head.

The composition of the organization staff depends on special circumstances Sometimes a good plan is to have a board of senior officers, presided over by the principal officer. All organization schemes are submitted for the approval of, and have to be passed by, this board, the members of which carry on the work in addition to their ordinary controlling duties. Under the board is the working staff, which carries out the works of organization, and which is recruited by drafting men into it from the ordinary branch of the service after they have served a few years and become thoroughly acquainted with the working of a sub-division.

This system is only suitable for districts in which the headquarters of the controlling officers on the board are all in one place. Each member looks specially after the working of the plans in his own division, and generally conducts the revisions in person.

An important duty of the organization branch is to collect and work up statistics. The business of collecting statistics and drawing general inferences is best done by a central institution of this kind, and much useful work would often be lost without a trained staff, whose special duty is to work up details collected in different parts of the country ; the "Bavarian tables," which have proved so useful, not only in Bavaria, but throughout Germany, are a case in point ; they would probably never have been constructed if there had not been a central organization department at Munich.

Speaking generally, the bent of the argument appears to be in favor of having this kind of work done by a special branch ; but not always, as circumstances may without doubt arise which render the alternative course advisable, as, for instance, when the aggregate area of forests requiring to be organized is so great that their organization could not be accomplished within a reasonable period by a necessarily limited staff, or when the methods to be employed are so simple that their execution does not require any special skill.

FORESTRY IN THE COLONIES AND IN INDIA.*

A circular containing questions relating to colonial timber was addressed by the Secretary of State for the Colonies to the administrative heads of the various British possessions in 1874, from the replies to which it appears that in none of the six Provinces of the Dominion had measures been taken to secure the replanting of cleared areas, or the afforestation by natural reproduction, notwithstanding an enormous and growing consumption.

In the Province of Ontario more than 87½ per cent. of the timber annually cut was exported, and, looking to the magnitude of the timber exports, it was remarkable that so little had been done to prevent the threatened exhaustion of the chief article of trade in the Province.

In Nova Scotia the amount of timber annually cut was estimated to exceed by 25 per cent. the amount which could be cut each year without permanent injury to the forests, while in Prince Edward Island the amount annually cut exceeded nearly seventeen times the quantity which would represent a prudent rate of consumption.

The timber resources of British Columbia were declared by local authorities to be practically inexhaustible, but it is probable that, should the whole strain of the demand be thrown upon British Columbia, a few years would make a very perceptible inroad upon the stock of native timber situated in accessible districts of the Province.

The importance of this trade to the commercial prosperity of the Dominion will be exemplified by the following table, compiled from materials contained in returns issued by the Board of Trade.

Comparative tables of money values of timber and corn (grain) exported to the United Kingdom during five years ending 1876.

Articles.	Value.					
	1872	1873	1874	1875	1876	Total.
	$	$	$	$	$	$
Timber and wood	4,218,661	5,220,296	5,706,567	4,205,045	5,282,657	24,633,226
Corn and grain......................	3,003,104	3,898,204	3,697,616	3,124,056	2,814,003	16,536,983

Timber and corn (grain) are the chief exports of the Provinces of the Dominion, but the value of the timber exports exceeded the value of the corn exports by more than one-third, and constituted nearly one-half of the total value of all the exports from the Dominion to the United Kingdom.

The returns exhibit, in a striking manner, the urgent need for some prompt and comprehensive action to stay the influences at work to destroy the indigenous forests, which constitute, in many instances, the principal natural riches of the colonies. There is a tendency in newly-settled countries to regard the timber as

* Schlich : In Proceedings of Royal Colonial Institute, vol. xxi., 1889-90.

48

a mere encumbrance to the land, and the finest timber is that first selected for destruction by fire, by ring barking, and other rude and wasteful methods in favour with settlers.

It is probably not possible in newly-settled colonies to put restraints upon the clearing of the most fertile soils, although it would seem to be advisable to leave belts for protection against the winds, and to enact that all the hills should be preserved in perpetual forest to protect the sources of the springs.

In many cases the reports of surveyors-general and other officials demonstrated the possibility of preserving, and even of restoring the forests, by the constitution of a small but energetic forestry department, but nothing worthy of notice had, up to the date of these returns, been done in the nature of forest conservation.

NEW ZEALAND.

The subject of forest conservation appears first to have engaged the attention of the Colonial Legislature in October, 1868, when a motion was made and agreed to that "steps be taken to ascertain the present condition of the forests of the colony."

In the course of a parliamentary debate in 1873, it was remarked, with reference to the Kauri wood, that extensive districts which were once covered with that wood were then totally destitute of it, and that its extermination progressed from year to year at such a rate that its final extinction was as certain as that of the natives of New Zealand. Another speaker maintained that "unless great care was taken, there would not be a Kauri tree in the colony in the next generation.*

As the result of the agitation of this question, an act was passed by the colonial Legislature in August, 1874, entitled " An Act to provide for the establishment of State forests, and for the application of the revenues derivable therefrom." The preamble recites that "it is expedient to make provision for preserving the soil and climate by tree planting, for providing timber for future industrial purposes, for subjecting some portion of the native forest to skilled management and proper control, and for these purposes constitute State forests."

The Act provides that an annual sum of £10,000 for thirty years is to be paid quarterly out of the Consolidated Fund into a special fund, to be called the "State Forests Account," and all receipts from State forests are to be paid into this account. The money is to be expended in managing and planting State forests and nurseries, and the establishment of schools for instruction in forestry. The department is placed under the supreme control of a minister of the Crown, who is to be assisted by a "conservator" and subordinate officers. Lands may, from time to time, be set apart as State forests on the recommendation of the superintendent or of the Provincial Council of any Province. Power is taken to set aside pastoral leases or licenses over lands so selected.

The Governor-in-Council may make, alter, and repeal by-laws and regulations—

Prescribing the duties of officers.

To regulate the form and issue of licenses.

*Kauri, *Dammara Australis*. A Conifer, the largest and most valuable tree in New Zealand. Attains a height of 120 feet and diameter of 10 feet to 15 feet Grows in Province of Aukland only. Exudes large quantities of resin, known as Kauri gum. Weight, 38 lb. to 41 lb per cubic foot; grows on clay soils. The above remarks as to the " Kauri " will apply to our most valuable hard-wood tree (the Black Walnut).

To control the management of the forests.

To determine the seasons for the cutting and removal of timber and bark.

To prevent waste and unnecessary destruction.

To prevent the danger and spread of fire.

To prohibit trespass and regulate access.

For constructing roads and tramways in the forests and charging of tolls.

The Act also provides for the punishment of offenders and for the application of money recoverable as penalties.

As a practical and comprehensive experiment in the direction of forest conservancy, the results were looked forward to with interest.

AUSTRALIA.

Australia proper consists of the colonies of New South Wales, Victoria, Queensland, South Australia, and Western Australia. The causes which determine the climate of Australia are remarkable in many ways. In the first place the northern parts of the country are situated in a tropical, and the southern parts in a temperate latitude. Secondly, between the two stretches the enormous central plain is daily heated in summer to a very high degree, the air expands, is lifted, and flows away on all sides, causing an indraught of moist sea air. This is forced to rise on reaching the high coast lands, which it moistens in various degrees. Owing, however, to the great distance from the shore to the centre of the country, the latter profits only at regular intervals by this, because the indraught is regularly stopped by the nightly radiation of the heat absorbed during the day, or the clouds are once more converted into vapour owing to the high temperature of the air.

Such is the heat of the interior during the summer that the air, if it moves at all, feels like a furnace blast. Sometimes, however, sufficient masses of clouds succeed in passing over the coast ranges, and, in such cases, floods of rain fall upon the inland country. The distribution of the rain differs considerably. The north coast has the advantage that the air drawn in from that side comes from the equatorial regions, the great reservoir of moisture.

Then the hills on the east coast are comparatively high, those on the west coast are lower, and along a portion of the south there are no mountain ranges at all. Thus it happens that the rainfall at the head of Spencer's Gulf is only 6 to 8 inches; at Adelaide, 20; Melbourne, 26; Portland, 32; Sydney, 48; Newcastle, 44; Brisbane, 49; and at Rockingham Bay, something like 90.

In every part, however, the rainfall decreases rapidly in passing inland, so that comparatively little falls on the inner slopes of the coast ranges.

The temperature depends on the situation and the rainfall. The northern part of the continent is tropical. Brisbane has a mean annual temperature of 6 degrees, Fahr.; Sydney, 63 degrees; Melbourne, 57 degrees; and Adelaide, 65 degrees.

The mean temperature in the interior is much higher than along the shore; it is said to rise as high as 130 degrees in the shade during summer.

South Australia was perhaps first in the field to introduce a separate forest law.

4 (F.)

In Victoria a new Land Act was passed in 1884, which provides, amongst others, for the following matters:

(1) The formation of State forests.

(2) The formation of timber reserves.

(3) The management of both.

4) The management and disposal of timber and other forest produce on the unalienated Crown lands not included in the State forests and timber reserves.

Under this Act the State forests can only be alienated with the consent of the Governor-in-Council. The timber reserves shall not be alienated in the first instance, but as the several parts become denuded of timber, they may be added to the pastoral or agricultural lands—in other words, thrown open to selection. The timber reserves are, therefore, only temporary reserves.

The forests generally are worked under the license system, regulated by rules made under the Act. There are licenses for felling, splitting, clearing under-growth, the erection of saw-mills, grazing, removal of wattle bark, etc. For each of these licenses certain fees are paid. Penalties are provided for breaches of the law, or any regulations issued under it.

The question is whether, and in how far, effect has been given to the policy which is indicated in the Act. Mr. Vincent, an expert and a trained forest officer of known ability, who served in the Indian Forest Departments since 1873, gives the following description of forest management in a report to the Governor of the colony, as existing in 1887.

The area of State forests and timber reserves then stood as follows :—

State forests ....664,710 acres,
Timber reserves...... ....690,732 "

Total......... ....1,355,442 acres.

Equal to 2,118 square miles, or about 2 per cent. of the area of the colony.

Mr. Vincent visited a number of the State forests, timber reserves, and other forest lands, and he draws a rather gloomy picture of their condition.

This is what he says, for instance, about the Wombat and Bullarook forest (area, 105,000 acres):—"This is said to have been originally a magnificent forest, chiefly of messmate or stringy bark, the timber being of the very best class—enormous quantities have been sent away to Melbourne, Sandhurst, and Ballarat—there were thirty-six saw-mills at work in 1884—the splitters have cut more timber than even the saw-millers—the good timber is now almost all worked out, except in certain localities in the southern half of the forest. In the portion which I visited there are only second-class trees, with a certain number of bigger ones, which have been left for some fault. There has been little or no repro-duction, the whole of the young trees have been burnt, and there are no middle-aged ones coming on to yield timber some twenty or forty years hence.

" The useless waste and destruction that have been going on in this forest for the past thirty years defy all description. The saw-mill fellers and the splitters have been allowed to go in and cut when and what they chose. Gener-ally the fellers took one log out of each tree, leaving the rest, which, although not quite so good as the butt-end log, still consisted of first-class timber. The splitters, as often as not, left trees to rot where they had fallen, without even taking out one log, on finding that the wood did not split well. Even if they did split, at least three-fifths of the timber in the trees was wasted. Subsequently, when the wood thus left on the ground was fired, a fierce blaze occurred, which killed or rendered useless almost as many trees as had been felled. The selection of the State forests has not been well made here, for some of the best forests have been left outside, and inferior growth taken up for the reserve.

" As a large increase in the consumption may be safely anticipated, taking into account the natural increase in the population, the present rapid extension of quartz mining, and the decrease of timber on private lands, there is likely to be a great scarcity of timber in the next ten or fifteen years. Already the mining community complain of the great increase in the price of firewood and timber, and the neglect which the large area of Crown lands in the vicinity of the mines receive. On some mines firewood costs now 30 to 40 per cent. more than it did five years ago, and there is a universal complaint that the timber now supplied for props, laths, etc., is very inferior and immature."

Mr. Vincent then sums up as follows :—" The immediate causes of this are the bad license system, the ill-arranged classification of State forests, timber reserves and Crown lands, the absence of professional foresters to direct operations, and the neglect to reserve the best natural forests. The officials in charge of the forests have often protested against the present license system, explaining that the forests were being rapidly ruined. They explain that they cannot protect the forests from theft, and yet no change is made. Why ? Because Parliamentary influence is brought to bear by the saw-mill owners and by the splitters, who are determined that no change shall be made in the present arrangements. Both these classes are powerful, the splitters especially. When an attempt is made by the foresters or the Secretary of Agriculture to do justice to the forests and to protect them, the persons affected organize deputations, questions are asked in Parliament, and concession after concession is made. There is little hope of the forests ever receiving proper treatment until the forest question is made a national one, and removed from the arena of party politics. The question is, are the electors prepared to allow the saw-millers and splitters to devastate the remaining forests, robbing them and their children of their supply of timber and firewood, and risking some of the climatic changes which are traceable to the destruction of forests ? Are they prepared to sacrifice a source of large and increasing revenue to the demands of a limited class ?"

It was suggested that the Victorian Government should secure the services of a fully competent forest expert, a man like those who introduced systematic forestry into India, who should be directed to go round the colony, see for himself, and then propose what, in his opinion, ought to be done. After all the passing of fine laws is not such a difficult thing. What is of much greater importance is the determination to carry the law into effect when once passed.

Under any circumstances the Government of Victoria should not fall a victim to the delusion that the formation of some limited plantations will make up for the loss of the natural forests. The all-important step to be taken is to gazette and demarcate on the ground a sufficient area of reserved State forests, and to provide for their systematic management, according to the approved rules of scientific forestry, and, in addition, to take what measures are desirable and practicable for the protection of the forest growth on the Crown lands, which are not included in the reserve State forests.

The following short abstract indicates what seemed to be required :—

(1) Engagement of a thoroughly competent forest expert to be the head of the Victorian Forest Department.

(2) Selection, demarcation, and legal formation of a sufficient area of reserved State forests, suitably distributed over the country, systematically managed and efficiently protected.

(3) Protection and disposal of forest produce on Crown lands not included in the reserved State forests.

If the Government makes up its mind to do this, all the details will settle themselves easily enough.

INDIA.

India has to provide an enormous population of 255,000,000 people with timber and firewood, and, apart from a certain amount of teak and fancy woods, that country can probably do little towards an increased export of timber.

There are certain reasons why State interference is more called for in the case of forestry than in most other branches of industry. Most of our valuable timber trees require long periods of time to ripen. Large-sized oak trees are from one hundred to two hundred, and even more, years old. The teak, which comes to England from India, is derived from trees which are on an average at least 150 years old. If forests are to yield a regular annual return of timber they require to have trees of all ages, and consequently a considerable accumulation of material, which has been produced in the course of a long period of time. To maintain the forests in that condition only a quantity equal to that which grows annually should be removed, and no more. If more is removed a reduction of the producing capital must ensue. As long as the estates are in the hands of private parties, they are at all times liable to be overworked, that is to say, more than the annual increment is taken out; and it is easy to see that in a comparatively short time the forests must cease to yield timber. Experience has proved over and over again that this is generally the result. If we are to make over to our children the forests in an unimpaired condition they must be treated in a systematic manner, and this can, as a rule, only be achieved for any length of time by State interference. But the mere theory of such is by no means sufficient. Nominal interference on the part of the State is the most disastrous of all. In that case the forests are looked at as common property, and everybody tries to get the most out of them and into his own pocket, the result being that they disappear faster than ever,

If the State, as such, has arrived at the conclusion that the maintenance under forest of a certain proportion of the area is essential or desirable, it must also, once for all, decide to do what is necessary to secure that area, and to see that it is managed in a systematic and orderly manner. There are various ways of doing this. Either the State establishes State forests by setting aside certain areas at its disposal for forest purposes, or it passes laws which empower it to supervise the management of communal and even private forests. The former alternative is much the best wherever it can be adopted, and this is the case in India and in most of the Colonies.

Practically, only India has really and honestly dealt with the forest question. Some of the Colonies are fairly in earnest, but too many have restricted their action to nominal measures.

India is situated between the 8th and 35th degrees of northern latitude, hence the southern half of it lies within the tropic. Its length, as well as its greatest breadth, is about 1,900 miles, leaving out of consideration the newly-acquired territory of Upper Burma. The area and population stand as follows :—

	Area in square miles.	Population. Total.	Per square mile.
British Territory without Upper Burma....................	912,000	202,000,000	221
Native States........	531.000	53,000,000	96
Total.......	1,468,000	255,000,000	170

The physical configuration is very peculiar. The country consists of three great sections :—

(1) The Himalayas.

(2) The Indo-Gangetic Plain.

(3) The Peninsula.

The Himalayan ranges stand out like a high wall on the north, separating India from the Thibetan high plateau. The great Indo-Gangetic plain runs along the southern edge of the Himalayas from Sind in the west to the Bay of Bengal in the east. To the south of this plain, and partly surrounded by it, lies the Indian peninsula, forming another plateau of moderate elevation. The contrasts of elevation which occur in these territories are greater than those in any other part of the globe. While the Himalayas reach a height of 29,000 feet, the plain of Hindustan, at the foot of the hills, rises only a few hundred feet above sea level; further south elevation increases again, since the peninsula shows a height ranging between two thousand and eight thousand feet.

Another peculiar fact is that India receives the drainage of both slopes of the Himalayas, which ultimately collects into the three great rivers, the Indus, Bramaputra, and Ganges. The first two rise in close proximity to each other at the back of the Himalayas; one runs towards the west and the other towards the east, until both break through the Himalayas—the former running through the Punjab and Sind to the Arabian Sea, and the latter through Assam and Lower Bengal to the Bay of Bengal. The Ganges drains the greater part of the south face of the Himalayas, finding its way, after uniting with the Bramaputra, into the Bay of Bengal. The highest part of the peninsula is situated along its western edge, in consequence of which the greater part of the drainage from this part of the country goes in an eastern direction into the Bay of Bengal.

It will be easily understood that in a country like India many different climates are found. As a matter of fact, they range from the driest in Sind to the wettest along the west coast of the peninsula, in Assam, Eastern Bengal, and Burma; and again from the hottest to an arctic climate in the highest regions of the Himalayas. Of these various climates the following four types may here be mentioned as most characteristic :—

(1) The climate of tropical India : Showing the highest average temperature ; the early arrival of the monsoon rains mitigates the summer temperature ; there is little or no cool season.

(2) The climate of North-western India : Showing the highest summer temperature, though the average temperature of the year is lower than in the

former region ; there are four or five cool and even cold months during winter, when the climate resembles that of South Italy.

(3) The climate of North-eastern India : Here humidity reigns supreme ; the extremes of temperature in summer and winter are moderated by the effects of the relatively large quantities of moisture in the air.

(4) The climate of the Himalayas : It is, according to elevation, more or less temperate, and even arctic, with frost, snow, and bitter winds in winter, and a moderate heat in summer.

The rainfall depends in the first place on a very simple set of phenomena. The extensive plains and table lands of India are in spring and summer heated to a much higher degree than the surrounding sea, while during winter the air overlying the sea is warmer than that over the dry land—in other words, sea breezes prevail during summer and land breezes during winter.

In spring, which shall here comprise the months of March, April, and May, the highest temperature is found over the centre of the peninsula (Nagpur-Hyderabad), the difference being from five to ten degrees compared with the temperature at the sea coast on the east or west, or at the foot of the Himalayas. The air in the centre expands, lifts the higher layers, causes them to flow away on all sides, and produces a centre of comparatively low pressure. Into this centre presses the heavier atmosphere from the surrounding country, principally from the sea on the south, east, and west, and from the dry table land of Beluchistan and Afghanistan on the west and north-west. As a general rule, the moist sea breezes gain the upper hand and bring a rainfall, ranging from three to six inches during this period. The north-western breezes, on the other hand, are dry, and known as the hot winds of the Bombay Presidency, the north-western provinces, and Centre India. With the advance of the season the sea-winds become stronger and stronger, and the air is then drawn from the most distant equatorial region, the great reservoir of moist air ; they now cause a copious rainfall, known as the south-west monsoon. The amount of rain differs, however, very considerably according to the configuration of the country ; in other words, according to the degree to which the clouds in their forward passage are forced to rise or sink again, owing to a rise or fall of the surface.

As long as the sea-winds are sufficiently strong to keep in check and even force back the north-western winds, all is well for India ; but occasionally the reverse occurs, that is to say, the north-west winds force back the sea-winds and proceed far into the Indian plain and the peninsula. If this ascendancy continues for some time, the rains fail, and scarcity, or even famine, is the result.

In September the monsoon commences to decline, and by degrees north-easterly winds replace the south-western and southern breezes. They are dry, except in part of Madras, where they bring heavy rains until December, and are known as north-east monsoon winds. Local rains of moderate extent are caused during winter, more especially in the Punjab and North-western Himalayas.

The total annual rainfall ranges from 4 inches in some parts of Sind to more than 500 inches in the Khasia Hills, and all intermediate grades are duly represented.

A country which shows such extremes of climate must necessarily show a most varied vegetation. The actual distribution of the forests is principally governed by the rainfall. Where that is favorable, production is great, and the forests are dense ; where it is unfavorable, production proceeds at a slow rate. Again, the nature of the rainfall governs the character of the forests. Where the rains are heavy, the country is generally covered with evergreen forests ;

where it is less copious, the forests are deciduous ; under a still smaller rainfall they become sparse, and more dry, untill they gradually end in desert. Consequently, the evergreen forests are found along the moist west coast of the peninsula, in the coast districts of Burma, Chittagong, and along the foot and lower slopes of the eastern Himalayas. The deciduous forests occupy the greater part of the peninsula and Burma away from the coast. Dry forests are found in Rajputana, and the Punjab, while deserts are the principal feature of Sind.* With rising elevation in the hills, the forests become gradually temperate, and then Alpine, until they disappear altogether on approaching the lower limit of the eternal snow.

These details on the great variety of climates prevailing in India are given, because some idea on the subject is necessary so as to understand the forest policy, which is indicated in the case of that country. The main issues of that policy depend on the following three points :—

    (1) Forests in relation to climate and rainfall.

    (2) The regulation of moisture, and

    (3) Forest produce required by the country.

The south-west monsoon must for ever be the main source of moisture in India, and the climate and rainfall of the Indian plain, and of the peninsula, are generally subject to other influences, in comparison with which the effects of forests must always remain small. On this account then, afforestation cannot be pushed in the case of India. It must, however, be mentioned that the shade and shelter of forests will be most gratefully accepted by man and beast in a hot country like India.

In a tropical climate like that of India, the evaporation from an area exposed to the full effects of the sun, is probably not less than four times that from an area which is covered by a dense growth of forest vegetation ; hence afforestation is of great importance wherever the rainfall is limited, or unfavorably distributed over the several seasons of the year.

Then, there is irrigation to be considered. No less than 30,000,000 acres of land are artificially watered in India by means of canals, wells, lakes, and tanks. Only three million acres depend directly on the melted snow of the Himalayas, and it will easily be understood of what importance it is to keep the areas which provide the remainder of the water properly sheltered. The larger the proportion of the catchment areas, whence the irrigation water comes, is shaded by forest vegetation, the more favorable and sustained will be the supply of water. On this account, then, forestry in India has an important mission to fulfil.

The mechanical action of forests in regulating the flow of water from hillsides also is not without importance in India, and cases are by no means rare, which show the mischievous effect of reckless deforestation. In this respect, none is more instructive than the case of the hills behind Hushiarpur in the Punjab. These, consisting of a friable rock, were safe until, some forty years ago cattle graziers settled in them and destroyed the forest and other vegetation. Since then a process of erosion has set it, which is carrying by degrees the hills into the plains, where they appear as huge sand-drifts which have already covered enormous areas of fertile cultivated land, and even destroyed part of the town of Hushiarpur. Such an evil can be avoided by preserving the natural vegetation on the land, but, if once started, special measures are required to meet it. In the

---

*Sind has some very valuable forests, which are situated on the banks of the Indus on land more or less regularly inundated.

first place, grazing must be stopped, at any rate that of goats and sheep, so as to allow a natural growth of plants, shrubs, and trees to come up; artificial sowing and planting must be done, preceded in bad cases by the construction of dams and dykes to steady the soil, until vegetation has once more laid hold of it. Mischief of this kind can be stopped and cured at a comparatively small sacrifice, provided it is taken in hand at an early stage; but if it has been allowed to grow for a series of years, the expenses of checking the evil may be beyond the means of the State.

Although forests are of considerable importance in India in respect of their action as regards the regulation of moisture, they are absolutely indispensable on account of the produce which they yield, since by far the greater part of India must rely on the timber and fuel produced in the country, apart from other produce. All the teeming millions of India use wood for their domestic firing, or, if such is not available, drier cow-dung, the latter being much to be deprecated from an agricultural point of shall k. At the same time, enormous quantities of timber are required for construction, boat-building, tools, agricultural implements, railways and other public works. If we add thereto a demand for many important items of minor produce, more especially cattle fodder in the drier parts of the country, it will easily be understood that at least 20 per cent. of the total area requires to be kept under forest. Even such an area would give only about half an acre per head of population, an allowance below that of most European continental countries.

The history of forestry in India is very instructive. According to the available evidence the country was in former times covered with dense forests. Then settlers opened out the country along the fertile valleys, but the destruction of the forest on a larger scale was carried out by nomadic tribes, who fired alike hills and plains as they moved from one pasture to another. This process is believed to have gone on for more than 700 years. Subsequently came British rule, and with it a more fierce destruction of the forests than before. Extension of cultivation became the order of the day, and before its march many of the remaining woods fell under the axe, no inquiry being made as to the ultimate result. Simultaneously with the extension of cultivation and the increase of population, the annual requirements of timber and fuel increased, while quickly multiplying herds of cattle roam far and wide over the remaining forests. Finally, railways came, and with their extension the forest disappeared with greater rapidity than ever, partly on account of the increased demand for timber used in construction and firewood, and partly on account of the fresh impetus given to cultivation on both sides of the line. I have watched this last process, and I can testify from personal experience how fatal railway extension is to forests which are not subject to proper control and protection.

For some time matters went smoothly enough in India, but then the shoe commenced to pinch. Difficulty was experienced in meeting the demands of timber for public works, sleepers had to be imported from foreign countries, and it was then recognized that a great mistake had been made in allowing the forests to be recklessly destroyed. Experience had definitely proved that the preservation and suitable management of a sufficient area as forests could not be left to private enterprise, and that the interference of the State had become a necessity in the general interest of the country.

The forest question commenced to attract attention in the early part of this century, in consequence of which a timber agency was established on the west coast of the Peninsula.

Next we find, in the year 1843, Mr. Conolly, collector of Malabar, planting teak on a large scale at Milambur, Dr. Gibson was appointed conservator of forests in Bombay in 1847.

In 1848, Captain Frederick Conyers Cotton caused the appointment of Lieutenant James Michael (now Major-General J. Michael, C.S.I.) as Forest Officer in the Anamalais, which post he retained for seven years. Dr. H. Cleghorn became connected with forest conservancy in Mysore in 1847, and he was appointed Conservator of Forests in Madras in 1856. He was on special duty with the Government of India about the years 1860-62 when he inquired into the forest matters in the north-western Himalayas and elsewhere. In the Central Provinces Colonel Pearson was the first Conservator who took up forestry in a business-like manner.

These gentlemen and others were the pioneers of forest conservancy in India. Their action, though localized, caused the matter to be discussed and kept before the public, and it led ultimately to the organiztion of a general department by Dr. D. Brandis (now Sir Dietrich Brandis, K.C.I.E.) The latter was appointed superintendent of Forests in Pegu in 1856 by that great administrator, Lord Dalhousie. Dr. Brandis was principally instrumental in saving the Burma teak forests from destruction by enterprising timber merchants—that is to say—estates which yield now a gross revenue of some £250,000 a year. In 1862, he was attached to the Government of India, and in 1864 appointed the first Inspector-General of Forests to that Government. He then set to work to establish the Indian Forest Department, and to introduce a systematic management of the forests. At first he devoted himself to the Provinces directly under the government of India; subsequently he was twice deputed to Bombay, and he totally re-organized the forest department in Madras in 1881-83, immediately before his final retirement from India.

The first duty of the new department was to ascertain the extent and character of the remaining forests and especially of that portion which still belonged to the Government. This inquiry was not of special difficulty, except in so far as a sufficiently trained staff was not available at the outset.

The next step was to take the State forests under protection and management, and now difficulties arose. There were no doubt some administrative officers who soon preceived that it was to the true interest of the people to preserve a suitable forest area, and who cordially assisted the new department, but the majority of the officers of the State failed for a long time to accept that view, principally because the idea of forest preservation was new to them, and they feared complications from the facts that the rights of government in the forests were in many cases ill-defined, and that the people claimed extensive rights by prescription, and on other grounds, in the areas which were the property of the State.

The first Indian forest law was passed in 1865; it provided that the Government might declare any land belonging to it a Government State forest, and that such declaration should not abridge any right held by private persons over such areas; but the Act did not provide power to inquire into and legally settle the rights of third persons in the State forests. Under this Act considerable progress was made in the preservation of the forests, wherever the population was limited and the forest areas extensive.

But where the reverse conditions prevailed, and where the rights claimed by the people, rightly or wrongly, were extensive, the benefits of the Act soon threatened to become abortive. Consequently fresh legislation was soon contemplated, and after years of discussion, a new Act was passed known as the Indian

Forest Act of 1878, followed by special Acts for Burma, Madras, and one or two other Provinces. Of these, the Burma Act is the best. Generally speaking, the enactments give power to the Government :—

(1) To declare any area belonging to the State, or over which the State has rights, to be a State forest.

(2) To demarcate such area, and to enquire into and settle, once for all, the rights claimed by third persons in or over such area ; to commute such rights if they seriously interfere with the maintenance of such forests ; and to prevent the springing up of new rights except by a Government grant.

(3) To provide for the proper protection and management of the State forests.

(4) To provide for the protection and management of Government forests not included in the reserved State forests.

(5) To provide for the preservation of private forests, which are of special importance to the community as a whole.

(6) To provide for the protection of forest produce in transit.

(7) To provide for the adequate punishment of persons breaking the forest law.

Passing over many other provisions, I shall only add that the Act is throughout permissive, that is to say, the Government may bring its provisions into operation or not, as may be required from time to time.

Under these laws an area of about 55,000,000 acres, which is just under 10 per cent. of the British territory, has been brought under the control of the Indian Forest Department ; thirty-three million acres are so-called reserved State forests, that is to say, areas which have been set aside and are managed as permanent forest estates ; while the remaining twenty-two million acres are as yet so-called protected or unclassed State forests, enjoying a limited extent of protection until it has been finally decided whether they are to be incorporated with the permanent State forests or not. Some fifteen million acres of additional forest lands are at the disposal of Government, which have not as yet been brought under the control of the Department.

It will be noticed that the area of State forests falls considerably short of 20 per cent. of the total area, the proportion which is believed to be that required to meet the demands of the country. There are however, as yet extensive forest lands in the hands of private persons, and although their extent and yield capacity is decreasing every year, a considerable portion is so situated, or of such a description, that it is not fit for permanent cultivation, and may be expected to yield always a certain amount of produce. Interference with these private forests will only be possible in cases of absolute necessity.

The bulk of the required produce must come from the State forests, and if they are to yield that, they must be managed in a careful and systematic manner.

Hence Sir Dietrich Brandis recognized at an early stage the paramount importance of providing a competent staff of officers. He obtained, as early as 1866, the sanction of Government to a scheme, under which every year a number of young Englishmen are selected, and trained in forest science and practice before they proceed to India to take their places as officers of the Forest Department. For many years these young men studied forestry in Germany and in France. Gradually the difficulties of studying in a foreign country and in a foreign language made themselves more and more felt, until it was decided to start, in 1885, an English forest school in connection with the Royal Indian Engineering College at Cooper's Hill. Under these arrangements, some 110 officers have been trained and drafted into the Indian Forest Department. At the present moment we have twenty-two forest students under instruction at Cooper's Hill.

These young men are destined to recruit the superior or controlling staff of the department. In addition, it was found necessary to let the future executive officers pass through a suitable course of training. Accordingly, an Indian forest school was started, in 1878, at Debra Doon, in the North Western Provinces, which has been gradually developed, so that it now turns out annually some thirty trained forest rangers. These are almost entirely natives of India; they enter the executive branch of the service, but those of special merit are eligible for promotion to the controlling staff.

The organization of the department may be shortly described as follows:— The Inspector-General of forests is the head of the department, and responsible to the Government of India. The department in each Province is presided over by a Conservator of Forests (or two, and even three in the large Provinces) who is responsible to the Local Government. He is assisted by deputy and assistant conservators, each of whom controls the management of the forests in a district or other part of a Province. Subordinate to this controlling staff are the executive officers, divided into various grades, and they in their turn are assisted by the protecting staff, consisting of foresters and guards, numbering many thousands.

In this manner a well-organized department has been built up during the last quarter of a century, which has under its charge an immense government property consisting at present of some 55,000,000 acres of forest lands. Some of the forests were taken in hand before they had been destroyed, but by far the greater part of the area was taken over in a reduced and even ruined condition. Although a quarter of a century is only a short period in the life of a timber tree, the effects of protection and systematic management are everywhere apparent. Economic systems of utilization have been introduced, a large proportion of the forests is successfully protected against the formerly annually recurring forest fires; young growth is allowed to spring up under the protection now afforded; sowings and planting are carried out when required; the forests are managed under carefully considered working plans; and all this without interfering with the acknowledged rights of the people, who receive every year enormous quantities of forest produce, either free of charge or at comparatively low rates. In many parts of the country the people have come to recognize the importance to themselves of the proper preservation of a suitable forest area, and this feeling is steadily extending.

What I have said above refers to British territory. Space does not permit my dealing with forestry in native States, beyond mentioning that of late years many native rulers have commenced forest conservancy in their States, with the assistance and advice of officers of the Indian Forest Department on lines similar to those followed in the British territory.

And now the question may well be asked, how about the cost of all this elaborate organization and the works of protection and improvement?

Well, on that hand, too, I can present you with what I consider satisfactory figures. The net surplus of the Indian Forest Department, after meeting all expenses, has been as follows since 1864:—

NET REVENUE OF INDIAN STATE FORESTS.

1864-67, average annual net revenue				£106,615
1867-72, " " "				133,929
1872-77, " " "				219,919
1877-82, " " "				243,792
1882-87, " " "				384,752

The annual net revenue during the period 1882-87 was nearly four times that of the period 1864-67, and although I am not in possession of the detailed figures for the years 1887-88 and 1888-89, I may state that the gross revenue realized in the latter year surpasses that for the period 1882-87 by about £300,000. Calculated for the whole area of the forests the revenue is as yet small, but there is little doubt, if any, that twenty-five years hence the net surplus will be four times the present amount, provided the Government of India perseveres in the forest policy as developed in the past. The growth of tress is of slow progress, and of all branches of the administration of a country the forest departments require to be more thoroughly guided than any other by the watchword, " continuity of action."

BURMA.

EXTRACTS FROM RULES RELATING TO RESERVED FORESTS.

1. Within a forest reserve no person shall poison water or set traps or snares, and no person shall hunt, shoot or fish without a license. Any person who in a reserved forest, in contravention of this rule, hunts, shoots, fishes, poisons waters, or sets traps or snares is punishable under section 25 of the Act with fine, which may extend to Rs. 50, or when the damage resulting from his offence amounts to more than Rs. 25, to double the amount of such damage.

2. Such license may be granted by the deputy commissioner or forest officer in whose local jurisdiction such reserve is situated. Provided that no such license may allow hunting or shooting during the season when forest fires most commonly occur, namely : from the 1st March to the 1st June, nor the hunting or shooting of pheasants, jungle-fowl, partridges, quails or hares, during the breeding season, namely : from the 1st March to the 1st September.

3. Such license to be in form provided, and a fee of Rs. 10 may be charged for the issue thereof,

4. Between the 5th day of January and the 15th day of June no person shall, within two miles of the boundary of a reserve, leave any fire burning unless he shall have taken the following precautions, namely:—

(a) He shall, at least one week before kindling such fire, have given notice of his intention to do so to the nearest forest officer.

(b) He shall have cleared of inflammable matter a belt of ground of not less than twenty feet in breadth around the place whereon he proposes to kindle such fire.

(c) He shall have kindled such fire at a time when no high wind is blowing in the direction of the reserve.

Any person who, in contravention to this rule, leaves any fire burning in such manner as to endanger a reserved forest, is punishable under section 26 of the Act with imprisonment for a term which may extend to six months, or with a fine which may extend to Rs. 500, or with both, in addition to such compensation for damage done to the forest as the convicting court may direct to be paid.

NOTES ON FOREST MANAGEMENT IN GERMANY.[*]

The following extracts are made from a work published by Sir Dietrich Brandis to facilitate the instruction of senior Forest students at the Cooper's Hill Royal Indian Engineering College, England, and deal with the subject of Forest Management in Western Germany. Though intended primarily for Indian Forest Officers, they are not without a bearing on the subject of Forestry in Ontario ;—

The character of modern forestry may be said to consist in this, that each portion of the forest is treated with special regard to the peculiar conditions of the locality and the requirements of the growing stock, while due attention is constantly paid to the systematic arrangements of the entire forest range. The working plans prepared at the present time are elastic, and they are carefully framed to adapt themselves to the circumstances of the case.

BADEN FOREST ADMINISTRATION ; BLOCKS OF THE WOLFSBODEN RANGE.

Block 1 (compartments 1–72.) Aha, 2,053 hectares, comprises the valley of the Aha stream, as well as the northern slopes of the range which separates the Alb valley from the Schluchsee, and the head waters of the Aha stream, 884 to 1,300 m. Spruce is the prevailing tree, with silver fir at lower elevations and beech near the top of the ridge. *Pinus montana* and Scotch pine in and near peat bogs at the head of the lake, and on the head waters of its feeders.

Block 2 (compartments 1–61.) Alb, 1,679 hectares, occupies the south-western slopes of the range mentioned under Block 1 above and opposite to St. Blassien, and a small area situated between the two branches of the Alb above their juncture, elevation 770 to 1,270 m. Spruce is the dominating tree, with silver fir here and there, occasionally up to one-fourth of the growing stock. Beech more abundant near the top of the ridge. Scotch pine in a few places on steep and dry slope, with a south-westerly aspect (compartments 7 and 8.) Very good natural reproduction on the piece situated between the two branches of the Alb, overlooking the northern branch, in compartments 40 and 44. Young poles of excellent growth, mainly spruce, with a little beech and silver fir, 430 cub. m. per hect., at 950 m. mean elevation in compartment, 28. Fine old forest though not completely stocked, spruce with 25 per cent. silver fir 100 to 150 years old, with 100 cub. m. per hect. in compartments 32 and 33, on opposite sides of the small valley leading to Muchleland. Also near the southeast end of the block, at a similar elevation part of compartments 7 and 8 stocked with spruce mixed with silver fir and beech, 70 to 150 years old, with 650 to 800 cub. m. per hect. A large extent, the greater part of compartments 14, 15, 23, 24 with portions of the adjoining compartment, nearly 100 hect. of mostly pure spruce 60 to 100 years old on the top of the hill called Botzberg, which overhangs St. Blassien, between 1,100 m. and 1,270 m. The forest is completely stocked, the soil is good, with a dense covering of moss, the climate very moist, but the trees are short. The growth is slow, and hence there is not so much timber upon the ground as there might otherwise be, 300–350 cub. m. per hect. The reason is the high elevation and severe climate. At a lower elevation the growing stock of the same age in such a locality would be much larger. According to old traditions the trees were habitually felled while the snow was on the ground, and this agrees with the remains of the stumps 2 m. high overgrown with moss, standing in these and other compartments. On the tops of such stumps the seed of the

[*]By Sir Dietrich Brandis K.C.I.E., Ph.D., F.R.S., late Inspector-General of Forests to the Government of India.

spruce has often germinated, and the young plant has sent its roots over the decaying stump into the ground. The stump has perished, and the tree stands now, as it were, upon stilts in the air supported by its roots.

Block 3 (compartments 1–4.) Kutterauer Halde, 90 hect., a small detached block, occupying the lower slopes of the Alb Valley, 2½ km. below St. Blasien, at an elevation of 700 to 840 m. under the village of Hochenschwand. Spruce, silver fir and beech, with a little oak.

Block 4 (compartments 1–3.) Schwarzathal, 616 hect., occupying the slopes on the right side of the Schwarza Valley, below the village of Hochenschwand. Spruce, silver fir and beech, with a little oak in the lowest part. More than half of this area has been acquired within the last fifteen years, chiefly by the purchase of private forests and pastures which now are planted up with pine, spruce and larch.

Blocks 5 and 6 (compartments 1–6), 179 hect., are two small detached blocks, Blasiwald and Dresselbach, lately purchased.

The total area as here stated, 4,617 hect. for Wolfsbolen, and 3,144 hect. for St. Blasien refers, it must be understood, to 1887, and comprises some pieces included within forest limits since 1885.

Utilization of Forest Produce.

With so large a forest area, the question naturally arises whether there is a sufficient demand for all the timber and wood annually produced. For charcoal there was formerly a very large local demand for the numerous iron and glass works which existed upon these hills. The names of places such as Althutte, Altglashutte, recall the former existence of this industry, which is now nearly extinct. The last considerable iron works belonged to the State, and these were closed in 1863; some glass works existed until 1877. The completion of works elsewhere using mineral coal, and favorably situated near railways and rivers, was too powerful, and the struggle of these small establishments, situated far from rivers and railways in out of the way and difficult mountainous districts, against such powerful opposition was hopeless, though the abundant water power available and the cheap and plentiful supply of charcoal gave them important advantages. Quite lately, I am informed, a prospect has arisen of selling large quantities of small wood for making charcoal which is wanted by large iron works in Switzerland. On the other hand, the construction of roads has greatly facilitated the export of forest produce. The road up the Wehra Valley to Todtmoos was built in 1848-49, and that from St. Blasien down the Alb Valley was completed in 1861, and it is now proposed to construct a road from the Schluchsee along the valley of the Schwarza to the Rhine, which will greatly facilitate the export of timber from the forests in the basin of the Schluchsee and in the Aha Valley. St. Blasien and the the country around has long been connected with the Rhine Valley by an old carriage road, but it ascended the plateau with a rise of nearly 300 metres and then continued with numerous descents and ascents to Walshut. For the export of timber, roads along the valleys with an even and gentle gradient are indispensable. The construction of convenient main roads has been supplemented by a system of well designed cart roads throughout the forest, and thus it has become possible as already mentioned, to sell the thinnings from some of the forest as hop-poles. The construction of these forest roads commenced in 1860 and every year new lines are added in order to complete the system laid down in the programme.

It is a remarkable fact that in this part of the Schwarzwald water carriage has never been used for timber to any large extent. The elaborate and most skilful arrangements which were formerly used, and are still in use to some extent, to facilitate the floating of timber on the Enz, Nagold, Murg, Wolfach and Kinzig rivers in the northern Schwarzwald, were unknown here. At first sight this seems difficult to understand. But the rivers of the southern Schwarzwald, particularly those here in question, the Alb and Wehra, have a much greater fall, and their bed in places is more narrow, and more obstructed by rocks and stones than that of the northern streams, in which floating has been chiefly practised.

The following is a sketch of existing arrangements for the utilization of the produce of these forests. For felling the trees, fashioning the timber, and bringing the logs to the roadside where they are sold, contractors (accordanten) here, as in the other State forests of Baden, are employed by the district forest officers. In each range four or five of these contractors generally find employment throughout the year. They are men of long experience in the business, picked, and to a great extent trained, by the Oberforster himself. Each works, according to circumstances, with from 8 to 15 timber cutters, the contractor being the foreman and working himself with the men, with whom he shares the profits arising from their operations.

As previously explained, by far the most important produce of these forests consists in the timber of spruce and silver fir. The following remarks will be mainly limited to them. The mode of utilizing beech wood offers no peculiar feature.

The thinnings in young forests are generally made between June and August. The poles to be taken out are marked by the forest guard, under the district forest officer's general direction. The forest guard is authorized to permit more experienced timber cutters to mark, under his supervision the poles to be thinned out. It has not having been possible to carry out thinnings in young forests as extensively as would be desirable, for the poles cut are in many cases, unsaleable. The spruce and silver fir forests of those portions of the Black Forest which adjoin the Rhine Valley and of the Vosges on the other side of that river, supply poles for the hop gardens of Alsace and Baden at lower rates. The forests of St. Blasien therefore can, in exceptional cases only, enter into competition with them. The present depressed state of the hop market makes it unlikely that hop-poles will be largely exported fr m these forests.

Mature timber is generally cut between April and the middle of June, and during the remaining months of summer and autumn the logs are prepared and brought to the roadside. As already mentioned, winter felling was customary, to a certain extent at least, in old times, but in those days charcoal was the main article produced, and the timber was at once cut up for the charcoal kilns into billets which were easily moved on sledges over the snow. At present, when large timber is the chief article produced, winter transport is not feasible, and during the last 40 years summer felling has become the general rule in these districts.

The first operation is to bark the trees. Of the spruce bark a considerable portion is sold for tanning, that of the silver fir is used as fuel. Timber of prime quality, whether spruce or silver fir is left in logs as long as possible and for such timber the contractor is paid for at the rate of 2 mark per cub. metre. It is classified under five classes, the first and second comprising those logs which at 18 m. (59 ft.) from the butt end have a diameter of 30 c.m. (12 in.)

and 22 c.m. (9 in) respectively, while logs of the third class are required to have a diameter of 17 c.m. at 16 m. from the butt. The rest, known as Sageklotze and Lattenklotze, is cut into convenient lengths, generally 5 to 6 m. and brought to the local saw mills, where it is converted into planks and battens For this class of timber the contractor is paid 1.60 mk. per cub. metre on delivery at the roadside.

A certain portion of the timber, chiefly of spruce, is in these forests used for splitting. At Bernau, a large village north-west of St. Blasien, and other places in the vicinity there is a thriving industry in sieves, tubs and other articles made of split wood. The workmen brought up to this special branch of the trade make a careful distinction between logs which split readily and those not so suitable for their work, and the former fetch a much higher rate at the sales. Thus, in the Alb Valley forests of St. Blasien, from which the coopers of Bernau draw their chief supplies, the three first classes of logs when fit for splitting, sold in 1885 for 21.2, 18.2, and 15.2 marks per cub. metre respectively, while ordinary logs fetched only 16.8, 14.4 and 12 marks. In 1886 the figures were 19.9, 16.5 and 14.1, for logs fit for splitting and 15.9, 13.8, 10.9 for ordinary timber.

The Oberforster of St. Blasien regularly employs one of the men from Bernau in order to mark the logs fit for splitting, and the logs thus marked are sold separately. I spent a day in the forest with this man in order to learn the characters upon which he relied. The first condition is that the log must be regularly shaped and clean, without knots and branches. Secondly, the fibres must not be much twisted and if twisted at all, the twist must go from right to left. On barked logs the twist is readily seen by the direction of the fine fissures on the surface of the wood. It can also be recognized, but less distinctly, on the bark of standing trees.

In order to drag the timber to the temporary depots at the roadside, where the sales take place, horses or bullocks are employed when the ground is level or nearly so. Down a slope, however, the timber is lowered by means of ropes. The tools used for this purpose are of the simplest description, a stout rope 140 feet long, a strong iron hook, with a ring to which the rope is attached (Seilhaken), and a kind of pick (krempe), the wooden handle 4 to 4½ feet and the iron 18 inches long.

On slopes large trees are always felled with top and branches down hill, and the logs are sent down top end forward, two men holding the rope, which is generally slung round a tree, while four or five men, all armed with picks, work alongside the log, lifting it over uneven places in the ground or other impediments, stopping it while the rope is slung over another tree lower down and generally directing its course. It is a fine sight to witness the speed and precision with which this difficult work is accomplished. Accidents very rarely happen.

In the latter portion of summer and in autumn, the timber intended for export is all carted down to the railway station. It chiefly goes westward, to Switzerland, Alsace, and France. This manifests itself in the higher rates realized at sales in the Wehra than in the Alb Valley division of the St. Blasien State forests range, which have to be carted nearly the same distance before they reach a railway station. In the Wehra forests first class logs in 1885 fetched 17.9 and in the Albthal forests 16.8 mk per cub. metre. In 1886 the figures were were 19.4 and 15.9.

5 (F.)

Small branches, as well as the underground wood, are unsaleable. The collection of small branch wood is permitted, and, where practicable, the condition is attached to the permission granted that what remains on the ground must be burnt.

Of the larger branches and of the poles obtained by thinnings, small quantities have as already mentioned, latterly been sold as hop-poles. A new demand for small wood has fortunately arisen through the erection of paper-stuff factories from wood, the first of which was built in 1874. There are four such factories now in the vicinity, and to these it is due that much of the small-sized spruce timber, for this is the only kind used by them at present, can be disposed of.

The yield of minor forest produce in these districts in insignificant. Here, as elsewhere in the Black Forest, the spruce was formerly extensively tapped for resin and in Wolfsboden the collection of resin in a few places, where the old trees formerly tapped are still standing, is still let out, but this will soon cease, as the old trees are being cleared away rapidly.

REGENERATION OF THE FOREST.

In some cases, for instance where a large proportion of old trees formerly tapped for resin are on the ground, it is necessary to clear and plant, but, as a rule, the system followed is to rely upon natural reproduction as much as possible, and this necessitates the gradual cutting of the mature stock. It has already been stated that the mean age at which the timber in these forests is cut is 120 years. When the time arrives for commencing cuttings in a compartment which has attained that age, the first operation is to clear away all soft woods and useless brushwood, and to cut out any advance growth that may be on the ground and that may not be suitable to form part of the young forest intended to be produced. After this follow in succession a series of cuttings more or less heavy, the beginning being made with the removal of all oppressed, damaged and diseased trees. These successive cuttings are generally continued in these ranges during a period of from 30 to 40 years. Thus in 1858, when a preliminary cutting had already been made, compartment 31 of Block 2 in Wolfsboden was stocked with a forest of two-thirds spruce, and one-third silver fir, containing 800 cub. metres per hectare, 11,432 cub. feet per acre. In 1864 the first heavy cutting removed 200 cub. metres per hectare (2,860 cub. feet per acre). Afterwards five successive cuttings were made, which only left 80 cub. metres (1,143 cub. feet per acre), in 1887. Meanwhile the ground has got well-stocked with young growth, and the remaining old trees will probably be cleared away by 1896. This long period of regeneration is necessary, because seeding years in the cold climate of these districts are scarce, in the case of spruce and silver fir every fourth or fifth year, and in the case of beech once in 8 to 10 years. But there is another object besides, viz. the great increase in timber of the trees left standing in a more isolated position after each successive cutting, hence the practice is to keep the most vigorous trees to the last. In the case of the silver fir, groups of younger trees or single trees are left standing after the series of cuttings has been completed; they form part of a new forest growing up, and will be cut, when the next rotation comes round, having then attained an age of 200 years or more.

FOREST RANGES FORBACH (I AND II) AND HERRENWIES.

The three forest districts to which the present remarks specially relate are the following. Area and actual yield are for 1886 :—

	Productive forest area, in hectares.	Annual yield in cubic metres.		
		Sanctioned.	Actual, 1886.	
			Total.	Per hectare.
Forbach, I............................	3,609	16,720	18,906	5.24
Forbach, II......	4,977	32,000	29,257	5.88
Herrenwies.........................	3,459	21,000	18,894	5.46

Of these, Herrenwies, which is entirely State forest, is situated on the head waters of the Raumunzach, and occupies a portion of the range west of the Murg. Forbach I comprises the communal forests of Forbach, Langenbach, Gausbach, and a few other villages, as well as the Heiligenwald (Saints' Forest) of 860 hectares (2,124 acres), which is the property of the Forbach parish church. In these forests the Oberforster has the management in his hands in the same manner as in State forests. In regard, however, to the disposal of the wood cut by him, the communal authorities, and the trustees of the church are at liberty to make their own arrangements. As a matter of fact, they gladly avail themselves of the Oberforster's services in this part of the business also. From the proceeds of these communal forests roads, school-houses, and other public buildings are built, and the income from these forests is sufficient to defray all other expenses of the municipalities, so that the members of these communities not only enjoy immunity from all local rates and taxes, but also pay no school fees for their children. Part of the firewood which forms the yield of these forests is distributed among the villagers, so that they have most of their fuel free. For timber and other wood they pay like other people, but the money yield of most of the communal forests in the range Forbach I is so considerable that a certain surplus is divided annually among the villagers. Thus the people of Gausbach and other villages have this year received a share of the surplus, amounting to 70 mk. (£3 10s.) each householder.

The study of this and other communal forests in the Grand Duchy of Baden, and in many other parts of Germany, will be found most instructive for those who have to deal with forest matters in India, for there is no doubt that the success of the endeavors which of late years have properly been made to stimulate the development of local self-government in the different Provinces of India will, to a great extent, depend upon the success which may attend the efforts to place the self-government of towns and villages upon a stable footing, by organizing a good management of such landed property as these communities possess, or may hereafter be able to acquire.

The income derived from the Heiligenwald has of late years been allowed to accumulate, until the amount had become sufficient for the erection of a new church, which has been completed at a cost of 350,000 mk. (£17,500).

Forbach II consists of the forests belonging to an ancient corporation (Murg Schifferschaft). The State has of late years purchased a large portion, about one-half, of the shares, and since that time (1886) the management of these forests has been intrusted to a State forest officer.

As already mentioned, the prevailing trees are spruce and silver fir, the latter being more abundant at lower elevations, while the spruce predominates in the upper portion. With them the beech is associated at all elevations, but in varying proportions, for while in some places it forms a large portion of the growing stock, it is almost absent in others. The Scotch pine is found in the granite region, chiefly upon dry, steep, rocky slopes with a southerly aspect, while in the sandstone region it occurs almost everywhere, sometimes scattered, and in other places forming an essential element of the growing stock. A remarkable feature here is the occurrence of mixed forest of Scotch pine and silver fir, the latter forming a kind of high underwood under the former.

Although these forests present great variety of soil and other conditions, yet upon the whole it may be said that in places the growth of the species mentioned is magnificent.

The three conifers attain a height of from 40 to 45 metres, the stems carry their girth well up to a great height, and are, as a rule, regularly shaped. Seed years occur frequently, and the reproduction is generally very good. A marked difference is, however, noticeable, especially at lower elevations, between slopes with a southerly and northerly aspect, the latter showing much better growth and more abundant reproduction The disease most frequently observed consists of the irregular swellings on the stem of the silver fir, commonly known under the name of cancer. Considering the enormous area of unbroken forest, on both sides of the Murg Valley, chiefly composed of conifers, it is remarkable how little damage by insects takes place. Storms and snows do some damage, but upon the whole it is insignificant. At times the pressure of the masses of snow is so heavy that large trees are bent down gradually and uprooted. There are not many species of subordinate importance, and those which occur are scarce upon the whole. Along the valley from Gernsbach to Forbach, and even higher up, the oak forms a fringe at the lower edge of the forest, and a few oak trees are seen scattered over the whole granite region. The hornbeam is found here and there, associated with the beech, and single specimens of the sycamore are now and then met with. On peat soil at high elevations, and on the top of the two chief hill ranges, a considerable area is stocked with the mountain pine, and in such places the birch is also common.

Of shrubs there is no great variety. It may be justly said that the forest is everywhere too dense and too well stocked for much subordinate vegetation. In old and dense forests, where there is not sufficient light for the young growth to come up, the ground is frequently covered with *vaccinium*, brambles are almost absent, and the wild raspberry is scarce.

From time immemorial these forests have been treated on the system of selection fellings (jardinage), and this system is still followed in the two Forbach ranges. Here, therefore, the character of the forest is extremely varied, trees of all ages standing on the ground together. Formerly, the practice was to select the finest and most accessible trees for felling. In this respect a great change for the better has now taken place, for the aim at present is to cut out all unsound and badly-shaped trees first, so as to leave more room for the young growth and the more vigorous trees A rational treatment of forest on the selection system cannot easily be brought under precise rules; the manager must consider the requirements of each plot separately, and this is being done at present in these forests.

In order to determine the annual yield of forests managed under the selection system it is necessary to measure the old timber over the entire area, and the work cannot be shortened by the examination of sample plots. Thus, at the renewal of the working plan for the Schifferschaft forests in 1886, all large trees (in diam. 15 c. m. = 6 inches and upwards) were measured on 85 per cent. of the total area. This was accomplished in two working seasons by two valuation officers, an immense and most difficult piece of work, considering the dense underwood of young growth which in most places covers the ground under the old trees. The volume of the smaller trees was estimated, and the total growing stock was thus determined at 1,912.244 cubic metres, exceeding the normal growing stock by 232,688 cubic metres. The rotation for this forest range was fixed at 120 years, and upon these data the annual yield was fixed at 32,000 cubic metres, or 6.43 cubic metres per hectare. The average growing stock was 382 cubic metres, but in many compartments the volume of timber exceeds 700 cubic metres per hectare.*

In 1886 the sanctioned yield of the Schifferschaff forests was not fully worked up to, whereas, in Forbach I, there was an excess of 2,200 cubic metres over the sanctioned yield, caused partly by the timber cut for road-making, partly by some extraordinary requirements of the village of Forbach.

In the Herrenwies range the plan is, not to continue the system of selection fellings, but gradually to introduce the system of felling by compartments (schlagwirthschaft), and in some portions of this range considerable progress has been made in this respect. Large areas are now stocked with uniform thickets up to twenty years old, while others are stocked with pole forests, so that in places a regular gradation of ages has been brought about. This has been accomplished by the gradual removal of the old trees, under the shelter of which the young growth had come up.

The rotation here, as in the two other forest ranges, has been fixed at 120 years, and it is intended that the period assigned to the cutting out of the old timber and the regeneration of the forest is eventually to occupy thirty to forty years.

Considering the enormous area of the forests, it is remarkable that all the timber in them can be sold. Underground wood, however, finds no purchaser anywhere in these forests, and only in the vicinity of the villages will people undertake to root up the stumps on taking away the wood without payment. The removal of tops and branches is free throughout these forests. The bark, at present at least, is not sold, but is removed free with the branch wood. The produce of thinnings formerly found a ready sale in the hop gardens of the Rhine Valley, but the cultivation of hops has of late years greatly diminished, and much of the small wood would remain unsaleable if numerous factories of paper pulp had not, about fifteen years ago, been established in the Murg Valley. This has opened a new demand not only for poles but also for small trees. At present the paper pulp factories have a decided preference for spruce, and pay more for clean stems without branch knots.

At Schonmunzach, already mentioned as the first village of Wurtemberg territory, is a large glass factory, which works with gas made from wood, and this factory consumes annually a very large quantity of small wood. It is not improbable that hereafter the inferior kinds of wood produced in the fire-protected forests of India, which at present are unsaleable, may be used for the pro-

---

* It would lead too far in these notes to discuss this question why the Murg Valley ranges here described, possessing as they do unusually favorable conditions for the growth of spruce and silver fir, have a smaller annual yield in material than the St. Blasien and Bonndorf Ranges in the southern Scharzwald.

duction of gas for iron smelting, and that thus it may be possible to revive the old iron industry in some parts of India. Indian forest officers will do well to visit the Schommunzach glass works.

The disposal of large timber is greatly facilitated by the numerous saw mills in the valley of the Murg and other rivers, and most of the timber from the Murg forest is now exported in the shape of beams and boards. Of the principal kinds of timber sold spruce and silver fir command the same rates, while Scotch pine generally fetches a somewhat higher figure.

Timber is cut during spring and summer. In forests like these with much young growth under the trees, great care is necessary and is used, so as not to injure the mass of seedlings and saplings on the ground. In dragging the timber much attention is paid to this, and as a further safeguard the branches of the standing trees are lopped before felling. This is done by men who climb the trees, with the aid of foot-irons, and who are paid at the rate of from 20 to 30 pfennige a tree. A skilful man can lop ten to fifteen trees, thus earning from 2s. to 4s. 6d. a day. Formerly the custom prevailed to lop the branches of trees standing over young growth intended to remain on the ground so as to diminish the shade. This practice, however, has of late years here been abandoned because it was found that the lopping of branches made the trees unsound. Through the whole forest are narrow dragging paths, some in their natural condition, others levelled and built up. The timber is brought to the edge of these dragging paths, and there the sale takes place. The logs are dragged generally by horses, with or without the help of a pair of wheels, according to the gradient of the path.

A system of carting roads, however, is being steadily extended over the whole forest, and the timber carts carry very heavy loads measuring up to ten or twelve cubic metres. Under the present practice the stems are brought out as long as possible, provided they are sound, for it is found that the proprietors of the saw mills pay higher rates when they can cut up the logs according to their particular requirements. For timber work on the dragging paths, as already mentioned, as well as for carting, horses are generally employed, and from spring to autumn the roads are full of large timber carts, on which huge logs, up to 30 m. in length, are carried down the valley.

COPPICE UNDER STANDARDS NEAR KIPPENHEIM, IN THE RHINE VALLEY.

The contrast between the unbroken forest of the Schwarzwald, interrupted only here and there by stretches of field, not very productive, and the rich plains of the Rhine valley bearing splendid crops, with luxuriant meadows on low ground and extensive vineyards on the hills, is exceedingly striking. In the Rhine valley, between Freiburg and Offenburg, one of the most fertile portions of Baden, there is not much forest, but what there is produces large quantities of most valuable material. Here are situated the forest districts of Kippenheim and Ettenheim, which comprise the State and communal forests situated in the civil district of Ettenheim. Kippenheim. where the Oberforster resides, is a large village, situated at the foot of the hills, which rise into the Schwarzwald. Like some of the other villages in this district, it has forests both in the outer hills and in the Rhine valley. The produce of these communal forests is sufficient in the case of this and other villages in the vicinity to cover all municipal expenses, so that the inhabitants have to pay no local taxes and no school fees. The forests in the plains are of special interest, and particularly those which belong to the Kippenheim forest district. They form a compact block six kilometers long and one to two kilometers wide, and are situated between the Rhine and the foot of the hills at an elevation of 170 metres. One hundred and forty-nine hectares of this area belongs to Government, and is known under the name of Kaiserwald; the rest belongs to six villages in the vicinity. Of these the best stocked and most instructive is that belonging to the village of Grafenhausen, 148 hectares, adjoining the Kaiserwald on the west side. Over the whole area of these forests the soil is not uniformly good. It always consists of a surface layer of loam with much vegetable mould, resting upon a stratum of more binding clayey soil under which there is a thick stratum of gravel. The thickness of the soil over lying the gravel varies. At the same time there are slight differences in level, and these circumstances are believed to have a considerable influence on the distribution of the trees of which the forest is composed.

The whole of this forest is treated as coppice under standards. In low situations and on moister soils, where soft woods prevail, the rotation of the coppice is 25 years, but most of the area is worked under a rotation of 30 years. The species of which these forests consist are ash, oak (*Quercus pedunculata*), alder, hornbeam, sycamore, and elm. Of subordinate importance are birch, aspen, maple, wild cherry, and willow. Over the greater part of the forest the ash prevails, and its proportion is being steadily increased by planting. In suitable localities this is also the case with the oak. In regard to the influence of the soil upon the distribution of the different kinds, it may be said that where the gravel is dry and near the surface the hornbeam prevails in the underwood and the oak succeeds particularly well, while in such localities the ash is apt to become stag-headed after the age of sixty years. Such places also generally have a somewhat higher level. Where the gravel is moist the ash, the hazel, and the elm prevail; where the gravel is wet the oak is wanting, while the alder, elm, and hazel abound. In such places, and on low ground generally, the rotation is only twenty-five years, because this is sufficient for the prevailing kinds, which, as already stated, are the alder, the elm, and the hazel. In such places, however, the proportion of the ash is being steadily increased by planting, and eventually the normal length of a thirty years' rotation will probably be established.

Coppice and Field Crops in the Siegen District.

The Sieg, which flows into the Rhine below Bonn, drains a large area of mountainous country, rising to near 700 m. (2,300 feet), the rocks of which, clay slate and grauwacke of the Devonian formation, are remarkable on account of the numerous veins of excellent iron ore which they contain. The oak is the principal tree, forming excellent high forest, more or less mixed with hornbeam and beech, up to the top of the highest hills, such as are rarely found in similar localities, while over an area of 57,000 hectares (140,800 acres), on both sides of the Sieg as well as on the headwaters of the Dill, a tributary of the Lahn River, a peculiar class of oak coppice prevails, known under the local designation of " hauberge." This area is situated in the circles (Kreise) of Siegen and Olpe, which form part of the civil district (Regierungsbezirk) of Arnsberg, belonging to the Province of Westphalia, and in some tracts belonging to the adjoining district of Coblenz and Wiesbaden. The following remarks relate chiefly to the circle of Siegen.

In the narrow valleys of this mountainous country are numerous factories, mines, and iron works, some of them of old standing, formerly worked with water-power, now mostly with water-power and steam, surrounded by well-watered meadows, with very limited areas of fields and gardens. The hills are well wooded, and, as already stated, almost exclusively stocked with oak. In the midst of the oak coppice, on the slopes and ridges, are numerous extensive fields with rye, the pale green of which in early summer contrasts strangely with the dark green color of the oak forest. These fields change their position from year to year, so that the traveller who visits these hills in two successive years finds the aspect of the landscape changed, though its general character remains the same. The high forest which covers the tops and ridges belongs to the State or to large private proprietors, but the vast areas of coppice which occupy the main portion of this tract of country do not belong to the State or to private proprietors, nor to town or village communities but to public corporations, commonly regarded as the remains of the old " Markgenossenschaften," which in the words of the late Sir Henry Maine, were " an organized, self-acting group of Teutonic families, exercising a common proprietorship over a definite tract of land, its mark, cultivating its domain on a common system and sustaining itself by the produce." The coppice is managed on rotation of from 17 to 20 years (19 years on an average), and the area assigned to each year's cutting is treated in this manner. Early in spring (March, April) all soft woods, birch, hazel, aspen, and others, as well as the most slender shoots of the oak coppice, are cut out, the operation proceeding from the bottom of the valley upwards. At the same time the poles intended to be peeled are cleaned by cutting off the lower branches. As soon as the season is sufficiently advanced for the bark to come off readily, generally in May or June, the poles are peeled standing, the operation being performed as follows :—From a cut made breast high the the lower portion of the bark is taken off downwards, while the upper portion is peeled upwards, the upper end remaining attached to the pole. In the case of high poles ladders are used, and weak poles are bent down in order to peel them. The naked poles remain standing until the bark is dry. Long strips remain hanging ; smaller pieces are tied up in bundles and are hung upon the poles. In the case of poles which have sprung from seed, either natural or planted, the rule is strictly observed to ring them close to the ground by a circular cut going through the bark only. The bark then comes off down to the girdle only, and this promotes the growth of coppice shoots from the stool.

The wood is cut as near the ground as possible, the cut being smooth and slanting without splitting and without injuring the roots. The poles over 5 cm. (two inches) diam. are cut one to two inches above the ground by means of two opposite cuts slanting upwards. Seedlings, whether natural or planted, not yet fit for peeling, remain standing so they may not be damaged by hoeing and when the corn is cut.

The wood is placed on the ground between the stools with the butt end down hill, and is removed as soon as possible without injuring the young shoots from the stools.

Immediately after the bark has been peeled and while the naked poles are still standing, the ground between the stools is worked up with a hoe of peculiar shape, the sharp edge indented, and is turned up in sods, which are gathered in heaps and when dry are burnt, with the aid of the small branch wood.

The burning takes place between July and September. The ashes are then evenly spread over the ground with a shovel, and the rye is sown broadcast.

The seed is worked into the ground with the aid of a peculiar kind of light plough without wheels, locally called" hainharch," drawn by cows or oxen, which are muzzled so as to prevent their browsing upon the young shoots of the oak coppice. The crop is always clean, without weeds. In August the harvest akes place, and the corn is cut with the sickle, so as not injure the young coppice hoots of the oak.

System of Management : Its Origin.

The management of the forest estates here described is entrusted to committees, elected by the shareholders for a period of six years. Each committee consists of a chairman and one or two members. The current duties are conducted by the chairman alone, but certain matters, such as the appointment of the guard who is intrusted with the protection of the estate, are by law and custom assigned to the full committee.

In order to maintain the coppice well stocked, cultural operations are regularly carried on in most of these estates. The old established practice is to dibble in acorns in lines about 2 m. apart, either with the seed corn in autumn, or in spring with the young crop, or in the second autumn into the stubble. Where sowings have been made the broom is cleared away when it threatens to choke the young plants. Cattle are excluded until the plants are sufficiently advanced to be beyond damage. Where it is not possible to keep the area closed so long, strong saplings 1½ to 2 m. high are are planted about 3 m. apart, and in order to provide a sufficiency of such plants suitable nurseries are established for each estate. The chief civil officer of the circle (Landrath), together with six shareholders who are elected for a period of six years by the whole body of hauberg associates, form a board of control for the management of these estates throughout the circle (Schoffenrath). This board appoints one or several forest officers, who have the supervision of the management of these estates as far as regards professional matters. The board also assigns the area to be subject to their inspection, and is empowered by law to decide all matters relating to these estates that may be referred to them. At present there is one forest officer (Hauberg Sachverstandiger) for the entire circle. His chief duty is to watch over the due observance of the treatment laid down by law, and generally by his advice and personal influence to promote the good management of these estates. All these matters are governed by a special law based upon old ordinances and customs existing in regard to these estates. The law which is in force at the present time was passed by both houses of the Prussian Parliament in 1879.

The system under which these forests are managed is very old. The oldest document preserved regarding it is of 1447, and a detailed account exists of 1553, from which it appears that in its main features the system then was the same as at present. The peculiar development of the system must be attributed to two circumstances, the requirements of the mining and iron-making industry, and the insufficiency of arable land in the district.

Formerly, all the iron works in this district were worked with charcoal, which the forests furnished, and for this purpose coppice was the simplest and most convenient mode of treatment. The poles, whether oak, birch, or other kinds, could readily be utilized for charcoal. The mines and the iron-works in this part of the country in former times were always owned by associations (genossenschaften) and in some cases these associations may have also owned the forest lands adjoining the works. In any case, the organization of the " Hauberg genossenschaften " has developed in a manner similar to that of the mining associations. At the same time, the population, though never dense, as compared with the plain country, nevertheless did not produce corn enough for their maintenance, nor was there sufficient litter for their cattle. In this manner the necessity arose to utilize the forest for the temporary cultivation of corn after the coppice had been cut over. These temporary fields furnished a large portion of the corn and straw which they required. In 1862 the total area of the circle, 64,653 hectares (159,800 acres) with a population of 48,479, consisted of 74½ per cent. forest (five-sixth hauberge) 10 per cent. meadows, and 13 per cent fields. Of the fields an average area of 6,940 acres was devoted to the production of corn, so that the addition of about 4,400 acres, which at that time was the aggregate area of coppice annually cut over and cultivated with rye, was an important addition to the corn-producing land. Even then, however, grain was imported largely, and now, with a vastly increased population, (77,674 in 1885) the importation of grain has largely increased. It is estimated that at present three-fourths of the corn consumed in the district has to be imported, hence the great importance f the system here described for increasing the corn producing area.

The custom of raising one or several crops of corn on forest land, and of letting the forest grow up again after the harvest, is an ancient custom in mountainous countries of all parts of Europe, nor is it limited to Europe, but is found in most other parts of the world. In India it is known as kumri in the south, as jhum in the east, as dhya in Central India, as khil in the north-west Himalaya, and as toungya in Burma. In Europe, however, the system has in so far developed, that the wood which grows up after the harvest is not all destroyed to furnish ashes for the field crops, but is otherwise utilized.

As the Siegen country was gradually opened up by railways, coal was imported, and the use of charcoal ceased. This diminished the value of the forest crop as far as the wood was concerned, but simultaneously the tanning industry, which was important as long ago as the fifteenth century, developed on a much larger scale, and at present bark is the most valuable produce of these lands. At Freudenberg, Siegen, and elsewhere in the circle, large tanneries exist which receive hides from all parts of the world and send away the leather prepared by them in all directions. The oak bark produced by these woods amounts to 85,000 cwt. a year.

COPPICE AND FIELD CROPS IN SOUTH GERMANY, AUSTRIA AND FRANCE.

Coppice, combined with corn crops, likewise occupies extensive areas on the sandstone (Bunter sandstein) of the Odenwald, the mountain range situated between the rivers Main and Neckar, in the Grand Duchies of Hesse Darmstadt and Baden. Here this kind of coppice is known under the name " hackwald." Further south, in the valleys of the Kinzig and Rench, on the gneiss and granite of the Schwarzwald, it is known as " reutfeld," and under the same name it is practised in Wurtemberg and some parts of Switzerland. Circumstances in some respects are similar to those existing in the Siegen district and on the Moselle. The arable land is limited and the forest area large. Hence the desire to utilize part of the forest land for the production of corn and straw, whenever the ground has been cleared by cutting the coppice.

In upper Styria there existed formerly and probably now to some extent exists a similar management of forests, chiefly consisting of alder, birch, aspen and sallow. At the age of twenty-five to forty the coppice was cut, the larger wood used or sold, while the branches were burnt, one crop of rye and a second crop of oats being taken, after which the forest was allowed to grow up again. Wessely, in giving an account of this system in 1853, states that it is disappearing, and that many of the areas formerly thus treated are gradually being converted into spruce forests.

The practice called schiffeln which prevails in the mountainous tracts of the Eifel is only a variety of this system, differing in this, that small brushwood only and no regular forest grows up on the land after it has yielded a cereal crop. Near Cochem, for instance, the land on the plateau of the Eifel which is not under the plough or kept as meadow land is either high forest, or coppice, or schiffel land, which is allowed to remain fallow for twelve years, between two crops of corn, and during that period gets covered with a dense matting of grass and bushes of broom and juniper. It is here the place to mention the system of *sartage*, which prevails in some mountainous districts of Belgium and France, particularly in the Ardennes, the continuation to the south of the " Hohe Venn." Sartage resembles the system here described, except that the coppice is worked under a longer rotation (twenty-four years), and that what is called open-air firing is more generally employed, that is, the sods of turf are not burnt in heaps, but small wood and branches are spread uniformly over the ground, and are fired during calm weather, with the needful precautions against spreading of the fire. The system is well described in that excellent work of Lorentz, " Cours elementaire de culture des bois," 4th edition, 1860, p. 424.

It is also treated in Bagneris' Elements of Sylviculture, English translation, 1882, p 125. Bagneris remarks, that the system is dying out in France. Zealous foresters, in Germany, as well as in France, have often condemned the system of combining coppice with field crops, as barbarous and indefensible. This, however, is not a correct view of the case. The system has certain positive advantages as far as the growth of the coppice is concerned ; moreover, in many districts it admirably adapts itself to the requirements of the population. With due care and with the aid of diligent sowing and planting, the coppice can under this system be maintained in excellent condition. On the other hand, where it is not carefully supervised, the system is wasteful and unprofitable. As a matter of fact, in some districts the altered circumstances of the people may perhaps eventually lead to a gradual extinction of the system, whereas in other districts it will be maintained and will continue to contribute materially to the well being of the population.

HIGH FORESTS AND FIELD CROPS.

The raising of cereal crops between two crops of high forest, or as an operation preparatory to the formation of new forests on waste lands, has been practised centuries ago in different parts of Europe. On the south-western portion of the mountain range which separates Bavaria from Bohemia, known as the " Bayrische Wald," a peculiar system of forest culture has existed since the fifteenth century. The forest which here chiefly consists of birch, is cut, a number of trees being left standing for seed. During one or two years rye, millet, potatoes and oats are raised on the ground, which had been fertilized by the ashes of the tops and branches. The birch seeds plentifully and regularly, and the ground soon gets covered with dense young growth, partly seedlings, partly coppice shoots. Where cattle have been kept out, the young forest is large enough to be cut and burnt after the lapse of twenty to forty years. Often, however, these areas, which are mostly private property and are known under the name of " Birken berge, Birken reuter," are indiscriminately opened to cattle.

In the large spruce forests on the mountains of upper Styria, during the first part of this century, the old wasteful system still existed of making wholesale clearances into which cattle were admitted immediately after cutting, no steps being taken to facilitate reproduction. When gradually the rapid development of the iron industry in those parts of Austria made wood (for charcoal) more valuable, one of the first measures to accelerate the regeneration of these forests, and thus to increase their productiveness, was to let out the clearances for cultivation, and to sow the spruce seed with rye. The stems were used for timber or charcoal, but tops, branches and trees without value were burnt. This system I found in existence in 1865 on the mountains west of Bruck. A large extent of well stocked forest has been raised in this manner.

On the mountains of the Odenwald, which have already been mentioned in connection with fields and coppice, a system of raising high forest with the aid of cereal crops, called Roderbau, has existed from time immemorial. The results of this system may be seen in the shape of excellently stocked forests (aggregating over 2,000 hect., equals 4940 acres) of Scotch pine, spruce, silver fir and beech up to 120 years old, with a mean maturity increment per hectare of 6 cubic metres, or 85.7 cub. ft. per acre.

GENERAL REMARKS ON THE SYSTEM OF HEAVY THINNINGS.

The system of heavy thinnings in high forests, combined with the raising of underwood, so as to produce as it were a forest consisting of two stories, the upper storey of trees which, like the oak and the Scotch pine, require much light. and the lower of shade supporting trees, such as beech and silver fir, has, as explained, first been applied in a methodical manner and upon a large scale to the oak and and beech in the Spessart about forty to fifty years ago. It is not impossible that the natural mixed forests of Scotch pine and beech in the Steigerwald, a beautiful forest-clad mountain range, situated east of the Spessart, may have given the idea of improving the growth of Scotch pine by means of an underwood of beech. To pure beech forest the principle was applied about 1830 in the Solling, a hilly country consisting of red sandstone, situated east of the Weser River. In these forests, which were burdened with heavy prescriptive rights of old standing, it was at that time found difficult to satisfy the requirements of right-holders in the matter of wood; and with the view of meeting immediate needs without at the same time impairing the productiveness of the forests, a new system of treatment was devised by Christian von Seebach, who at that time had the control of those forests under the Government of the former kingdom of Hanover. The period of rotation was 100 to 120 years, but all compartments which had attained that age had been gradually cut and renewed, and it became necessary to commence cuttings in forests 70 to 80 years old, of which, fortunately, there was a large area. In the areas taken in hand, about three-fifths of the trees were cut, the ground got covered with a dense growth of self-sown seedlings, with a few coppice shoots, and thirty to forty years later the crowns of the trees left standing had again closed and had formed a complete canopy. In this manner a portion of the crop was cut by way of anticipation, and what remained had more vigorous growth through the greater space given to the trees, the ground remaining all clothed with what may be termed the ground floor or the lower story of the forest. Subsequently, as the canopy of the older trees became complete, the undergrowth gradually went back, and most of it died. The forest, after it had attained the full age prescribed by the term of rotation, was then cut by means of successsive fellings, and renewed by self-sown seed in the usual manner.

Of late years this system of heavy thinnings has been somewhat overdone in various localities, and a great deal has been written upon the subject. It would lead too far on the present occasion to enter further into this matter.

TREATMENT.

The rotation in this forest is 120 years, divided into six periods of twenty years each, but it is part of the general system of treatment followed here to allow selected vigorous oak and Scotch pine trees to remain on the ground when a piece of forest is cleared which had attained maturity, and thus to produce larger timber than the ordinary rotation would yield. This method is of old standing. In a portion of compartment Laubchesbusch (4) of the "Oberwald," there is a beech forest dating from about 1808, with a number of Scotch trees at that time 60 years old, which were allowed to remain standing when the original forest was cleared. The result, vigorous and well grown Scotch pine, 60 years older than the beech forest which surrounds them, proves the excellency of the arrangement. The old Scotch pine trees standing over the young growth of oak in Kaisertanne (2a) have already been mentioned. In this case the expediency of the measure is somewhat doubtful, the Scotch pine being rather aged (140 years), and, as a matter of fact, a portion of these old trees have already become dry, and had to be removed. In compartment Scheerwald (7) (Oberwald), which is under renewal at present, and where the last clearance is expected to be made in 1890, after a period of regeneration of about fifteen years, it is intended to hold over fifty to one hundred stems (oak and Scotch pine) per hect., the young growth consisting of oak and beech, with groups of silver fir and maple, planted chiefly where stumps have been rooted up.

In the regeneration of these forests, night frosts are one of the chief difficulties and it may here be mentioned that this is felt throughout the tract with comparatively dry climate, which extends from the foot of the Taunus range to the Rhine. In the Rhine valley, near Darmstadt, I am told that there is hardly a month in spring and summer when night frosts do not occur, and it is not impossible that there is a connection between the frequency of night frosts and a comparatively dry climate. This circumstance has, to some extent, influenced the treatment of these forests. Species which are readily damaged by frost, such as the beech and silver fir, can here only be raised under cover, and even the oak greatly profits while young by a certain amount of shelter. The combination of field crops with sylviculture, the system of partial clearances with underwood, and the method of allowing older trees to stand among the young forest, all these measures have a special value in this district, where a young forest growth is so much exposed to damage by night frosts.

Great stress is justly laid in this forest district upon the early cutting out of brushwood, of soft and inferior woods, and of woods which have served their object in acting as nurses to the more valuable kinds. A considerable amount of pruning also is done, always with the saw. Thinnings are commenced early, and under ordinary circumstances, are repeated once in ten years. The peculiar treatment of these forests, which results in mixed forests, consisting of different species, necessitates much attention to these operations, whereby the development of the more valuable kinds is generally promoted. Fortunately, the vicinity of the town makes it possible, as a rule, to sell nearly all the small wood which is the result of these operations. At times, however, the market gets overstocked, and these operations have then to be delayed.

PROVISIONS OF THE BADEN FOREST LAW ON THE SUBJECT OF PASTURE.

32. In high forests, pasture is only admissable where the young growth has attained the age of thirty-five years in deciduous, and of thirty years in coniferous forests.

In coppice woods, pasture is not permitted unless the young growth in hard wood is twenty-five, and in soft wood, twelve years old. Where the forest is mixed, the age of the dominating kind, and in cases of doubt, the age of the hard wood, decides the point.

33. Forest pasture can only take place between the months of May and October.

34. Before sunrise and after sunset, cattle are not allowed in the forests : exceptions may be made in those cases where, on account of the distance, the cattle must remain in the forest. In such cases, however, they must be kept during the night in sheds, or within a ring fence.

35. Unless proprietors of cattle have a right to use certain paths, the line of road to be used in going to the pasture grounds and watering places, is indicated by the forest officers.

36. Sheep and goats are not admitted into the forests. In special cases, exceptions may be made by the forest authorities, with the consent of the forest proprietor.

37. Each head of horned cattle must be provided with a bell.

38. Each village community is obliged to employ one or several herdsmen for their cattle, as may be necessary.

The members of a village community may not drive their cattle themselves into the forests, nor may they send them with a herdsman of their own, separate from the herd of the village. Where a right of pasture does not belong to a village, but to an association of proprietors, they must entertain a common herdsman for their cattle. Single proprietors having rights of pasture entertain their own herdsmen.

FOREST DESTRUCTION: ITS CAUSES AND RESULTS.*

FORESTS OF UNITED STATES.

In his book, "The Earth as Modified by Human Action," Mr. George P. Marsh devotes considerable space to the effects upon the earth's surface and the conditions of human life which have followed the removal of forests both in Europe and America. Some of the more pertinent paragraphs are herewith appended :—

It is, perhaps, a misfortune to the American Union that the State Governments have so generally disposed of their original domain to private citizens.

Within the memory of almost every man of mature age timber was of so little value in the northernmost States that the owners of private woodlands submitted, almost without complaint, to what would be regarded elsewhere as very aggravated trespasses upon them. Persons in want of timber helped themselves to it wherever they could find it, and a claim for damages for so insignificant a wrong as cutting down and carrying off a few pine or oak trees was regarded as a mean-spirited act in a proprietor. The habits formed at this period are not altogether obsolete, and even now the notion of a common right of property in the woods still lingers, if not as an opinion at least as a sentiment. Under such circumstances it has been difficult to protect the forest, whether it belonged to the State or to individuals. Property of this kind is subject to plunder as well as to frequent damage by fire.

It is evidently a matter of the utmost importance that the public and especially land-owners be aroused to a sense of the dangers to which the indiscriminate clearing of the woods may expose, not only future generations, but the very soil itself.

Some of the American States, as well as the Governments of many European colonies, still retain the ownership of great tracts of primitive woodland. The State of New York, for example, has in its north-eastern counties a vast extent of territory in which the lumberman has only here and there established his camp, and where the forest, though interspersed with permanent settlements, robbed of some of its finest pine groves, and often ravaged by devastating fires, still covers far the largest proportion of the surface. Through this territory the soil is generally poor, and even the new clearings have little of the luxuriance of harvest which distinguishes them elsewhere. The value of the land for agricultural uses is therefore very small, and few purchases are made for any other purpose than to strip the soil of its timber. It is desirable that some large and easily accessible region of American soil should remain, as far as possible, in its primitive condition, at once a museum for the instruction of the student, a garden for the recreation of the lover of nature, and an asylum where indigenous tree and humble plant that loves the shade, and fish and fowl and four-footed beast may dwell and perpetuate their kind, in the enjoyment of such imperfect protection as the laws of a people jealous of restraint can afford them. The immediate loss to the public treasury from the adoption of this policy would be inconsiderable, for these lands are sold at low rates. The forest alone, economically managed, would without injury, and even with benefit to its permanence and growth, yield a regular income larger than the present value of the fee.

*Marsh ; The Earth as Modified by Human Action.

The collateral advantages of the preservation of these forests would be far greater. Nature threw up those mountains and clothed them with leafy woods, that they might serve as a reservoir to supply with perennial waters the thousand rivers and rills that are fed by the rains and snows of the Adirondacks, and as a screen for the fertile plains of the central counties against the chilling blasts of the north wind which meet with no other barrier in their sweep from the north pole. The climate of northern New York even now presents greater extremes of temperature than that of southern France. The long-continued cold of winter is more intense, the short heats of summer even fiercer than in Provence, and hence the preservation of every influence that tends to maintain an equilibrium of temperature and humidity is of cardinal importance. The felling of the Adirondacks woods would ultimately involve, for northern and central New York, consequences similar to those which have resulted from the laying bare of the southern and western declivities of the French Alps, and the spurs, ridges and detached peaks in front of them.

It is true that the evils to be apprehended from the clearing of the mountains of New York may be less in degree than those which a similar cause has produced in southern France, where the intensity of its action has been increased by the inclination of the mountain declivities, and by the peculiar geological constitution of the earth. The degradation of the soil is perhaps not equally promoted by a combination of the same circumstances in any of the Atlantic States, but still they have rapid slopes and loose and friable soils enough to render widespread desolation certain if the further destruction of the woods is not soon arrested. The effects of clearing are already perceptible in the comparatively unviolated region of which I am speaking. The rivers which rise in it flow with diminished currents in dry seasons, and with augmented volumes of water after heavy rains. They bring down larger quantities of sediment, and the increasing obstructions to the navigation of the Hudson, which are extending themselves down the channel in proportion as the fields are encroaching upon the forest, give good grounds for the fear of irreparable injury to the commerce of the important towns on the upper waters of that river, unless measures are taken to prevent the expansion of "improvements" which have already been carried beyond the demands of a wise economy.

In the Eastern United States, wherever a rapid mountain slope has been stripped of wood, incipient ravines already plough the surface, and collect the precipitation in channels which threaten serious mischief in the future.

There is a peculiar action of this sort on the sandy surface of pine forest, and in other soils that unite readily with water, which has excited the attention of geographers and geologists. Soils of the first kind are found in all the Eastern States; those of the second are more frequent in the exhausted counties of Maryland, where tobacco is cultivated, and in the more southern territories of Georgia and Alabama. In these localities the ravines which appear after the cutting of the forest, through some accidental disturbance of the surface, or, in some formations through the cracking of the soil in consequence of great drought or heat, enlarge and extend themselves with fearful rapidity.

In Georgia and in Alabama, Lyell saw "the beginning of the formation of hundreds of valleys in places where the primitive forest had been recently cut down." One of these, in Georgia, a soil composed of clay and sand produced by the decomposition *in situ* of hornblendic gneiss with layers and veins of quartz, "and which did not exist before the felling of the forest twenty years previous," he describes as more than fifty-five feet in depth, three hundred yards in length, and from twenty to one hundred and eighty feet in breadth. He refers to other cases in the same States "where the cutting down of the trees, which had prevented the rain from collecting into torrents and running off in sudden land floods, has given rise to ravines from seventy to eighty feet deep."

Similar results often follow in the north-eastern States from cutting the timber on the "pine plains" where the soil is usually of a sandy composition and loose texture.

WOODLANDS IN EUROPEAN COUNTRIES.

In 1862, Rentzsch calculated the porportions of woodland in different European countries as follows :—

Norway	66.00	per cent.
Sweden	60.00	"
Russia	30.90	"
Germany	26.58	"
Belgium	18.52	"
France	16.79	"
Switzerland	15.00	"
Sardinia	12.29	"
Neapolitan States	9.43	"
Holland	7.10	"
Spain	5.52	"
Denmark	5.50	"
Great Britian	5.00	"
Portugal	4.40	"

In many places peat is generally employed as a domestic fuel, hence, though Norway has long exported a considerable quantity of lumber, and the iron and copper works of Sweden consume charcoal very largely, the forests have not diminished rapidly enough to produce very sensible climatic or even economical evils.[*]

FORESTS OF GREAT BRITAIN.

The proportion of forest is very small in Great Britain, where, on the one hand, a prodigious industrial activity requires a vast supply of ligneous material, but where, on the other, the abundance of coal, which furnishes a sufficiency of fuel, the facility of importation of timber from abroad, and the conditions of climate and surface combine to reduce the necessary quantity of woodland to its lowest expression.

With the exception of Russia, Denmark and parts of Germany, no European countries can so well dispense with the forests, in their capacity of conservative influences, as England and Ireland. Their insular position and latitude secure an abundance of atmospheric moisture; the general inclination of surface is not such as to expose it to special injury from torrents, and it is probable that the most important climatic action exercised by the forest in these portions of the British Empire, is in its character of a mechanical screen against the effects of wind. The

---

[*] Railway ties, or sleepers, are largely exported from Norway to India, and sold at Calcutta at a lower price than timber of equal quality can be obtained from the native woods.

From 1861 to 1870, Norway exported annually, on the average, more than 60,000,000 cubic feet of lumber.

Since 1872 the quantity of the annual exportation of timber from Norway and Sweden has steadily increased, and in 1881 it was so large that it might well excite the grave anxiety of all friends of the primeval forest.

6 (F.)

due proportion of woodland in **England** and Ireland is, therefore, a question not of geographical but almost purely of economical expediency, to be decided by the comparative direct pecuniary return from forest growth, pasturage and plough land.

In England, aboriculture, the planting and nursing of single trees has, until comparatively recent times, been better understood than sylviculture, the sowing and training of the forest. But this latter branch of rural improvement now receives great attention from private individuals, though not, so far as I know, from the National Government, except in the East Indian provinces, where the forestal department has assumed great importance. Many laws for the protection of the forest, as a cover for game and for the preservation of ship timber, were enacted in England before the 17th Century. The Statutes I Eliz. c. xv., XIII Eliz. c. v., and XXVII Eliz. c. XIX., which have sometimes been understood as designed to discourage the manufacture of iron, were obviously intended to prevent the destruction of large and valuable timber, useful in ordinary and naval architecture, by burning it for charcoal. The injury to the forges was accidental, not the purpose of the laws.

In Scotland, where the country is for the most part broken and mountainous, the general destruction of the forests has been attended with very serious evils, and it is in Scotland that many of the most extensive British forest plantations have now been formed.

FORESTS OF FRANCE.

The preservation of the woods was one of the wise measures recommended to France by Sully, in the time of Henry IV., but the advice was little heeded, and the destruction of the forest went on with such alarming rapidity, that, two generations later, Colbert uttered the prediction: "France will perish for want of wood." Still, the extent of wooded soil was very great, and the evils attending its diminution were not so sensibly felt, that either the Government or public opinion saw the necessity of authoritative interference, and in 1750 Mirabeau estimated the remaining forests of the kingdom at seventeen millions of hectares (42,000,000 acres).

In 1860 they were reduced to eight millions (19,760,000 acres) or at the rate of 82,000 hectares (202,000 acres) per year.

In a country and a climate where the conservative influences of the forest are so necessary as in France, trees must cover a large surface and be grouped in large masses, in order to discharge to the best advantage the various functions assigned to them by nature. A large part of its territory is mountainous, sterile, and otherwise such in character or situation, that it can be more profitably devoted to the growth of wood than to any agricultural use.

The conservative action of the woods in regard to torrents and inundations has been generally recognized by the public of France as a matter of prime importance, and the Government has made this principle the basis of a special system of legislation in the protection of existing forests, and for the formation of new. The clearing of woodland, and the organization and functions of a police for its protection are regulated by a law bearing date June 18th, 1859, and provision was made for promoting the restoration of private woods by a statute adopted on the 28th July, 1860. This latter law appropriated 10,000,000, francs to be expended, at the rate of 1,000,000 francs per year, in executing or aiding the replanting of woods.

In 1865 the Legislative Assembly passed a bill amendatory of the law of 1860, providing, among other things for securing the soil in exposed localities by grading, and by promoting the growth of grass and the formation of greensward over the surface.

In 1863, France imported lumber to the value of twenty-five-and-a-half millions of dollars, and exported to the amount of six and a half millions of dollars. The annual consumption of France was estimated in 1886 at 212,-000,000 cubic feet for building and manufacturing, and 1,588,500,000 for firewood and charcoal. The annual product of the forest soil of France does not exceed 70,000,000 cubic feet of wood fit for industrial use, and 1,300,000,000 cubic feet consumed as fuel. This estimate does not include the product of scattered trees on private grounds, but the consumption is estimated to exceed the production of the forests by the amount of about twenty millions of dollars.

The timber for building and manufacturing produced in France comes almost wholly from the forests of the State or of the communes.

FORESTS OF ITALY.

According to statistics, Italy had 17.64 per cent. of woodland in 1872, a proportion which, considering the character of climate and surface, the great amount of soil which is fit for no other purpose than the growth of trees, and the fact that much of the land classed as forest was then either very imperfectly wooded, or covered with groves badly administered, and not in a state of progressive improvement, might advantageously be doubled.

Taking Italy as a whole, we may say she is eminently fitted by climate, soil and superficial formation for the growth of a varied and luxuriant arborial vegetation. In such a country the promotion of forestal industry was among the first duties of her people.

The denudation of the central and southern Appenines and of the Italian declivity of the western Alps began at a period of unknown antiquity, but it does not seem to have been carried to a very dangerous length until the foreign conquests and extended commerce of Rome created a greatly increased demand for wood for the construction of ships and for military material.

The eastern Alps, the western Appenines, and the maritime Alps retain their forests much later; but even here the want of wood, and the injury to the plains and the navigation of the rivers by sediment brought down by the torrents, led to legislation for the protection of the forests by the Republic of Venice at various periods between the fifteenth and nineteenth centuries,[*] by that of Genoa, as early at least as the seventeenth, and both these Governments, as well as several others, passed laws requiring the proprietors of mountain land to replant the woods.

Although no country has produced more able writers on the value of the forest and the general consequences of its destruction than Italy, yet the specific geographical importance of the woods, except as a protection against inundations has not been so clearly recognized in that country as in the States bordering it on the north and west. It must be remembered that the sciences of observation did not become knowledges of practical application till after the mischief was

*According to Hummel, the desolation of the Karsts, the high plateau lying north of Trieste, one of the most parched and barren districts in Europe, was owing to the felling of its woods centuries ago to build the navies of Venice. "Where the miserable peasant of the Karst sees nothing but bare rock swept and scoured by the raging Bora, the fury of this wind was once subdued by mighty firs which Venice recklessly cut down to build her fleets."—*Physiche Geographe*, p. 32.

already mainly done and even forgotten in Alpine Italy, while its evils were just beginning to be sensibly felt in France when the claims of natural philosophy as a liberal study were first acknowledged in modern Europe. The former political condition of the Italian peninsula would have effectually prevented the adoption of a general system of forest economy, however clearly the importance of a wise administration of this great public interest might have been understood. The woods which controlled and regulated the flow of the river-sources were very often in one jurisdiction; the plains to be irrigated or to be inundated by floods and desolated by torrents in another.

Action under a single government can alone render practicable the establishment of such arrangements for the conservation and restoration of the forests, and for the regulation of the flow of the waters as are necessary for the full development of the yet unexhausted resources of that fairest of lands, and even for the maintenance of the present condition of its physical geography.

FORESTS OF GERMANY.

Germany, including a considerable part of the Austrian Empire, from character of surface and climate, and from the attention which has long been paid in all the German States to sylviculture and forestry, is in a far better condition in this respect than its more southern neighbors; and though in the Alpine Provinces of Bavaria and Austria the same improvidence which marks the rural economy of the corresponding districts of Switzerland, Italy, and France has produced effects hardly less disastrous, yet, as a whole, the German States must be considered as in this respect the model countries of Europe. Not only is the forest area in general maintained without diminution, but new woods are planted where they are specially needed, and though the slow growth of forest trees in those climates reduces the direct pecuniary returns of woodlands to a minimum, the governments wisely persevere in encouraging this industry. The exportation of sawn lumber from Trieste is large, and in fact the Turkish and Egyptian markets are in great part supplied from this source.

As an instance of the scarcity of fuel in some parts of Bavaria, where, not long since, wood abounded, the fact may be mentioned that the water of salt springs is, in some instances, conveyed to the distance of sixty miles, in iron pipes, to reach a supply of fuel for boiling it down.

The Austrian Government has made energetic efforts for the propagation of forests in Tyrol and on the desolate wastes of the Karst. In 1866 upwards of 400,000 trees had been planted on the Karst, and great quantities of seed sown. The results of this important experiment are said to be encouraging, (*Chronique Forestiere* in the *Revue des Eaux et Forets, Feb. 1870.*) Later accounts state that the Government nurseries of the Karst supplied between 1869 and 1872, 26,000,000 young forest trees for planting, and that of 70,000 ash trees planted in the Karst scarcely one failed to grow.*

FORESTS OF RUSSIA.

Russia, which we habitually consider as substantially a forest country, which has in fact a large proportion of woodland, is beginning to suffer seriously for want of wood. Jourdier observes:—" Instead of a vast territory with immense forests, which we expect to meet, one sees only scattered groves thinned by the

* For information respecting the forests of Germany, as well as other European countries, see the very valuable *Manuale d'Arte Forestale* of Siemoni, 2d. edizione, Firenze, 1872.

wind or by the axe of the *moujik*, grounds cut over and more or less recently cleared for cultivation. There is probably not a single district in Russia which has not to deplore the ravages of man or fire, those two great enemies of Muscovite sylviculture. This is so true, that clear sighted men already foresee a crisis which will become terrible, unless the discovery of great deposits of some new combustible, as pit coal or anthracite, shall diminish its evils."

Hohenstein, who was long professionally employed as a forester in Russia, describes the consequences of the general war upon the woods in that country as most disastrous and as threatening still more ruinous evils. The river Volga, the life artery of Russian internal commerce, is drying up from this cause, and the great Muscovite plains are fast advancing to desolation like that of Persia.

ECONOMY OF WOODLANDS.

The action of the forest, considered merely as a mechanical shelter to grounds lying to the leeward of it, might seem to be an influence of too restricted a character to deserve much notice, but many facts concur to show that it is a most important element in local climate. Experience, in fact, has shown that mere rows of trees, and even much lower obstructions, are of essential service in defending vegetation against the action of the wind. Hardy proposes planting in Algeria, belts of trees at the distance of 100 metres from each other, as a shelter, which experience has proved to be useful in France.

In the report of a committee appointed in 1836 to examine an article of the forest code of France, Arago observes :—" If a curtain of forest on the coast of Normandy and of Brittany were destroyed, these two Provinces would become accessible to the winds from the west, to the mild breezes of the sea. Hence a decrease of the cold of the winter. If a similar forest were to be cleared on the eastern border of France, the glacial east wind would prevail with greater strength, and the winters would become more severe. Thus the removal of a belt of wood would produce opposite effects in the two regions."

It is thought in Italy that the clearing of the Appenines has very materially affected the climate of the valley of the Po. It is asserted in *Le Alpi che cingono l'Italia* that :—" In consequence of the felling of the woods on the Appenines, the sirroco prevails greatly on the right bank of the Po, in the Parmesan territory, and in a part of Lombardy ; it injures the harvests and the vineyards, and sometimes ruins the crops of the season."

According to the same authority, the pinery of Porto, near Ravenna—which is twenty miles long, and is one of the oldest pine woods in Italy—having been replanted with resinous trees after it was unfortunately cut, has relieved the city from the sirroco to which it had become exposed, and in a great degree restored its ancient climate.[*]

The local retardation of spring so much complained of in Italy, France, and Switzerland, and the increased frequency of late frosts at that season, appear to be ascribable to the admission of cold blasts to the surface, by the felling of the forests, which formerly both screened it as by a wall and communicated the

---

[*]The following well attested instance of a local change of climate is probably to be referred to the influence of the forest as a shelter against cold winds. To supply the extraordinary demand for Italian iron, occasioned by the exclusion of English iron in the time of Napoleon I, the furnaces of the valleys of Bergamo were stimulated to great activity. "The ordinary production of charcoal not sufficing to feed the furnaces and the forges, the woods were felled, the copses cut before their time, and the whole economy of the forest was deranged. At Piazzatorre there was such a devastation of the woods, and consequently such an increased severity of climate, that maize no longer ripened. An association, formed for the purpose, effected the restoration of the forest and maize flourishes again in the fields of Piazzatorre." Report by G. Ross, in Il Politecnico, Dicembre, 1861, p. 614.

warmth of their soil to the air and earth to the leeward. Caimi states that since the cutting down of the woods of the Apennines, the cold winds destroy or stunt the vegetation, and that, in consequence of "the usurpation of winter on the domain of spring," the district of Mugello has lost its mulberries, except the few which find, in the lee of buildings, a protection like that once furnished by the forests.

Influence of the Forest on the Flow of Springs.

It is an almost universal and I believe well-founded opinion that the protection afforded by the forest against the escape of moisture from its soil by superficial flow and evaporation, insures the permanence and regularity of natural springs, not only within the limits of the wood, but at some distance beyond its borders, and thus contributes to the supply of an element essential to both vegetable and animal life. As the forests are destroyed, the springs which flow from the woods, and consequently the greater watercourses fed by them, diminish both in number and in volume. This fact is so familiar throughout the American States and the British Provinces that there are few old residents of the interior of those districts who are not able to testify to its truth as a matter of personal observation.

The hills in the Atlantic States formerly abounded in springs and brooks, but in many parts of these States, which were cleared a generation or two ago, the hill pastures now suffer severely from drought, and in dry seasons furnish to cattle neither grass nor water.

Almost every treatise on the economy of the forest adduces facts in support of the doctrine that the clearing of the woods tends to diminish the flow of springs and the humidity of the soil.

Marchand cites the following instances:—

"Before the felling of the woods within the last few years, in the valley of the Soulce, the Combe-es-Mounin and the Little Valley, the Sorne furnished a regular and sufficient supply of water for the iron works of Unterwyl, which was almost unaffected by drought or by heavy rains. The Sorne has now become a torrent, every shower occasions a flood, and after a few days of fine weather, the current falls so low that it has been necessary to change the water wheels, because those of the old construction are no longer able to drive the machinery, and at last to introduce a steam engine to prevent the stoppage of the works for want of water."

"The spring of Combefoulat, in the commune of Seleate, was well known as one of the best in the country; it was remarkably abundant, and sufficient, in the severest droughts, to supply all the fountains of the town; but as soon as considerable forests were felled in Combe-de-pré Martin and in the valley of Combefoulat, the famous spring, which lies below these woods, has become a mere thread of water, and disappears altogether in times of drought."

"The Wolf spring, in the commune of Soubey, furnishes a remarkable example of the influence of the woods upon fountains. A few years ago this spring did not exist. At the place where it now rises, a small thread of water was observed after very long rains, but the stream disappeared with the rain. The spot is in the middle of a very steep pasture inclining to the south. Eighty years ago, the owner of the land, perceiving that young firs were shooting up in the upper part of it, determined to let them grow, and they soon formed a flourishing grove. As soon as they were well grown, a fine spring appeared in place of the occasional rill, and furnished abundant water in the longest droughts. For forty or fifty years this spring was considered the best in the Clos du Doubs.

A few years since, the grove was felled, and the ground turned again to a pasture. The spring disappeared with the wood, and is now as dry as it was ninety years ago."

IMPORTANCE OF SNOW.

The quantity of snow that falls in extensive forests far from the open country, has seldom been ascertained by direct observation, because there are few meteorological stations in or near the forest. According to Thompson, the proportion of water which falls in snow in the northern States does not exceed one-fifth of the total precipitation, but the moisture derived from it is doubtless considerably increased by the atmospheric vapour absorbed by it, or condensed and frozen on its surface. Though much snow is intercepted by the trees, and the quantity on the ground in the woods is consequently less than in open land in the first part of the winter, yet most of what reaches the ground at that season remains under the protection of the wood until melted, and as it occasionally receives new supplies, the depth of snow in the forest in the latter half of winter is considerably greater than in the cleared fields. Measurements in a snowy region in New England in the month of February, gave a mean of thirty-eight inches in the open ground and forty-four inches in the woods, but the actual difference between the quantity of snow in the woods and that in the open ground in the latter part of winter, is greater than the measurements would seem to indicate. In the woods the snow, which remains constant, is consolidated by a pressure, while in the open ground, being blown off, or thawed several times in the course of the winter, it seldom becomes as densely packed as in the woods, except in the bottom of valleys or other positions where it is sheltered both from wind and sun.

The water imbibed by the soil in winter sinks until it meets a more or less impermeable or a saturated stratum, and then, by unseen conduits, slowly finds its way to the channels of springs, or oozes out of the ground in drops, which unite in rills, and so all is conveyed to the larger streams, and by them finally to the sea.

IMPORTANCE OF SUMMER RAINS.

In countries like the United States (and Canada) where rain is comparatively rare during the winter and abundant during the summer half of the year, common observation shows that the quantity of water furnished by deep wells and by natural springs depends almost as much upon the rains of summer as upon those of the rest of the year, and, consequently, that a large portion of the rain of that season must find its way into strata too deep for the water to be wasted by evaporation.

According to observation at one hundred military stations in the United States, the precipitation ranges from three and one-quarter inches at Fort Yuma in California, to about seventy-two inches, at Fort Pike, Louisiana, the mean for the entire territory, not including Alaska, being thirty-six inches. In the different sections of the Union it is as follows :—

Northeastern States	41	inches.
New York	36	"
Middle States	$40\frac{1}{2}$	"
Ohio	40	"
Southern States	51	"
S. W. States and Indian Territories	$39\frac{1}{2}$	"
Western States and Indian Territories	30	"
Texas and New Mexico	$24\frac{1}{2}$	"
California	$18\frac{1}{2}$	"
Oregon and Washington Territory	50	"

The mountainous regions, it appears, do not receive the greatest amount of precipitation.

The average downfall of the Southern States, bordering on the Atlantic and the Gulf of Mexico, exceeds the mean of the whole United States, being no less than fifty-one inches, while on the Pacific coast it ranges from fifty to fifty-six inches.

INCREASED DEMAND FOR LUMBER.

With increasing population and the development of new industries come new drains upon the forests from the many arts for which wood is the material. The vast extension of railroads, of manufactures and the mechanical arts, of military armaments, and especially of the commercial fleets and navies of Christendom, within the present century have incredibly augmented the demand for wood, and but for improvements in metallurgy and the working of iron, which have facilitated the substitution of that metal for wood, the last twenty-five years would have almost stripped Europe of her last remaining tree fit for these uses.

Let us take the supply of timber for railroad ties. According to Clavé, France had in 1862, 9,000 kilometres of railway in operation, 7,000 in construction, half of which is built with a double track. Adding turn-outs and extra tracks at stations, the number of ties required for a single track is stated at 1,200 to the kilometre, or, as Clavé computes, for the entire net-work of France, 58,000,000. This number is too large for 16,000 + 8,000 for the double track half way = 24,000, and 24,000 × 1,200 = 28,800,000. Gandy states in 1863, that 2,000,000 trees had been felled to furnish the ties for the French railroads, and as the ties must be occassionally renewed, and new railways have been constructed since 1863, we may probably double this number.

The United States had in operation on the first of January, 1872, 61,000 miles, or about 97,000 kilometres of railroad. Allowing the same proportion as in France, the United States railraods required 116,400,000 ties. The number of ties annually required for these railways was estimated at 30,000,000. The annual expenditure for lumber, buildings, repairs and cars was estimated at $38,000,000, and the locomotive fuel, at the rate of 19,000 cords of wood per day, at $50,000,000.

The walnut trees cut in Italy and France to furnish gunstocks to the American Army, during the late civil war, would alone have formed a considerable forest.

The consumption of wood for lucifer matches is enormous, and thousands of acres in extents are purchased and felled, solely to supply timber for this purpose. The United States Government tax, at one cent per hundred, produced $2,000,000 per year, which shows a manufacture of 20,000,000,000 matches. Allowing nothing for waste, there are about fifty matches to the cubic inch of wood or 86,400 to the cubic foot, making in all upwards of 230,000 cubic feet, and as only straight grained wood, free from knots, can be used for this purpose, not less than three or four thousand well-grown pines are required.

Add to all this the supply of wood for telegraph poles, wooden pavements, wooden wall tapestry paper, shoe-pegs, wooden nails, and wood-pulp and other recent applications which ingenuity has devised, and we have an amount of consumption for entirely new purposes, which is really appalling. Wooden field and garden fences are very generally used in America, and some have estimated the consumption of wood for this purpose as not less than that for architectural uses.

Fully one-half our vast population is lodged in wooden houses ; and barns and country out-houses of all descriptions are almost universally of the same material.

The consumption of wood in the United States as fuel for domestic purposes, for charcoal, for brick and lime-kilns, for breweries and distilleries, for steam-boats, and many other uses, defies computation, and is vastly greater than is employed in Europe for the same ends. For instance, in rural Switzerland, cold as is the winter climate, the whole supply of wood for domestic fires, dairies, breweries, distilleries, brick and lime-kilns, fences, furniture, tools, and even house-building and small smitheries, exclusive of the small quantity derived from the trimmings of fruit trees, grape vines, and hedges, and from decayed fences and buildings, does not exceed *two hundred and thirty cubic feet* or less than two cords a year, per household. The annual consumption of firewood by single families in France has been estimated at from two and a half to ten Paris cords of 134 cubic feet.

The report of the Commissioners on the Forests of Wisconsin, 1867, allows three cords of wood to each person for household fires alone. Taking families at an average of five persons, we have eight times the amount consumed by an equal number of persons in Switzerland for this and all other purposes to which this material is ordinarily applicable. It has been estimated that in the cold climate of Sweden, 144 solid, or 200 loose cubic feet of pine or fir are required per head of the population. The consumption in Norway is about the same.

Evergreen trees are thoughtlessly destroyed in immense numbers for the purpose of decoration and on festive occasions. Thrifty young groves of evergreen of considerable extent have been completely destroyed in this reckless way.

France employs 1,500,000 cubic feet of oak per year for brandy and wine casks, which is about half her annual consumption of that material ; and it is not a wholly insignificant fact that, according to Rentzsch, the quantity of wood used in parts of Germany for small carvings and for children's toys is so large that the export of such objects from the town of Sonneberg alone amounted in 1858, to 60,000 centner, or three thousand tons weight.

In an article in the Revue des Eaux et Forets for November, 1868, it is stated that 200,000 dozens of drums for boys were manufactured per month in Paris ; this is equivalent to 28,800,000 per year, for which 56,000,000 drumsticks are required. The consumption of matches in France is given at 7,200,000,000.

EFFECTS OF FOREST FIRES.

Only trees fit for industrial uses fall before the lumberman's axe, but the fire destroys, almost indiscriminately, every age and every species of tree. While, then, without fatal injury to the younger growths, the native forest will bear several "cuttings over" in a generation—for the increasing value of lumber brings into use, every four or five years, a quality of timber which had been before rejected as unmarketable—a fire may render the declivity of a mountain unproductive for a century.

Aside from the destruction of the trees and the laying bare of the soil, and consequently the free admission of sun, rain, and air to the ground, the fire of itself exerts an important influence on its texture and condition. It cracks and sometimes even pulverizes the rocks and stones upon and near the surface ; it consumes a portion of the half decayed vegetable mould which served to hold its mineral particles together, and to retain the water of precipitation, and thus loosens, pulverizes and dries the earth; it destroys reptiles, insects, and worms, with their eggs and the seeds of trees and of smaller plants; it supplies, in the ashes which it deposits on the surface, important elements for the growth of a new forest clothing as well as of the usual objects of agricultural industry; and by the changes thus produced, it fits the ground for the reception of a vegetation different in character from that which had spontaneously covered it. These new conditions help to explain the natural succession of forest crops, so generally observed in all woods cleared by fire and then abandoned. There is no doubt, however, that other influences contribute to the same result, because effects more or less analogous follow when the trees are destroyed by other causes, as by high winds, by the woodman's axe, and even by natural decay. *

When the forest is left to itself the order of succession is constant, and its occasional inversion is always explicable by some human interference. It is curious that the trees which require most light are content with the poorest soils, and *vice versa*. The trees which first appear are also those which propagate themselves farthest to the north.

The birch, the larch, and the fir, bear a severer climate than the oak or the beech.

The difficulty of protecting the woods against accidental or incendiary fires is one of the most discouraging circumstances attending the preservation of natural and the plantation of artificial forests. In the spontaneous wood the spread of fire is somewhat retarded by the general humidity of the soil, and of the beds of leaves which cover it. But in long droughts the superficial layer of leaves and the dry fallen branches become as inflammable as tinder, and the fire spreads with fearful rapidity, until its further progress is arrested by want of material, or more rarely, by heavy rains, sometimes caused, as many meteorologists suppose, by the conflagration itself.

---

* Trees differ in their power of resisting the action of forest fires. Different woods vary greatly in their combustibility, and even when the bark is scarcely scorched, trees are, partly in consequence of physiological character, and partly from the greater or less depth at which their roots habitually lie below the surface, differently affected by running fires. The white pine, *Pinus Strobus*, as it is the most valuable, is also perhaps the most delicate tree of the American forest, while its congener, the northern pitch-pine, *Pinus Rigida*, is less injured by fire than any other tree of our country. Experienced lumbermen maintain that the growth of this pine was even accelerated by a fire brisk enough to destroy all other trees.

UNITED STATES CONSULAR REPORTS ON EUROPEAN FORESTRY.

There is a great deal of useful information respecting European systems of forestry contained in a volume entitled "Forestry in Europe," published at Washington, Government Printing Office, 1887, from which the following extracts have been made :—

COPY OF CIRCULAR.

DEPARTMENT OF STATE, }
WASHINGTON, November 30, 1886. }

To the Consular Officers of the United States:

GENTLEMEN :— *You are instructed to prepare a report covering the following questions on Forest Culture and Forest Preservation. I would ask you to devote especial attention to the practical phases of the question, that your replies may serve as a basis for framing forestry legislation in this country, where the subject is of great and increasing importance.*

1. Areas under forests, distinguishing, where possible, between State and private areas.

2. Common forests, if any, and privileges of the population in them. If pasture is permitted, how are the trees, etc., protected ?

3. Organization and functions of government forest bureaus.

4. Revenues from government forests, cost of maintaining or managing forests ; profits of forest cultivation.

5. Forest planting and culture; methods; bounties, if any; schools, their organization and courses of study.

6. Destruction of forests, causes and results.

7. Reclamation of sand dunes, or waste places by tree planting.

8. Sources of lumber supply ; trade in lumber, bounties on importation, if any, and customs duties.

9. Give the names of three reliable sellers of seeds and shoots in your district.

10. Transmit to the Department copies and translations of the forest laws of the district in which you reside. (The general laws should be forwarded by the Consul-General ; the local laws, by the Consul.)

I am, Gentlemen,

Your Obedient Servant,

JAMES D. PORTER,

Assistant Secretary.

WEIGHTS, MEASURES AND CURRENCY.

Cental	Equals	220 pounds.
Centner meter	"	221.5 pounds.
Florin	"	35.9 cents.
Franc	"	19.3 cents.
Hectare	"	2.471 acres.
Joch	"	1.42 acres.
Kilogram	"	2.2046 pounds.
Mark	"	23.8 cents.

AUSTRIA HUNGARY.—REPORT OF CONSUL-GENERAL JUSSEN.

GOVERNMENT CONTROL.

The forest laws of Austria prescribe and control not only the culture of the forests belonging to the imperial domain, but also all woodlands which are the property of municipalities, private corporations or private individuals, and are based upon the theory of paternal government.

If the law as it stands is enforced not a tree can be cut nor a load of dry leaves gathered in a forest which is situated in Austria except in accordance with certain rules and restrictions, and although there may be much in these laws which may serve for framing future forestry legislation in the United States, the greater portion of the enactment is in direct conflict with the American idea of home government and property rights.

The Austrian Empire is unusually rich in forest lands. There is no lack of dense woods in any of its Provinces, except in Dalmatia and Istria and in the territory near Trieste, and the culture of forest lands may be called exemplary, especially in Bohemia, Moravia, Upper Austria, Silesia and Salzburg.

The yield of these vast forests, although it is said to be on the decline, still far exceeds the home demands, and large quantities are exported.

AREAS UNDER FORESTS, PUBLIC AND PRIVATE.

The latest statistics place the total area of the productive land of the Empire at 28,406,530 hectares ; of these total numbers of hectares 9,227,061.20 hectares are forest lands, and these again are divided into imperial (State), municipal and private forests, as follows :—Imperial forests, 952,689.96 hectares ; municipal forests, 1,297,238.21 hectares. The private forests, therefore, cover about 32 per cent. of the total area of the productive land of the Empire.

COMMON FORESTS AND PRIVILEGES OF THE POPULATION IN THEM.

As common forests of the Empire only the woodlands belonging to the several cities and villages can properly be denominated. The residents of these cities and villages undoubtedly enjoy certain privileges as to the use of these forests, by virtue of the local laws and regulations. I am not in a position, however, to have access to these local regulations, which undoubtedly differ in the different communities, but are one and all subject to the general law on forest culture and preservation hereinafter cited. This general law, if strictly enforced, furnishes the means of ample protection against any injury that may possibly threaten these common forests by the wasteful or careless exercise of any privilege granted by local enactment.

Organization and Functions of Government Forest Bureaus.

The cultivation and preservation of the forests of the Empire of Austria and the administration of the laws with reference thereto are entrusted to the Ministry of Agriculture. The right of appeal, however, in certain contested cases to the Ministry of the Interior is reserved.

Under the supervision of the Minister of Agriculture the several Provincial presidents (*statthalters*) are authorize l to execute the forest laws and regulations, and as next in authority to these *statthalters* the several district captains are empowered to enforce the laws in question, and to exercise a general authority, supervision and control over all the subordinate officers charged with the execution of the forest laws, and forest police regulations. This subordinate class of forestry officers is composed of two classes :—

1. The officers who have entered the service permanently, after passing the requisite examination, and are in the line of promotion, like officers of the regular army.

2. The volunteer officers who for the sake of pursuing their studies and adding practical experience to theoretical knowledge, accept the position in the forest service as an honorable distinction, but receive a salary in proportion to the extent of their field of action and responsibility.

This latter class, however, like the first, must have passed certain examinations, proving their qualifications before they can enter the service as such volunteers.

The professional and regular forest officers in the Empire are classified as follows :—

A. Forest Inspectors.

2. Chief forest counsellors (called oberforesträthe).

5. Forest counsellors (or forestnäthe).

7. Chief forest commissaries (called oberforestcommissäre).

B. Forest Technicists.

Forest inspection commissaries (called forestthechniker).

C. Forest Wards Belonging to the Category of Servants.

Forest wards, class I, salary per annum, 500 florins.

Forest wards, class II, salary per annum, 400 florins.

Forest wards, class III, salary per annum, 300 florins.

The forest inspectors are charged with the duty of superintending the execution of all forest laws, of examining the condition of the forest, fostering and furthering instructions in forest culture and acting as adjuncts to the Statthalter.

From early spring until late in the fall the forest inspector should visit and inspect the forests in his district and make a report of each inspecting tour to the Statthalter.

The Statthalter may also order the forest inspector to make special inspecting tours in addition to the regular tour.

The forest inspector is required to inspect the offices of the district captains with reference to forest affairs.

The instructions to forest inspectors contain sufficient points, elaborately presented, to fill a moderate-sized pamphlet, and the gist of the whole matter is that the forest inspector acts as a paternal adviser, and if need be as an imperative commander to all owners of forests in the empire, as well as a superintendent of imperial forests.

He controls and commands private owners as to the manner and order in which they should cut their timber, as to the necessity of replanting, the preventing of waste, the preservation of timber against floods, and as to the danger and injury threatening from insects, as to the fitness and capacity of the subordinate forest inspectors and hunters and forest wards to be employed by these owners; in short, there is not a single act of ownership which the holder of the titled deeds of woodland could possibly exercise over his own domain which is not directly under the control, and which does not require the approval of the forest inspector.

In the light of these instructions it is not at all paradoxical to say that the owner of forest land in Austria must exercise extraordinary care not to be guilty of trespass upon his own lands. There can be no question, however, that this paternal control has achieved most excellent practical results, though it is said that the discipline of forest officials has been lax, and that the laws and instructions have not been enforced with uniform strictness. The forest technicists and forest wards are the subordinate officers of the forest, instruments by which the duties above enumerated and imposed upon the forest inspectors are practically performed.

Forest Register (Waldkataster).

In pursuance of a decree of the Ministry of Agriculture, under date of July 3, 1873, the respective forest officers are required to keep a forest register of each district, which specifies the number of acres covered by forest, its condition, state of growth, etc.

In connection with this register maps are prepared and kept open for inspection at the offices of the district captains upon which the condition and extent of the several forests in the districts are shown.

At the close of each year a report about the progress of forest culture, etc., is to be made to the Ministry of Agriculture, which report is to be published in the *Landes Zeitung*.

The total number of forest officers of all grades, public and private, employed in Austria, reaches the respectable figure of 31,826.

Revenues from Government Forests—Cost of Maintaining or Managing Forests—Profits of Forest Cultivation.

On the point of the profits of Government forests there are absolutely no statistics published in the Empire, so far as I have been able to ascertain, except those given in the budget under the head of forest revenues and expenditures. The last budget published places :—

The forest revenues, p. a. florins	3,951,650
The forest expenditures, p. a. florins	3,546,240
Profit of State forests	405,410

These net proceeds of an area of government forest land, containing 952,689.96 hectares, certainly seem very inconsiderable, but in order to estimate the true value of these forests to the Empire, their influence upon the climate, the rainfalls, and the consequent benefit to agricultural land, as well as to the health of the population, should be taken into consideration.

A direct benefit also results to the population from the employment of numerous officers attending to the cultivation and preservation of these forests, all of whom are paid and supported by the profits derived from the culture.

It cannot be contended, therefore, that the people are taxed in order to support a small army of forest officers, who are actually producers, earning more than they expend.

I find on an examination of the meagre statistics to which I have had access that the profits of forest land culture have materially increased during the last fifty years.

The Kataster (Real Estate Register) shows that in Lower and Upper Austria the net profit per joch was estimated at 1.41 florins in the year 1830, while in the year 1880 this estimate rose to 2.62 florins per joch, an increase of almost 100 per cent., an incontrovertible proof that the forest laws of Austria, which were passed in 1852, have been of great practical benefit to forest culture.

This benefit is proven, not only by the increased net proceeds of a given area of forest lands, but also by the growth and greater extent of the area itself.

FOREST PLANTING AND CULTURE METHODS.—BOUNTIES IF ANY.—SCHOOLS, THEIR ORGANIZATION AND COURSE OF STUDY.

The method of forest planting and culture prevailing in Austria are quite particularly prescribed in the forest laws. There are no bounties paid in the Empire for planting or replanting of forests.

SCHOOLS.

The schools for forest culture were transferred in 1878 from the Minister of Agriculture to the Minister for Culture and Education, but all organic order and appointments of professors are made by the Ministers of Culture and Education with the concurrence of the Minister of Agriculture.

While there are undoubtedly numerous provisions of the forest culture law which cannot be applied or enforced in the United States, the system inaugurated in Austria to fit and educate young men for the duty of enforcing this law seems beyond all question worthy of imitation to the fullest extent.

These Austrian schools for forest culture consist of :—

A. University (hochschule).

B. Middle or preparatory schools.

C. Elementary or lower schools.

The university (hochschule) is situate in Vienna; it was founded in October, 1875.

Its aim and purpose is the highest possible scientific education in land and forest culture, All expenditures are borne by the State. The semesters (terms) are limited to six—that is, complete instruction is not perfected under six semesters.

The Students.

The students are either ordinary?or extraordinary hearers. The ordinary hearer must produce a testimonial as a 'graduate of a gymnasium (college) or high school (*oberrealschule*)—a testimonial which would also admit the students to any university.

Whoever does not possess the qualification of an ordinary hearer may be admitted as an extraordinary hearer if he is eighteen years old and has that degree of preparatory education which will enable him to understand the lectures.

Guests may be admitted to single lectures on notice of the the dean (rector). All hearers are subject to the dicipline regulations of the university.

Immatriculation, Tuition Fee, and Laboratorium Tax.

The immatriculation fee is five florins for all hearers. The ordinary hearers pay a tuition fee of twenty-five florins at the beginning of the semester (term).

Extraordinary hearers pay 1.50 florins (per week) for each lecture.

Ordinary hearers, if poor, may, as a reward for great diligence, be released from the payment of tuition fees if the college of professors so decides.

The laboratorium tax is five florins for fifteen hours.

Certificates of Attendance.

The attendance at lectures is certified to at the end of each semester. In case of non attendance, the fact is stated on the certificate. The certificates are to be delivered to the dean for examination.

Examinations and Testimonials.

The examinations are public and conducted under the supervision of the dean (rector).

In deciding the degree of succes in examinations, not only the written school examination, but also the labor in the laboratory and the authenticated studies in chambers are to be taken into consideration.

Every ordinary hearer has the right to be admitted to the state examination if he so desires.

Regular and full diplomas are only issued to ordinary hearers.

Extraordinary hearers can claim only a testimonial certifying to their attendance at lectures, good conduct and general progress in their studies.

Terms of Examinations for Diplomas in Forest Culture.

First Group.

1. Physics with climatology.
2. Chemistry.
3. General and special botany.
4. Mineralogy and geology.
5. Mathematics.

6. Geodesy.
7. Mechanics.
8. Geometry.
9. National economy.

Second Group.

1. Forest culture.
2. Forest felling with forest technology.
3. Forest preservation with forest zoology.
4. Forest laws
5. Forest yield, regulation and management.
6. Forest statistics.
7. Forest engineering.

The examinations are both oral and in writing, during which the use of, and reference to, books and memoranda are not permitted.

Only ordinary hearers who have performed the three years' course in the university are admitted to examinations for diplomas.

If the student desires to enter the service of the State as a forest officer, he must subject himself to, and pass two State examinations, after he has obtained his university diploma.

The subjects of the first State examination are the following;—Physics, climatology, chemistry, botany, geology, higher mathematics, geodesy, and national economy.

The second State examination embraces the subjects of culture, use and yield of forest lands, calculations on values of forests, forest machinery, and forest laws.

These State examinations are conducted orally and in public. The State issues diplomas to the successful candidates.

MIDDLE OR PREPARATORY SCHOOLS FOR FOREST CULTURE.

Three of these preparatory schools have been established in the Empire, one at Eulenberg, another at Weisswasser, and the third at Lunberg.

The conditions of admission to the Eulenberg school are the following :—

1. The applicant must be a graduate of a lower gymnasium (under gymnasium, or *unterrealschule*, preparatory college.)

2. He must have served with good success for two, or at the very least, one year, as the apprentice of a forest official.

3. He must not be less than seventeen and not more than twenty-four years old.

4. Must be in perfect health and vaccinated.

5. Must furnish security as to means required for instruction, clothing and support.

6. Must pass a preliminary examination by the teachers of the school.

The scholars, whose numbers shall not exceed twenty to twenty-five per annum, reside at the institute.

7 (F.)

The branches taught embrace mathematics, field engineering, drawing, natural history, forest culture, forest laws, business coorrespondence, office routine business, and hunting.

The conditions of admission to the other two middle schools are of about the same character, and nearly the same branches are taught there, all calculated to fit the student for admission at the university at Vienna.

In all these schools excursions are made by the scholars under the guidance of the teachers, for the purpose of combining practical illustration with theoretical knowledge, in the branches of natural history, forest culture, preservation and valuation.

Examinations take place at the end of each semester (term).

ELEMENTARY (NIEDERE) SCHOOLS FOR FOREST CULTURE.

The Ministry of Agriculture has established four of these lower schools, one in Tyrol, one in Styr, one in Galicia, and one in Agglsbach.

Course of Study :—Mathematics, geometrical exercises, field engineering, measuring of wood and timber cut and standing, measuring of arth and excavations, writing, drawing, natural history, geology, mineralogy, zoology, game as distinguished from other animals.

Practical Works :—Felling timber—numbering, measuring and piling same, planting and replanting forests, draining and irrigation, protection against insects and fires, charcoal making, sawing lumber and hunting.

The scholars are also required to construe and explain the most important provisions of the forest laws and to commit them to memory.

They are also taught the use, value, etc., of all building material, viz., wood, lime, bricks, stone, sand, etc., and are instructed in the building and clearing of forest roads, and the securing of the banks of forest streams, and repairing fissures in same, etc.

As teachers in these elementary schools experienced forest officers are detailed.

The discipline in these schools as regards the conduct and studies of the scholars, as well in school as in chambers, is very strict.

No scholar is permitted to absent himself from the institution without leave the side arms and guns intrusted to the scholars for practise, must be cleaned in the presence of the teachers and delivered to their care; all tools used by them must be cared for in the same manner.

If an offence against the regulations is repeated three times, dismissal follows.

Strict moral conduct is enforced, and the scholars are continually under the direct control and supervision of one of the teachers, who is also charged with the duty of visiting the scholars in their rooms.

All moneys belonging to the scholars must be deposited with the teachers, who supply the depositors with the amount actually needed from the deposit funds, and the parents are advised of this regulation.

The regulations of discipline are too voluminous to be cited here in full. They also differ somewhat in the different schools, but on the whole they are framed in a strict military spirit, which looks upon obedience to rules of conduct as a first requisite to a successful course of study.

A young man who has graduated from an elementary to a middle forest school, and from that to the university, or high school of forest culture, who has obtained his diploma at the latter, and has also passed the two State examinations, may be said to be thoroughly fitted for his profession, and besides undoubtedly clean, healthy, robust, and thoroughly manly in a physical as well as in a moral sense.

Extracts from General Forest Law of Austria in Force since January 1, 1853.

Cultivation of Forests.

Sec. 1. Forests are distinguished as (a) State or Imperial forests under the control of the State authorities. (b) Common forests, belonging to the city and country communities. (c) Private forests, belonging either to private individuals, or to corporations, or to orders, monasteries, benefices or prebends.

Sec. 2. No forest can be withdrawn from cultivation and used for other purposes except by consent. This consent can only be granted with reference to State forests by the proper authority, and if questions of strategy or military defence arise the concurrence of the Ministry of war is required.

With reference to common and private forests the consent of the district authorities is required, and all parties interested are to be heard on the application, and in case of conflict of interests the matter is to be submitted to the proper civil judge.

The arbitrary use of forests for other purposes is punished by a fine of five florins per joch. (1 joch equals 1.42 acres).

The area thus converted to improper use must be replanted within a certain time, to be fixed by experts. In case of default the fine is again imposed.

Sec. 3. Newly-cleared tracts of State or common forests must be replanted within five years. A longer time may be allowed for the replanting of private forests, according to circumstances, and in pursuance of the provisions of section 20.

Sec. 4. No forest should be devastated; that is, so treated that the cultivation is either jeopardized or made impossible. If the cultivation has only been jeopardized a fine will be imposed in accordance with section 2, and the replanting is to be enforced. If, however, cultivation has been impossible, a fine up to ten florins per joch will be imposed.

Sec. 5. A cultivation which exposes neighboring forests to injury from winds is prohibited. A strip of woods at least twenty Vienna klafter wide must be left, when such danger exists, along the margin of the neighboring woods until the same is in full growth. In the meantime this wind-cloak can only be thinned.

Sec. 6. On sandy soil and on steep mountain slopes the timber can only be cut in narrow strips or thinned out, and must be immediately replaced. The woods upon the summits of mountains must only be thinned.

See. 7. On the shores of large rivers or lakes, if the shores are not composed of rocks, and on the slopes of mountains where land slides are possible great care is to be exercised, and roots can only be dug if the fissure is immediately repaired.

Sec. 8. Violations of sections 5, 6 and 7, are punished with a fine of from 20 to 200 florins. Damage accruing to others to be paid by offender.

Sec. 9. Provides for cultivation of common forests, and for limitation and official control of grazing and other privileges and uses.

Sec. 10. Grazing is not permitted in young timber, where it might injure the growth, and no more cattle are to be driven into any woods than can find sufficient food within the area. Herdsmen must be employed, and the cattle shall graze together, and not isolated, as much as possible. The driving of cattle to the place of grazing to be done with due regard to the preservation of the forest; if necessary a circuitous route is to be taken.

Sec. 11. Bedding of dry leaves and moss must be gathered only with wooden rakes, without scratching up the soil. In young timber no gathering of bedding is permitted.

Sec. 12. From felled trees all the branches may be cut; from standing trees selected for future cutting, the lower two-thirds of branches may be cut. The young shoots between the strong branches must be preserved. From trees which are not to be felled immediately the branches must be cut between the months of August and March, excepting only during severe frost. No climbing vines are permitted.

Sec. 13. The gathering of bedding can only be permitted on the same ground every third year. Young shoots, may, however, be gathered with permission of owner.

Sec. 14. Provides for regulating the exact time within which parties possessing the privilege may gather bedding. Time to be fixed by the owner.

Sec. 15. Provides for the different marks on timber to be felled.

Sect. 16. Where the preservation of young timber requires it the cutting and transporting of timber must take place in the fall or in winter during snow fall. Generally timber may be cut also in spring or summer, but in such case it must be taken out of the woods before the next ensuing spring.

From trees felled in the green leaf the bark is to be peeled at once; from these felled in the late fall the bark must be taken in strips before the next spring. The stumps must not be left too high. In felling trees, hewing and transporting timber, all injury to standing trees is to be avoided. The same rule obtains with reference to the gathering and transportation of bedding, which must be removed out of the woods within three months.

Sec. 17. All products of the woods must be removed on the road designated by the owner. The time of removal as agreed between owner and purchaser of timber to be requested; if not so requested owner may give fourteen days' notice and dispose of products if notice is not complied with.

Sec. 18. Provides that forest officials (political authorities) shall decide all differences and disputes. Owners of forests who violate regulations to be fined for each offence from 20 to 200 florins.

Sec. 19. Provides that the State, in case of necessity, can take possession of forests for the purpose of protection against avalanches, land slides, etc. If claims for damages arise they are to be settled according to law.

Sec. 20. Provides manner of proceeding for the purpose of taking such possession, examination of experts, etc.

Sec. 21. As a rule no partition of common forests can be made.

Sec. 22. For the purpose of insuring the proper cultivation of forests all owners of forests of sufficient dimensions (which dimensions are prescribed by the authorities) are required to employ only such forests officers as are considered qualified by the government.

Sec. 23. The political authorities are charged with the general superintendence of all forests in their respective districts.

Transportation of Forest Products.

Every land owner is required to permit forest products to be transported across his land if no other outlet is convenient, or if other transportation is too expensive.

The transportation must be conducted with proper care, and all damages accruing must be paid.

The political authorities decide whether such transportation across lands of third parties is necessary and also fix amount of damages, from which decision as to damages an appeal may be taken to the courts.

Sec. 25. Provides for jurisdiction with reference to transportation over public roads, etc.

Sec. 26. The transportation of wood by means of rafts and the building of booms require special permission by the authorities of the district. If the use of private waters is required proceedings must be had according to section 24.

Secs. 27 to 43. Refer to the regulation of rafting, marking of timber transported by rafts, use of rivers and other waters, public and private, for rafting, etc.

Forest Fires and Damages by Insects.

Sec. 44. The greatest care must be exercised in igniting fires or in using combustible materials in or near the forests. If dangers arise in consequence of neglect to use such care the offenders must pay all damages, and may according to circumstances, be either prosecuted under the general criminal code or fined from 5 to 40 florins, or imprisoned from one to eight days.

Sec. 45. Every person who finds a deserted and unextinguished fire in or on the edge of a forest is required to extinguish it if possible. If any person observes a forest fire he is held to give notice to the next inhabitants in the direction of the road which he travels. These parties so notified are required to give notice to the nearest local authorities and to the owner of the forest or to his forest officials.

Sec. 46. All surrounding villages can be required by the owner of the forest, or his forest officials, or by the local authorities to extinguish the fire. The posse must at once repair to the place of the fire with the necessary fire-extinguishing apparatus. The local authorities and the forest officials must accompany the posse.

Sec. 47. Unconditional obedience is to be paid to the superior officer commanding the posse. The other local officials must preserve order among the firemen and cause the execution of the orders. After the fire has been extinguished the place where it occurred is to be guarded from one to two days, or longer, if necessary, and the necessary number of men must be furnished for this purpose.

Sec. 48. Local officers who neglect to perform their duty will be fined from 5 to 50 florins, and all persons who refuse to obey their orders will be punished by a fine of from 5 to 15 florins or by imprisonment from 1 to 3 days.

Sec. 49. Damages to property of third parties, caused by extinguishing these fires, are to be paid by the parties for whose benefit the posse was called, unless this third party was protected against still greater loss by the efforts of the posse.

Sec. 50. The damages caused to forests by insects are to be closely watched. The owners of forests and their employees are required, in case they cannot succeed in preventing the spread of such damage to adjacent woods, or on their own grounds, to notify the political authorities at once, or in default thereof to pay a fine of from 5 to 50 florins. Every person is authorized to give such notice.

Sec. 51. The political authorities, with the assistance of experts, must at once take the proper measures to prevent this damage by insects. All owners of forests whose woods may be in danger are bound to render assistance and to submit to the order of the authorities, who are herewith authorized to enforce their orders.

The expense shall be borne by the owners of the forests in proportion to the dimensions of their respective tracts.

Forest Preservation Service.

Sec. 52. Provides for the organization of forest guards to be attached to the forest administration service. These guards, whether employed by the State, by communities, or by private individuals, to take the oath of office. (Form of oath general, with reference to performance of duties in preserving forests as law requires).

Sec. 53. These sworn guards to be regarded as public guards, with all rights guaranteed to public officers by law, and authorized to carry the usual arms. Every persons is required to obey their orders given in the line of their duties.

Sec. 54. The guards shall use their arms only in case of self-defence. To wear uniforms.

Sec. 55. The guards are authorized to order suspicious persons to leave the forests and to confiscate all tools used for gathering forest products if the parties carrying them in the forest cannot give a satisfactory explanation.

Sec. 56. Confiscation of forest products in possession of suspicious party in the forest.

Sec. 57. Offender who are strangers to the guards are to be arrested; offenders known to the guard are to be arrested only in case they attack or abuse him, or if they have no fixed home. Persons arrested to be delivered at once to the competent authorities.

Sec. 58. In case the offender was caught in the act and took flight he may be pursued beyond the forest and the stolen product attached.

Miscellaneous.

Chapter five contains an enumeration of minor forests offences not hereinbefore particularly mentioned, and fixes the punishment.

These offences are : Gathering of loose wood and twigs, marking and barking of trees, using climbing irons, boring into trees, appropriating bark from felled timber, exposing the roots of trees, cutting or tearing off limbs or twigs or leaves, digging or cutting out young trees, gathering twigs for brooms, gathering tree juice of all sorts, gathering tree seeds or sponges or rotten wood or digging out roots, gathering bedding of all sorts, especially if gathered with hoes or iron rakes ; taking away earth, clay, turf, stones, and other minerals, or cutting sod, or mowing grass and herbs.

This chapter also provides for proceedings and estimate for damages to forests by cattle. Chapter six provides for mode of procedure and proper tribunal to fix damages. Chapter seven provides for proceedings on appeal.

KINGDOM OF PRUSSIA ; REPORT OF CONSUL WAMER OF COLOGNE.

FOREST AREA IN THE PRUSSIAN MONARCHY.

The total area of the Prussian Monarchy amounts to 35,479,536 hectares.* Of this amount 8,124,521 hectares are forests, being an equivalent of 23.33 per cent. of the total area. It may be stated that this estimate includes all land devoted to the culture of wood.

The apportionment of the forests is as follows :—

(a) 29.4 per cent., equivalent to 2,374,039 hectares belong to the State.

(b) 11.9 per cent., equivalent to 983,727 hectares belong to the Communes.

(c) 1.5 per cent., equivalent to 122,759 hectares belong to institutions.

(d) 2.1 per cent., equivalent to 170,063 hectares belong to corporations.

(e) 55.1 per cent., equivalent to 4,473,933 hectares belong to private individuals.

Under the same heads the Rhenish Province and the district of Cologne have the following area respectively :—

RHENISH PROVINCE.		COLOGNE.	
	Hectares.		Hectares.
(a)	143,284	(a)	11,766
(b)	321,019	(b)	7,358
(c)	7,149	(c)	1,773
(d)	15,303	(d)	1,201
(e)	342,687	(e)	98,284

The forests of Prussia stretch from the Baltic coast over the mountains of the Sudeten, Hartz, Thuringia, Teutoburg, Meissner, Taunus, Rhön, and the slate mountains of the Lower Rhine.

According to a rough estimate, 4,043,800 hectares of forest area are level, 2,089,500 hectares are hilly, and 1,991,200 hectares are mountainous.

GOVERNMENT SUPERVISION OVER COMMUNAL FORESTS.

Although the Communes are left free to manage the Communal forests, the State government reserves for itself certain rights over the general administration in order to prevent any mismanagement or abuses. For instance, in Westphalia and the Rhineland, which embrace this consular district, the communities and public institutions are left free to administer their own forests, but at the same time the government gives certain instructions regarding the culture and utilization of the forests, which the local authorities are bound to carry out without any alteration on their part not first consented to by the government. Whether it is considered best that the Commune should appoint the officials intrusted with the supervision of the forest is left to the discretion of the government. In leaving the election of the forest officials to the Communes, they are to elect

*Consul Wamer says :—Considering the vast amount of technical knowledge required to fully comprehend the whole system of forest culture in Prussia, I have found it extremely difficult in preparing a report on this subject, and I am greatly indebted to *Oberforstmeister* (head forest master) von Wurmb, chief of the forest department of the Government district of Cologne for such information.
One hectare is equivalent to 2.471 acres.
The total area of forest in the Empire of Germany is 13,900,611 hectares, or 25 per cent. of the total land area.

such persons whose qualifications are approved by the government, to whom the election is submitted for consideration and confirmation. It is the duty of the government, either by virtue of its office or for some special reason, to examine into any changes made in the management of the Communal forest, and to proceed against all adverse administration by assuming special supervision or by instituting any other judicious precaution.

As technical organs for the supervision of the Communal forests, the government can make use of its foresters, who are generally bound to report to government any wrong done to the Communal forests that may come to their knowledge. The technical supervision is conducted by its technical foresters, namely, by the *oberforstbeamte* (head forest officers), and *forstmeister* (forest master).

Each *forstmeister* has a special geographical district allotted to him, who has not only to superintend the State forests, but also, at the request of the government department of the interior, of which he is a technical member, to examine the management of all the Communal forests situated within his particular district. The *oberforstbeamte* (head forest officer), has general supervision over the administration of all the Communal forests situated within the government district of which he is the head, and by whom all orders are issued. The privileges of the Communes in their forests consist in their having sole benefit of all the income or any other profit derived from the forests. The use of the pastures, as well as the straw and grazing, is usually permitted to the consumers whenever such use is very necessary, but is so far restricted that the condition and value of the forests and the maintenance of the pastures may not suffer thereby. The Communes are bound to bring all sand dunes and waste lands under forest cultivation as soon as it is shown it can be profitably done. On all sales of Communal forest area the permission of the Government must first be obtained.

ORGANIZATION AND FUNCTIONS OF GOVERNMENT FOREST BUREAUS.

The State forest administration is under the Ministry for Agriculture, Domain, and Forest. The chief direction of forest affairs is divided into four heads:—

1. The central direction: Forest Department in the Ministry for Agriculture, Domain, and Forests.

2. Local direction: Inspection and control by the district government under the Department of Taxes, Domain, and Forest.

3. District administration by the chief forester (*oberforster*), respectively the Bureau of Receipts and Disbursements.

4. Forest preservation and special superintendents over the management of the subordinate foresters (the so-called *forstschutzbeamten*).

The revision of all forest accounts is done at the so-called *Ober-Rechnungskammer* (Head Bureau of Accounts) of the Ministerial Department. The entire organization is based upon the division of the State forests into so-called *oberforstereien* (forest districts). Every principal forest district is an independent administration, for whose administration a separate finance is kept, and the chief forester, who is the responsible administrator of the finance, submits all the accounts through the forest treasurer of his district to the Finance Department of the government, for auditing.

The duty of the *oberforster* (head forester) is to watch and take care of the rvation of his forest district and to make his administration useful in every possible way. It is, therefore, the duty of the *oberforster* to possess the most exact knowledge of the working of the district confided to his care, and not to neglect visiting the forest daily, if possible.

The *oberforster* is an independent officer, and is alone responsible for the duties and salaries of his assistants.

The *oberforster*, on having passed the scientific examination required by the State, is appointed by the Minister of Agriculture, Domain, and Forest, and receives a definite salary, with the right of pension. His rank is that of a government assessor. The extent of each forest district varies. There are 679 forest districts in Prussia, and the average size of each district is 3,496 hectares.

The following table shows the area of State forests in the different Provinces, also the number and the average size of each *oberforsterei* :—

Province.	Forest area.	Oberforsterei.	Average size of oberforsterei.
	Hectares.	Number.	Hectares.
East Prussia	359,241	74	4,855
West Prussia............................	273,174	47	5,812
Brandenberg	369,510	72	5,132
Pomerania	170,619	42	4,062
Posen	162,029	28	5,787
Silesia.....	151,325	34	4,451
Saxony......	169,480	55	3,081
Schleswig	30,111	16	1,882
Hanover	235,074	104	2,260
Westphalia.....	57,189	19	3,009
Hessen-Nassau	253,003	146	1,733
Rhine Province.........................	143,284	42	3,412
Total.....................	2,374,039	679	45,476

There is a treasurer for each *oberforsterei*. He is an independent officer and is alone responsible for the administration of his bureau. The government appoints him, and he is required to give bond for the faithful discharge of his duties.

The foresters under the supervision of the chief forester (*oberforster*), are of two classes, namely, those who protect and attend to the practical management of the forests, the so-called *forsters* (foresters), and *waldwarter* (forest attendant), and the assistant foresters, the so-called *forsthulfsaufscher*. The immediate head of the chief forester is the district government, especially its department of finance, and whose organs for the administration and supervision of the forests and the finances are in the person of the *forstmeister* (forest master), and *oberforstmeister* (head forest master). The former has control over a certain number of the chief forests within the government district, and the latter over all in the government district.

The *forstmeister* resides at the seat of the district government as a technical member of it. He has to personally inspect every part of the district at least three times a year, assist in carrying out the regulations of work and in adjusting the finance, in controlling and fixing the annual plans of culture and the felling of the forest, subject to the supervision of the *oberforstmeister*. Further, he has to examine all the work done in the forest and its protection, inspect the book of the *oberforster*, and the accounts of the treasurer, check the forest cash account and the inventories, and inspect once in every five years all the forest boundaries of every forest district within his district and report as to their condition. As a member of the district government he has to work out all business matters which directly concern his inspection district, except in cases where the work is specially provided for. All reports of the *oberforster* to the government must be sent through the hands of the *forstmeister*.

The *forstmeisters* are appointed by His Majesty the Emperor and King, on the proposal of the Minister for Agriculture, Domain, and Forest, of the *oberforsters* who have distinguished themselves by their superior technical education and business management of forests. They have not to pass any special examination for this promotion beyond the forest scientific examination originally passed and required by the government for *oberforsters*.

Forstmeisters rank as *regierungs rathe* (government councillors). The number of *forstmeister*s at present in Prussia is 92, which is on an average of about 6 to 7 *oberforstereien* (chief forest districts) to each *forstmeister*. This estimate, however, does not include the communal forests of Westphalia and the Rhineland nor the 80 royal *oberforstersien* under the supervision of 26 *oberforstmeisters* who are the directors of the whole administration of the entire government district, and, as such, are the superior officers of the *forstmeister*.

Accordingly, there is one *oberforstmeister* for each government district, who is, by virtue of his office, a member of its department. The *oberforstmeister* is selected out of a number of the most capable *forstmeisters*, who is proposed by the Minister for Agriculture, Domain and Forest, with the sanction of the State Ministry, and appointed by His Majesty, the Emperor and King. The *oberforstmeister*, having the entire forest administration of the government district, has to make annually, in conjunction with the *forstmeister*, an inspection tour and to see that the management of the forest is properly carried out. He has, under the direction of the government district president the appointment and arrangement of the pay of the forest police according to the general instructions issued by the Minister. He has, further, the regulating of the general business, the preparing of the budget, the super-revision and approval of the annual felling and cultivation plans, the distribution of the means for the cultivation of the forests, and the disposition of the funds set apart for the entire district.

The Ministry for Agriculture, Domain and Forest contains in its department for forest the central direction for the entire state forest administration, consisting of an *oberland-forstmeister* (head State forester) a ministerial director and four forest technical ministerial councillors, whose departments of business are arranged according to the Province. The general regulations for the maintenance and utilization of the State property, consisting in forests, are fixed by the Minister, who also takes care that they are properly executed.

REVENUES FROM GOVERNMENT FORESTS; COST OF MAINTAINING OR MANAGING FORESTS; PROFITS OF FOREST CULTIVATION.

The estimated revenue and expenditure of the State forests for the years 1886-87, according to official statement, are given as follows:

Gross receipts	M.	56,070,000
Ordinary expenses	M.	31,062,200
Surplus	M.	25,007,800
Extraordinary expenses		2,450,000
Net income	M.	22,537,800

FOREST SCHOOLS.

In Prussia there exist three kinds of forest schools.

(a) Two preparatory forest schools for *forster* and *forstschutzbeamte*. The pupils, from 12 to 17 years of age, receive at these schools an elementary education and practical instruction in forestry under the direction of a *forster*. These schools are intended to take the place of the apprenticeship of two or three years which the student, on the completion of his elementary education elsewhere, would otherwise be obliged to serve at an *oberforsterei* under the direction of the *oberforster*. The advantage of the former is that it combines the elementary education with practical forest instruction.

(b) Two forest academies, one at *Eberwalde* and the other at Minden, under the department of the Minister for Agriculture, Domain and Forest, and the immediate supervision of the *oberlandforstmeister*, one of the chief state forest officials in the Forest Department of the Ministry. These academies are intended to give a scientific education and to fit students for the forest administration service, that is to say, for the higher forest career, from *oberforster* upward.

The term of study is two years and embraces the following branches:

A. Fundamental Science

1. Physics, including meteorology and mechanics.
2. Chemistry, organic and inorganic.
3. Mineralogy.
4. Geognosy and geology.
5. Botany.
    (a) General botany.
    (b) Anatomy, physiology and pathology of plants.
    (c) Special forest botany.
    (d) Anatomical and miscropical demonstration.
6. Zoology.
    (a) General zoology.
    (b) Special zoology particularly with respect to the different kinds of forest animals and birds.

7. Mathematics.

(a) Repertory and practice in arithmetic, planimetry, trigonometry and stereometry.

(b) Principles of analytical geometry.

(c) Principles of high analyses.

8. General political economy, particularly with respect to forest affairs.

B. Branch Science.

1. History and literature of forest affairs.
2. Forest statics.
3. Forest planting.
4. Forest preservation.
5. Forest technology.
6. Forest valuation, wood measuring, forest survey.
7. Forest statistics.
8. Forest administration, particularly with respect to the organization of forest affairs in Prussia.
9. Forest administration.
10. Redemption of forest claims.

C. Adjunct Science.

1. Jurisprudence, Prussian civil and penal code.
2. Forest road construction.
3. Game law.

As aids to study, these academies have extensive collections relating to forest and natural science, botanical gardens, seed collections, etc. Each academy is under the direction of an *oberforstmeister*. The lectures are given by scientifically educated foresters and special professors. The student, before he is admitted to these academies, must produce a diploma showing that he has passed the course of studies required at a German gymnasium or at a Prussian technical school of the first class. He must be under 25 years of age. have a good character and show that he possesses the necessary means for studying.

EFFECTS OF FOREST DESTRUCTION.

The destruction of forests is caused mostly by parcelling off large forest estates, which leads to a careless felling of the trees and little·disposition to restore the loss. An eminent authority on forestry science, Dr. Otto von Hagen, in writing on this subject makes the following observations :

"The forest is a trust handed down to us from past ages, whose value consists not alone in the income derived from wood, but also in the importance which it exerts, through its influence on climate and rainfall, on land culture. Its importance is not merely a question of the present day or of the present ownership, but is also a matter which concerns the future welfare of the people. This is a truism beyond contradiction, but nevertheless it is daily disregarded by those who are indolent and selfish.

" When such evils reach the stage of common danger, and this is in a great measure already the case, it then becomes a duty to interfere by legislation. Neither the decrease of the wood production nor the difficulty at times to meet the demand for wood, nor the rise in the price can confer upon the State the right to interfere with the freedom of private ownership or of private administration of forests, but this right and duty would devolve upon the State in case that any injury is done to the welfare and existence of the inhabitants of a certain locality resulting from the destruction of the forest. How entire districts which flourished in the past have been reduced to poverty and want through forest destruction, has been seen in Prussia, where large tracts of lands have suffered under such calamities.

" By stripping the beeches of their forests in the seventeenth and eighteenth centuries, the sea coasts have become exposed to all winds and storms. Fields, once fertile, have been transformed into waste sand dunes, and whole villages, whose agricultural people formerly prospered, have ceased to exist.

" In the middle and eastern Provinces light and undulating soil has been replaced by small or large sand hills, and places where forests once stood and served to carry off stagnant moisture, have been turned into marshes. In the western mountainous Provinces the fertile forest soil, the waste product of thousands of years of the trees, has disappeared. It has been dried up by the sun and wind, and washed into the valleys by rain and snow-water, and left the mountains bare and unfertile, whose soil is scarcely capable of supporting any vegetation save heath and broom-grass.

" The rich meadows in the valleys have vanished, they have been again and again, after every rainstorm, washed and torn by the water rushing from the mountain tops. The high moors which have been formed by the destruction of the forest, emit at all times of the year vapors and fogs which kill vegetation far into the land. Thus the soil becomes directly impoverished, and the climatic conditions change and become worse. Instances of the injurious effect upon the culture of the soil caused by the destruction of the forests can be seen to a smaller or larger extent throughout Prussia."

A BRIEF RETROSPECT.

Early in the fourteenth century, in the more thickly populated sections of Switzerland, the people appear to have been forced, through apprehension of a deficiency in their wood supply, to take some measures for the preservation of their forests. In the year 1314 Zurich forbade its foresters (vorsters) to "fell, raft or sell wood from the Sihlwald." In 1339 Schwyz issued a prohibition against charcoal burning, and in 1438 Freiburg decreed that no wood should be cut in the environs of the city. In Entlebuch it was forbidden in 1471 "to draw wood from forests situated high up in the mountains," and in 1592 Berne called attention to the need of economy in the use of wood. Finally similar decrees became general, but while serving to preserve forest areas they proved a hindrance to the progress of agricultural and vine-growing interests. Zurich, for instance, in 1563, forbade the establishment of any new vineyards, and the prohibition was kept in force up to the beginning of the eighteenth century. At that period the dread of a deficiency of wood became so general that it was even forbidden to purvey or export any of it from one village to another. Contemporaneously with these prohibitions were issued others forbidding the pasturage of cattle, sheep and goats in the forests. The old law generally ran in some such homely text as this: Whoever keeps a cow at home in summer is allowed to drive no goats, and nobody more than he actually requires for his house-keeping.

But spite of all these precautions and prohibitive measures the lack of combined action became painfully apparent. Moreover the individual owners were refractory, resented interference, and held on to their woodlands, so that, in fact, to-day the comparatively small forest area belonging to the State is what has principally been acquired by direct purchase, by inheritance or by the suppression of monasteries, as in the Bernese Jura, in Thurgau and in Schaffhausen.

With the advent, however, of the eighteenth century, Swiss forestry took on, in an official sense at least, a more active existence.

In 1702 Zurich, always foremost in the work, appointed a commission to devise a general forestry system. In 1825 Berne followed suit, and later Freiburg, Lucerne and Schwyz took action in the same direction. From this time on the several cantons managed their own forestry matters as they wished, and entirely independent of each other up to ten years ago, when the imperative needs of combined action having become apparent the matter was taken in hand by the federal authorities, whose attention had been called to the pressing demand for a legislative action to arrest the destruction of forests especially in the higher mountain regions. Accordingly on the 24th of March, 1876, a law was passed establishing federal control over the forests in all the mountain regions of Switzerland, embracing eight entire cantons, viz., Appenzell, Glarus Graubunden, Schwyz, Tessin, Unterwalden, Uri and Valais, and parts of seven others, viz., Berne, 41.48 per cent.; Freiburg, 32.70 per cent; Lucerne 53.50 per cent; St. Gallen, 76.17 per cent; Waadt, 22.98 per cent; Zug, — — per cent; Zurich, 6.86 per cent.

ZURICH FOREST SYSTEM.

As will be observed from the foregoing, Zurich has always evinced an actual and especial interest in forestry matters, and the result is that her forestry system at the present day is a model one, and is so regarded throughout Switzerland. Her forestry law, which has been in operation in its present form for over a quarter of a century, is so complete in every detail as to form a report in itself and it is therefore translated and incorporated in full herewith.

1. ORGANIZATION.

1. Cantonal, township, and corporation forests shall be subject to the control of the government forestry system. Private forests come under the same provision, in so far as the safety of the others or regard for a common danger renders necessary.

2. According to article 49 of the law pertaining to the organization of the government council, supreme control of the forestry system is vested in the direction of the interior. A yearly sum of 8,000 francs will be allowed it for the cost of management, and for the interests of forest culture, as, for instance, in the award of premiums for distinguished services, establishment of a course of instruction for foresters etc.

3. The canton is divided into four forestry districts, the limits of which shall be fixed by the council.

4. The cantonal forestry board shall consist of one overforest master and four district forest masters. The council is authorized to furnish an adjunct thereto. In said board is vested the duty of superintending all forestry affairs. The maintenance of the cantonal forest under control of the director of finance is also transferred to it; the duties of its members will be especially determined by official instructions from the council.

5. Only those who shall have passed a government examination, as prescribed by the council, and have been declared competent by the direction of the interior, shall be employed as forestry officials.

6. The overforest master, the district forest masters and the adjunct, shall be chosen by the council on the simple, though not binding, nomination of the direction of the interior.

The term of service of the over and district forest masters shall be for three years. The adjunct shall be chosen for a period to be fixed by the council. Retiring officers are eligible for re-election.

7. The overforest master receives a salary of 3,500 francs. When travelling on official business, his cash outlays will be reimbursed. The further sum of 1,000 francs is allowed for clerk and office expenses.

8. District forest masters receive an annual salary of 2,200 francs. While on official journeys on forest service, they receive a daily allowance of 10 francs, and while on official journeys, in the interest of cantonal forests, a daily allowance to be determined by the council.

9. The council shall fix the sum to be paid to the adjunct out of the appropriation provided in article 2.

10. The daily allowances, when involving cash outlays, shall be paid from the cantonal forestry fund, or from the appropriation provided in article 2, depending on whether they concern business connected with the cantonal or non-cantonal forests.

11. Forestry officials are required to give bonds in the amounts to be determined by the council.

12. Each corporation shall elect a board of overseers of not less than three members, for a period of three years, and shall give notice of such election to the directors of the interior.

13. The employment of foresters is obligatory upon (a) the canton for all forests directly or indirectly belonging to it, (b) all forest-owning townships, and (c) wood corporations. Townships and corporations are directed to appoint an over-forester. Several townships or corporations may unite on one and the same person for this purpose. Forestry officials appointed by townships and corporations, are, at the same time, subordinate to the cantonal forestry officials in matters pertaining to cantonal forests.

14. Townships, corporations, and private owners are to pay the salaries of forestry officers appointed by them. Where the forests of a township or corporation are so small that such salary does not amount to 100 francs, then the township or corporation in question shall unite with one or several neighboring townships or corporations to appoint a forester in common. The proper method of procedure in such cases shall be determined by the direction of the interior.

15. Cantonal foresters shall be chosen by the direction of finance on the simple, though not binding, nomination of the overforest master. The choice of township and corporation overforesters and foresters is vested in the board of overseers, which may, for this purpose, be increased to six or eight members. The term of service for overforesters and foresters shall be three years. Elections shall always be held after the renewal of the board of overseers. Retiring members are eligible for re-election. This period of service takes effect in individual cases from the first election held after the promulgation of this law.

16. The providing of foresters for private forests is left to the owners. But, in case a forest district is adjacent, the owner may decide upon the appointment

of a forester and be present at his election, at which the minority must submit and assume its proportional share of the salary. The proportional voice in voting, as well as in paying, shall be determined on a ratio of the area represented. Private individuals may, with the consent of the township or corporations, transfer to the latter's foresters the care of their forests, in which case they shall arrange with said township or corporation for what they are to pay.

17. Applicants for the position of overforester must furnish proof of their competency in the form of an essay to be submitted to the overforest master. Special instructions as to the nature of such essay will be furnished in an order from the direction of the interior. As conditions of eligibility as forester, active citizenship, a good physical constitution, and a knowledge of reading, writing, and arithmetic will be required.

18. Elections of overforesters and foresters by boards of overseers of townships and corporations, are subject to examination and confirmation by the direction of the interior. To this end, certificates, stating the manner of election, name, age, and former employment of the candidate, and the annual salary pertaining to the position, shall be forwarded, through the Statthalter's offices, to the direction of the interior. The examination by the latter covers in part the validity of the election, and in part the existence of the lawful qualifications, and it is ordered that confirmation be withheld where a candidate has previously been convicted of serious violations of, or misdemeanors against, forestry regulations. After confirmation, the newly-elected candidate is ordered to be sworn (oaths are no longer administered, the " hand vow," as it is called, having been substituted), which duty is to be performed by the Statthalter's office. Private owners appointing foresters must have them sworn by the Statthalter's office.

19. Sworn forestry employees stand, in regard to the performance of police duties, on an equal footing with police employees. The same official credit is, consequently, to be accorded to their reports, made under the provisions of article 96 and the following article of this law, as would be accorded to the same, if made by the police officers.

20. It shall be the duty of foresters in the cantonal, township and corporation forests to attend a course of instruction on the subject of forestry, to be provided by the direction of the interior, and imparted by the forest masters. They may be required by the direction of the interior to attend a second course, when a previous examination shall have proved unsatisfactory. They receive their service instructions from the direction of the interior. Foresters in private forests shall be allowed to participate in the courses of instruction referred to.

21. The consent of the direction of the interior is necessary whenever the overforest master, forest masters, or foresters, in cantonal forests desire to fill any other official position, or follow any other pursuit in conjunction with their position as stated. Overforesters, and foresters in township and corporation forests, cannot at the same time be members of their election boards. Before entering upon any other township office or service, they must procure the consent of the direction of the interior.

22. The following of any business in wood, or manufactured wooden-ware, or of any industry in which wood is the leading material, is unconditionally forbidden for all persons in the cantonal, township and corporation forestry service.

FORESTRY IN FRANCE.*

THE WOODS AND FORESTS OF FRANCE.

In 1876, the last year for which anything like complete details are available, the total wooded area of France, exclusive of isolated trees, such as those growing in parks and on road-sides, which were not planted for the sake of the timber they produce, amounted to 35,464 square miles, or a little more than 17 per cent. of the entire area of the country. The proportion in other European countries is as follows, viz.:—

	Per cent.
Russia	40
Sweden	34
Norway	29½
Germany	26
Turkey	22
Switzerland	18
Greece	14
Spain, Belgium, and Holland, each	7
Portugal	5
The British Isles	4
Denmark	3½

The average of all European States taken together, is 29½ per cent. The population of France being 181 per square mile, it follows that the area area of woodland per head is about three-fifths of an acre.

Some changes, which will be noted in a subsequent chapter, have taken place in the area of the State forests since 1876, but in that year the woods and forests were owned in the following proportions by the different classes of proprietors, viz.:—

	Square miles.	Per cent
The State	3,734	10.7
Communes and sections of communes	7,949	22.4
Public institutions	124	0.3
Private proprietors	23,657	66.6
Total	35,464	100.

and these figures may be taken as fairly representing the actual position at the present time.

Forests are not so exhausting to the soil as agricultural crops. In the case of the latter, the entire plant, except the roots, which are sometimes also taken, is removed, whereas with a crop of trees, the leaves, flowers and fruit, which are far richer in nutritive elements than the wood, are annually returned to the soil, and thus serve to maintain its productive power, as well as, by their protective action, to keep it in a good physical condition. Hence forests can flourish on comparatively poor soil; some kinds of trees, notably most of the conifers, being able to grow on ground that would be quite incapable of producing a series of

*By Major F. Bailey, R. E.   Vol. XI. of the "Transactions of the Scottish Arboricultural Society."

remunerative agricultural crops; and it is, therefore, generally speaking, out of place to keep rich fertile valleys under forests, which ought rather to be maintained on ground which cannot be profitably cultivated. In well populated districts, matters naturally tend to settle themselves in this manner; the better classes of ground being brought under the plough, while every acre of the rest of the country is kept wooded, in order to meet the domestic and agricultural wants of a dense population. But it is otherwise in less favored localities. Here vast areas might be devoted to the production of wood; but while, from the nature of the case, the local consumption is, in such places, very small, the absence of communications frequently renders export very difficult. Hence wood has but a very small value, and the forests tend to disappear gradually before the excessive grazing to which they are subjected; for the population of such regions, being unable to make its living by agriculture, is, generally speaking, driven to adopt a pastoral life.

Forests grow in France at all altitudes up to about 9,000 to 9,500 feet above the sea, a much larger proportion of them being found at low than at high levels. Thus it has been calculated that, if the country were divided into altitude-zones of 200 meters each (656 feet), the lowest zone would contain 36 per cent. of the forests, while the highest would not contain more than .04 per cent. of them; the fifth zone (2,600 to 3,300 feet) would, however, on account of the extensive plateaus existing at this level, contain more than the fourth. Forests situated at high altitudes do not produce so much wood, and are, therefore, not so profitable as those grown lower down; consequently the private owners, who have done their best to preserve their woods in the plains and low hills have, in the majority of cases, allowed the mountain forests they once possessed to be destroyed by over grazing Hence it arises that, while at altitudes below 4,000 feet, the proportion of State and communal forests is comparatively small, hardly any private woods are found above the level of 6,000 feet, such forests as exist there being, generally speaking, maintained by the State or communes in the public interest, as a protection against avalanches and the formation of torrents. The private forests are then, taken as a whole, more favorably situated than those which belong to the State and the communes, both as regards soil, climate, means of export, and proximity to the markets. It has been calculated that the distribution of the forest area by zones of altitude is thus proportioned:—

Altitude.					Forests under the forest dept.		Private and communal forests not under the forest department.	Total.
					State.	Communal.		
	M.	M.	Ft.	Ft.	Per cent.	Per cent.	Per cent.	Per cent.
Plains	0 to 200	=	0 to	656.	41	5	45	36
Low hills	200 to 500	=	656 to	1,640	32	48	25	31
Mountains above.......	500	=	above	1,640	27	47	30	33
					100	100	100	100

It is said that if the trees could be grouped together, so as to form a series of pure forest, the proportion of the total area which would be occupied by each species would be as follows :—

	Per cent.
Oak (Q. sessiliflora and Q. pedunculata)	29
Beech	19
Hornbeam	12
Silver fir	7
Scotch pine	4½
Evergreen oak (Q. ilex)	4
Maritime pine	3
Spruce	3
Larch	2
Other kinds	16½
Total	100

The small number of species which enters to any important extent into the composition of the French forests is very remarkable. Thus it appears that oak, beech, and hornbeam occupy 60 per cent. of the tree covered area, more than one half of the remainder being taken up with six other species ; but many other kinds are disseminated throughout the forests in various proportions according to circumstances. As a matter of course, however, the trees are not grouped together in the above manner, and, neglecting blanks, the crop on the ground is actually constituted somewhat as follows :—

Pure forests—	Per cent.	
Broad-leaved (oak or beech)	15	
Coniferous (silver fir, pine, spruce, or larch)	13	
		28
Mixed forests—		
Broad-leaved (oak, beech, and hornbeam)	52	
Broad-leaved and coniferous (beech and silver fir, or oak and pine)	18	
Coniferous (silver fir and spruce)	2	
		72
Total		100

Or separating the broad-leaved and the coniferous forests from those which consist of a mixture of the two, we have :—

	Per cent.
Broad-leaved forests, pure and mixed	67
Coniferous forests, pure and mixed	15
Broad-leaved and coniferous forest	18

The State forests show a smaller proportion of pure crops than are found in those of the communes, but they also comprise a very much larger proportion of forests in which the crop consists of a mixture of broad-leaved and coniferous species. The first of these differences is due to the circumstance that a mixture, which is always desirable from cultural considerations, has been systematically maintained in the State forests from a remote period, whereas this has not always been the case in the communes. The second difference is chiefly accounted for by the fact that those parts of the State broad-leaved forests, where, from various causes, the soil has become much deteriorated, have frequently been planted up with conifers, which are the only kinds likely, on account of their capacity to grow on poor soil, to succeed under such conditions; but these are in such cases, only intended to act as nurses to broad-leaved species, which are subsequently to be raised under their shelter. But little work of this kind has yet been accomplished in the communal forests from want of the needful funds. The private forests resemble those of the communes rather than those which are State property but a further comparison in this respect between them and the other classes of forests need not be made at present.

Many circumstances combine together to influence the nature of the vegetable growth, which characterizes any particular locality.

Thus, a "limestone soil," which is one containing more than four or five per cent. of carbonate of lime, is usually marked by a rich and varied vegetation; while on a silicious soil the flora is much more simple and uniform, the undergrowth being often formed of bilberry (*Vaccinium myrtillus*), broom and heather. Forty-four per cent. of the French forests are on limestone. But the principal forest trees are not much affected by the chemical composition of the soil, the two deciduous oaks, the beech, the hornbeam, silver fir, spruce fir, the larch, being classed as "indifferent" to it The ever-green oak, however, shows a preference for limestone; and the Scotch pine flourishes best on a silicious soil; but the maritime pine will not grow on limestone. The climate, which varies with the latitude, altitude, amount and distribution of the rainfall, proximity, or otherwise of the sea, and other conditions, is the principal factor in determining the distribution of trees, each of which finds its home in the locality which best suits its temperament. The hot region of the south, the temperate regions of the north and centre, and the mountains, are each characterized by the spontaneous vegetation to which they are adapted. Thus, in the south, are found the evergreen oak and the maritime pine; while the spruce, the silver fir, and the larch inhabit the mountains; and the five other species mentioned, grow chiefly in the temperate regions. The physicial condition of the soil also exercies an important influence on the growth and local distribution of trees; for example, *Quercus pedunculata*, and the hornbeam, will grow on moist soil, which does not suit either *Quercus sessiliflora*, the beech or the evergreen oak.

During the entire course of their development, trees of all kinds require light; but during the early stages of their existence, some of them must be completely in the open, without any cover at all; while for others, various degrees of shade are necessary. This quality of the young plants is, generally speaking, in direct relation to the abundance of the foliage of the adult tree from which they spring.

Those which, when young, require much light, such as the larch, the pines and the oaks, are called the "robust," or trees of light cover, while others, which will not stand exposure such as the beech and silver fir, are called "delicate," or trees of heavy cover. The spruce and the hornbeam are classed intermediately between kinds of light and heavy cover. This is a very important question to the forester not only with reference to the method to be adopted for raising a crop of any particular kind of trees, but also with regard to their coppicing power, their effect on the soil, and other matters. Trees of light cover, generally speaking, coppice better than those of heavy cover, but the latter have a much greater effect than the former in improving the soil.

It is estimated that the 35,464 square miles of woods and forests yielded the following produce in 1876, viz., 17,896,227 loads (50 cubic feet) of wood of all qualities, 321,741 tons weight of tanning bark, 2,556 tons weight of cork, and 31,539 tons weight of resin; the whole being valued at £9,471,017. The average production of wood was therefore 39 cubic feet per acre; and the gross revenue, omitting that on minor produce, which was very small, was equal to 8s. 4d. per acre.

But in addition to this, it is calculated that the isolated trees, not grown for the sake of their timber, and vines yield together three and one-half million loads per annum, valued at £1,000,000; so that the total production of wood in France is raised to about twenty-one and one-half million loads, and the value of the wood, bark, and resin to about £10,500,000. This gives the amount of wood and the money value of the forest produce per head of the population as 29½ cubic feet, and 5s. 9d. respectively.

Of the twenty-one and one-half million loads of wood produced, about four million loads were timber and the rest firewood. The latter sufficed for the national requirements, but the former was far from doing so; for the imports of wood of this class exceeded the exports by 2,062,432 loads, valued at £6,408,000 —that is to say, that it was less than two-thirds of the amount required The question of foreign timber supply is, therefore, a very important one, even for France, which has 17 per cent. of its area under forest.

FORESTS MANAGED BY THE STATE FOREST DEPARTMENT.

The forest law of 1827, which is still in force, confirmed the previous legislation, under which all woods and forests which form part of the domain of the State, all those which being the property of communes or sections, or of public institutions, are susceptible of being worked under a regular system, and finally all those in which the State, the communes, or public institutions possess a proprietary right jointly with private persons, are administered directly by the State Forest Department in accordance with the provisions of the forest law.

The areas thus administered at the commencement of 1885 were as follows, viz.,

	Hectares		Square miles
State forests......................	1,012,688	=	3,910
Communes, sections, and public institutions	1,967,846	=	7,598
Total	2,980,534	=	11,508

These figures, which include the dunes, represent about 5½ per cent. of the entire area of France, and nearly one-third of the total wooded area. An additional 144 square miles of barren land had, up to the end of 1884, been purchased by the State in connection with the project for the consolidation of bare and unstable slopes on the great mountain ranges; and this area is also administered by the department under the forest law. About 40 per cent. of the State forests are situated in the plains, while the rest of them, together with nearly the whole of the communal forests, are found in about equal proportions on low hills, up to an altitude of 1,700 feet, and on the higher mountain ranges. About one half of them stand on limestone rock, 92 per cent. of their entire area being actually under wood.

The principal object of the following pages is to sketch in a brief and summary manner the system of management adopted for these forests, so that some general idea may be formed of what the business of the French forest department consists in, and what the results of their labors have been, up to the latest date to which information is available under each head. The organization of the professional staff of the department, and the manner in which it is recruited, will then be explained

STATE FORESTS.

The forests now belonging to the State owe their origin to one or the other of the following sources :—They either formed part of the ancient royal domain, as it was consituted at the time of the ordinance of 1669, or of the sovereign domains united to France since that year ; or else they were ecclesiastical property confiscated at the time of the revolution in 1790, or they have been more recently acquired by purchase, legacy or gift. About one-half of them are ancient roya domains.

The State forests were formerly of much greater extent then they are at present. In 1791 they covered an area of 18,166 square miles, which was reduced to 3,792 square miles in 1876, the reduction being almost solely due to sales effected for the benefit of the exchequer ; but the loss of territorty after the war of 1870 was the cause of a diminution of 374 square miles. The records show that, between 1814 and 1870, 1,362 square miles of State forests were sold for nearly twelve and one-quarter million pounds sterling, or about £14 per acre; but since 1870 no such sales have taken place, and since 1876 the area has been somewhat increased by purchases and otherwise. It now includes 33 square miles of forest owned jointly with private persons, and 450 acres are temporarily held by the families of some of Napoleon I.'s generals, whose right will in the course of time either lapse or be commuted. The remainder of the area is owned absolutely by the State, but the enjoyment of the produce does not belong exclusively to the treasury, for, as will be explained hereafter, certain groups of rightholders participate in it.

In the next section, the principal points of laws relating to the communal forests, and of their management by the State Forest Department, will be brought to notice ; while in the subsequent sections of this chapter the work of the department in connection with the State and the communal forests will be briefly treated of in such a manner as to bring out and compare the results obtained in the two classes of forests.

FORESTS BELONGING TO COMMUNES, SECTIONS AND PUBLIC INSTITUTIONS.

The territory of France is divided into 39,989 communes or village communities, of which about one-third are forest proprietors. Certain groups or sections of the inhabitants have, however, rights and own property, apart from the commune in which they reside, and these are also owners of considerable areas of woodland. Those forests belonging to communes or sections, which are susceptible of being worked on a regular system, are managed directly by the State Forest Department for the benefit of their owners, the principal features of this management being as follows, viz.: The laws relating to State forests are, generally speaking, but with certain exceptions, applicable to them ; they cannot be alienated or cleared without the express and special sanction of the government in each case; they cannot be divided up among the members of the community ; the annual sales of produce are effected by the State forest officers, and the money realized is paid directly by the purchasers into the communal treasury; before the sale takes place the quantity of timber and firewood required by the inhabitants for their own use is made over to them usually standing in the forest, and it is subsequently worked out by a responsible contractor; three-quarters only of the total annual yield is available for distribution or sale, the remaining quarter being left to accumulate, and thus form a reserve fund or stock of timber from which exceptional necessities either in the way of wood or money can be met; the distribution of firewood is made according to the number of heads of families having a real and fixed domicile in the commune; the entry of goats into the forest is absolutely prohibited, while the grazing of sheep is only permitted temporarily, and under exceptional circumstances, with the special sanction of the government in each case; no grazing of any kind can be carried on in the forests, except in places declared out of danger by the forest officers who have the power to limit the extent to which it can be practiced with reference to the quantity of grass available; the forest guards are chosen by the communal authorities, subject to the approval of the forest officer, who delivers to them their warrants; the State defrays all expenses of management, including the officers' salaries, the marking of trees, notifying of sales, office charges, and the prosecution of offences; the State is reimbursed by the payment from the communal treasury of a sum equal to 5 per cent. on the sales of principal produce, including the value of the wood, made over to the inhabitants; but this payment, which forms a first charge on the forest revenue, can never exceed the rate of one franc per hectare (about four pence an acre) of the total area thus managed; the communes pay the guards' salaries, the taxes, and all charges for the maintenance and improvement of the forest, including planting, sowing, and road-making, as well as those for extra-ordinary works, such as demarcation, survey, and the preparation of working plans. In all this the forest officers are bound by law, to act on the principle that they are managing the property for the benefit of its owners, who must be consulted through their representatives, the mayor and municipal council, in all matters affecting their interests, and whose wishes must be acceded to when they are not opposed by the legislation, or contrary to the recognized principles of scientific forest management.

The principal public institutions are hospitals, charitable associations, churches, cathedral chapters, colleges, and schools; and the forests belonging to them are subject to administration by the State Forest Department on precisely the same terms as are those of the commune and sections.

Of the area of 7,598 square miles shown as being thus managed on behalf of these bodies at the commencement of 1885, about 100 square miles belong to public institutions, and about 7,500 square miles to communes, including sections. Of the remainder of their forests, about 410 square miles owned by the latter and about 27 square miles by the former are managed respectively by the communes themselves under the municipal laws, and by the administrative councils of the institutions.

Changes in this respect frequently take place; for every year a certain number of applications to free forests from the restrictions which State control involves are granted, while in other cases the owners demand or consent to their imposition. The records show that sanction has, since the year 1855, been accorded to the clearing of thirty-five square miles and to the alienation of forty square miles of the forests belonging to these bodies, but it is probable that the permission has not, in all cases, been acted on.

For the sake of convenience the forests belonging to communes, sections and public institutions will in future be spoken of collectively as "communal forests."

DEMARCATION AND SURVEY.

Up to the end of 1876 the work of demarcation had made good progress in the State forests, only 13 per cent. of which then remained to be completed, while 30 per cent. of the communal forests had still to be dealt with. The demarcation is indicated by dressed-stone pillars, with intermediate ditches or dry-stone walls, according to the custom and resources of each locality. The ground is usually resurveyed after the demarcation has been completed, and at the end of 1876 about three-fourths of the State forests and one-half of the communal forests had been thus re-surveyed and mapped, the prevailing scale being $\frac{1}{5000}$ (12 2/3″ = 1 mile) and $\frac{1}{10000}$ (6 1/3″ = 1 mile). Pending the completion of this work, the old maps are used for such of the forests as have not yet been resurveyed. In the communal forests the work of demarcation and survey is less advanced than it is in the State forests, because the charges for such work have to be defrayed from the communal treasury, and the needful funds are not always forthcoming.

SYSTEMS OF CULTURE.

The climate of France is singularly favorable to the natural regeneration of forests, which is, generally speaking, relied on—planting and sowing being only resorted to in the comparatively rare instances in which success cannot otherwise be achieved, such cases including, of course, the stocking of extensive blanks.

There are two main systems of culture—one known as "high forest," and the other as "coppice."

A high forest, which is usually destined to produce timber of large size, is one composed of trees that have been raised from seed, its regeneration being effected by means of seed, generally speaking, self-sown. There are two methods of treating the forest in order to produce this result. In one of these the trees of each age-class are grouped together, and are subjected to periodical thinnings, until the time arrives for regeneration, which is effected by a series of fellings, the first being a more or less light thinning, intended to promote the formation of seed and the springing up of the young seedling plants. The seed-felling, as this is called, is followed at intervals by a series of secondary fellings, usually three or four in number, which are made in order to meet the gradually increasing requirements of the young growth in the way of light; and ultimately the remainder of the old stock is removed by a "final felling." In this manner the marketable stems are gradually cut down and disposed of, the young crop being left to go through the same stages as its predecessor, and so on throughout successive generations of trees. In the selection method (known as *jardinage*), on the contrary, the trees of all ages are mixed over the whole area of the forest; there are no regular thinnings of the kind made under the first method; and the annual cuttings are effected by taking marketable trees here and there within a certain area of the forest, the blocks composing which are successively treated in the same manner, so that the entire forest is worked over within a fixed period of time. When treated by the first method, the forest is grown under very artificial conditions; for the aged classes are never in nature found thus grouped together; but by the selection method, on the contrary, a more or less near approach to a natural forest is obtained.

In the coppice system the regeneration is principally effected by means of coppice shoots.

There are two methods of treatment, *simple coppice*, in which there are no reserve trees, and the crop is clean-felled over successive portions of the forest:

and *coppice under standards*, in which standard trees are selected and reserved. with a view to their remaining throughout several generations of coppice shoots, generally at least three, but often four or five. Many forests are now undergoing conversion from the system of coppice to that of high-forest.

The following statement shows the extent to which the two systems were applied in the State and communal forests in 1876, since which year no important changes have taken place. The areas are given in square miles:

	High forest	Under conversion.	Coppice.	Pastures.	Total.
State forests	1,648	1,121	740	225	3,734
Communal forests	2,229	54	4,808	92	7,183
Total	3,877	1,175	5,548	317	10,917

It will be seen that there is a marked difference between the State and communal forests in this respect. In the former nearly three-quarters of the total area are either now under high-forest or under conversion to that system; while in the latter two-thirds of the total area are under coppice, and less than one third is either under high forest or under conversion.

High forest being usually destined to produce large timber, the trees must be left standing until they have attained a considerable age; and the capital, both in timber and money, which is locked up in it is therefore much larger than that in a forest under coppice. Other conditions being equal, the quantity of wood produced annually is, however, much the same under both systems; but owing to the greater value of the produce obtained from the high-forest, its money revenue is greater than that of the coppice, while on the other hand, it is found that coppice yields a higher rate of interest on its smaller capital value than than high forest, and on this account it is a more suitable system for adoption by communes. Coppice possesses, also, a further advantage for them, in that it yields for the use of the inabitants timber and other produce more varied in kind and dimensions than are obtainable from high-forest, and it thus satisfies their requirements, which are chiefly in fuel and small-sized timber, much better than forests managed under the latter system. But even in cases where the conversion of communal coppice to high-forest is deemed advisable, it is always found difficult to reduce the annual fellings to the quantity necessary in order to allow the growing stock to accumulate to the required extent; while the small size of the greater part of these forests renders them unsuited to the treatment which they would have to undergo in order to effect their conversion. The coppice system, including coppice under standards, is therefore in vogue in almost all communal broad-leaved forests, such high-forests as the communes possess being found chiefly in mountainous regions, and being composed of coniferous trees, which will not coppice. The area of communal forest shown as under conversion, consists principally of tracts in which the coniferous trees are spontaneously taking possession of the ground and driving out the broad-leaved species. It follows from what has been said above, that the State alone can, generally speaking, raise broad-leaved high-forest on a large scale, or undertake the conversion of coppice to high forest.

A further difference between the systems of culture generally adopted for the State and the communal forests may be noted, viz., that whereas in the former less than one-fifth of the high forest is treated by the selection method, three-fourths of the communal forest are so treated. In mountainous regions, where, as has just been said, the greater part of the communal high-forest is found, the selection method possesses incontestable advantages, in consequence of the continuous cover which it affords to the soil; but although the respective merits of the two methods, as applied to coniferous forests situated in such regions, are much disputed at present, there has of late years been an undoubted tendency to return to selection, which has for some time past fallen into discredit, and, taking the State and communal forests together, somewhat more than one-half of the total area of their high forest is now treated in this manner.

Two variations of simple coppice are sometimes practised, (1) that known in the Ardennes as *sartage*, in which, after the wood has been cut and removed, the twigs and chips are burnt on the ground, in order that their ashes may give to the soil sufficient manure to permit of the growth of a crop of cereals during the year immediately following the cutting. This system, which, as carried out in France, seems to be practised rather for the sake of obtaining the crop of corn han as a method of forest culture, is gradually dying out. It is not adopted in the areas under the State forest department. (2) That known as *furetage*, in which instead of clean cutting the coppice, those shoots only are taken which have attained to certain fixed dimensions, the operation being repeated annually, or after intervals varying from two to five years. *Furetage* prevails chiefly in the valley of the Seine, in the forests fiom which the fuel supply of Paris is drawn; but it is also employed in the mountainous districts of the south, in the case of forests maintained for the protection of steep slopes, which it is undesirable to denude entirely.

It it is impossible here to enter into anything like full details regarding these sylvicultural questions. To study them completely, as they are taught and practised in Frence, reference must be made to the books on the subject, among which may be mentioned " The Manual of Sylviculture," by G. Bagneris, (translated into English by Messrs. Fernandez and Smythies), Ryder & Son. London; and " *Le traitement des bois en France,*" by C. Broillard, Berger-Levrault, Paris.

WORKING PLANS.

Working plans or schemes, will, in course of time, be prepared for all forests administered by the forest department. The law provides that all these forests shall be subjected to the provisions of such plans, and that no fellings which are not provided for therein, and no extraordinary cuttings, either from the communal reserve, or in the blocks destined to grow from coppice to high forest, shall be made without the express sanction, in each case, of the government, by whom all plans must be approved before they can be adopted.

Subject to due provisions being made for the exercise of rights of user, the working plan provides for the management of the forest in the way that will best serve the interests of the proprietor. Unlike an agricultural crop, which ripens and is gathered annually, trees take many years to grow to a marketable size, the actual period that they require being dependant not only on their species and the natural conditions under which they are grown, as climate, soil, etc., but also on the use to which they are to be put. Thus a coppice being required to yield wood of small size only, may be cut every twenty-five to forty years, whereas a high forest, which is destined to produce large timber, must stand for a much longer time. It would be excessively inconvenient if the entire crop of such a forest were felled only once in every 100 or 150 years; and it is chiefly to avoid this that a working plan is required, which prescribes the arrangement necessary in order to allow of the produce being taken out annually, without intermission and in equal quantities, so that a regular and sustained income may be drawn from the forest. For example, a simple coppice thirty acres in extent, of which the crop is to be felled at the age of thirty years, might either be entirely cut down at one time, and then allowed to grow up again for thirty years, or, which would be found much more convenient, it might be divided into thirty one-acre compartments, each of which is to be felled in succession, so that by taking one plot each year, the whole area would be worked over in thirty years. The working plan must then, in the first place, prescribe the age at which the trees are to be felled, with reference to the average number of years that they take to arrive at maturity, or to attain the required size, and it must then fix the yield, or the amount of wood to be annually removed, this quantity being expressed either in the form of an area to be cut over, or a number of cubic feet of wood to be taken out. But in the case of a high forest managed under the selection method, it is sufficient to fix the number of trees of a minimum size to be cut out annually.

The provisions of a working plan vary according to the nature of the forest to which it relates. In the case of the simple coppice instanced above, the first thing to do would be to obtain a map showing the principal features of the ground, such as the edge of the plateau, the stream, and the road. The area would then be broken up, for purposes of examination and description, into temporary plots, each plot comprising a portion of forest more or less homogeneous in its composition. This study of the crop would enable the area to be divided into the thirty permanent compartments above alluded to, and it would also determine the order in which they should be numbered, so that the older portions might be cut first. It is evident that if one of these be cut every year the series of compartments will, after the lapse of thirty years, contain forest of all ages, from one to thirty years; and if the annual felling be invariably made in the oldest compartment, it is evident that the age of the crop cut will always be thirty years.

To make a working plan for a regular high forest, to be treated by successive thinnings, is not quite such a simple matter. If the forest is of great extent, it is, first of all, divided into two or more series or sections, each of which is dealt with separately. After the examination and description of the temporary plots, the section is divided into a number of equal compartments called *affectations* and when the ground has once been completely worked over the crop on each of these will always be, within certain limits, in the same stage of development, and subjected to the same kind of treatment. Thus, if the trees are to be felled at the age of 120 years, and there are six compartments, the sixth may contain the young growth, aged from one to twenty years; the fifth young poles from twenty-one to forty years old, and so on; the first containing the old trees which are to be felled. The compartments having been formed, each of them is then sub-divided into compartments usually corresponding in number with the years over which the fellings within it are spread (twenty in this case), and while the trees are being cut in the first compartment, clearings and thinnings, of various recognized degrees are going on in the compartments of the others, until each in its turn arrives at the age at which the trees are to be removed; and it is clear that in this case also the forest will ultimately contain a due proportion of trees of all ages, from one to 120 years, which is an essential condition.

The working plan prescribes the order in which all this is to be done, and it lays down the number of cubic feet of timber of the oldest class which are to be taken out annually from the first or oldest compartment, so that the entire stock on it may be removed within the first period of twenty years, windfalls and dead or dying trees being always taken first ; each of the remaining compartments is similarly dealt with when its turn to be felled arrives. The quantity of wood to be removed by thinnings cannot be prescribed by the working plan, as they must be made to the extent which is judged necessary in order to develop the trees which are left. The forester's art is to do this skilfully, and ultimately to remove the old trees in such a manner that they may leave behind them a young self-sown crop to take their place, and so on throughout successive generations.

For a high forest to be managed under the selection method the arrangement is different. Here it is, of course, equally necessary that all the age-classes should be represented in due proportion, but instead of the trees or poles of each class being grouped together in separate compartments, all classes are mixed indiscriminately over the entire area of the forest, and there is thus no necessity for the formation of *affectations*, or compartments, of the kind just described. After the main features, such as the streams, ridges and roads, have been laid down on the map, the temporary plots, and the descriptions of them are made as before. The forest might in the present case be divided into three sections, the upper of which being on the crest of the hill, is required to be kept as dense as possible, and will not be dealt with in the working plan, as dead or dying trees alone will be removed from it. Suppose that the annual yield of the central section which is 150 acres in extent, has been fixed with reference to the estimated rate of growth and degree of completeness of the stock, at 50 cubic feet per acre, and the trees of marketable girth within it contain on an average 100 cubic feet of timber, it follows that the number of such trees which may be removed annually from the section is $\frac{150 \times 50}{100} = 75$. Theoretically this number should be taken one here and one there over the whole area ; but this would be very inconvenient, so the forest is divided into twelve or any other convenient number of equal or nearly equal blocks, from each of which, in succession, the entire number of trees is to be cut ; after taking windfalls, the choice falls on the ripest trees, those which are dead or dying being selected first. The section below the road is in another zone of vegetation ; it is 100 acres in extent, and its annual yield is calculated at the rate of 60 cubic feet per acre. Suppose, then, that the trees of marketable girth contain on an average 110 cubic feet of timber, the number of such trees to be cut annually $\frac{100 \times 60}{110} = 54$. The section will then be divided into blocks, from each of which in succession the entire number of trees is taken. In this manner each zone of altitude may be dealt with on its own merits, while at the same time, the annual fellings being localized, are easy to supervise, and the wood can be disposed of more readily and more profitably than if the trees had been felled here and there over the entire area. The working plan for a forest under conversion would, of course differ from any of the above ; but this somewhat complicated question will not ne dealt with here. It is only by an arrangement similar to one of those above briefly sketched that a permanent annual yield of a particular class of produce can be assured, and that the forest can be secured against the risk of gradual extinction.

A special branch of the forest department is charged with the preparation of working plans, which are not made by the local officers, except in the case of small forests, the plans for which they can frame without interference with their ordinary duties ; but they undertake the revisions, which are made every ten or fifteen years in order to guard against errors, and to allow for changes in the rate of growth, or other causes of disturbance. Pending the preparation of such regular plans the forest department draws up provisional rules, which must accord with local usages, where these are not opposed to the recognized principles of sylviculture. Up to the beginning of 1877 regular working plans had been completed for more than two-thirds of the total area of the State forests, and for somewhat less than one-half of the communal forests. The work progresses more slowly in the latter than in the former, because in their case the funds have to be provided by the communes, and the money is not always available; but as a matter of course the most important forests were taken in hand first, and these have for the most part been completed.

The question of working plans has only been dealt with above in an extremely superficial manner. In order to gain anything like a complete idea of the systems pursued in France the following works, should, among others, be studied, viz., "Amenagement des Forets," by C. Broillard, Berger-Levrault, Paris, 1878, and "Amenagement des Forets," by A. Puton. A translation of the latter work has appeared in vols. VIII. and IX. of the "Indian Forester."

PRODUCTS OBTAINED FROM THE FORESTS.

The yield in wood of various classes having once been fixed by the working plan it is the business of the department to realize it as nearly as circumstances will permit.

As to tanning bark, all that the felled trees or poles will yield is utilized. Cork bark is taken from the living trees, which will not bear the removal of a too large proportion of their protecting covering, and hence care has to be taken not to overwork them. Resin is collected on a large scale in forests of maritime pine (*Pinus maritima*), which only yields it freely on the hot and damp coasts of the south-west.

The yield of minor produce, such as grass, moss, litter, and other things, being small, and details regarding it not being available, this class of products cannot receive more than a passing mention. Neither can account now be taken of the numerous advantages which the forests undoubtedly render to the population, but which cannot be expressed in the bulk or weight of the products drawn from them.

The latest available statement of yield relates to 1876, in which year the state and communal forests taken together gave 5,620,663 loads (50 cubic feet) of wood, or an average of about 40 cubic feet per acre ; also 59,742 tons of tanning bark, 292 tons of cork bark, and 1,967 tons of resin.

The yield of wood per acre of the State forests somewhat exceeded that of the communal forests ; but while, in explanation of this, it must be said that the greater extent to which grazing is practiced in the latter affects their wood production unfavorably, it must also be admitted that a large proportion of their produce is made over to the inhabitants for their own use, and that this is estimated at a low figure, so as to reduce as far as possible the charges against them on account of management by the forest department ; and the apparent difference is largely due to the latter cause. Of the total yield in wood 1,364,846 loads were timber and 4,255,817 loads were firewood ; and as might be expected from what has been said before regarding the different systems of culture adopted, the State forests give the larger proportion of timber, one-third of the wood from them being of that class, while in the case of the communal forests, the proportion of timber was only one-fifth. A still more striking result would follow a comparison of the nature of the produce obtained from the State and from private forests ; and since timber is a more useful and valuable product than firewood, the advantage to the country from this point of view, of considerable areas of forest land being owned by the State is apparent, and the more so when it is remembered that France does not grow more than two-thirds of the amount of building timber that she consumes.

The communal high forest is for the most part situated in the mountains, and is composed of coniferous trees, which explains the fact that the greater part of the timber derived from the communal forests is fir and pine, whereas only about one-third of that coming from the State forests is of those kinds.

SALES AND EXPORT.

Principal produce (wood, bark and resin).—With the exception of the produce made over to right-holders, and of that delivered to the inhabitants of the communes from their forests for their own consumption, as well as of comparatively small quantities of timber cut in the State forests for the war department and admiralty, the whole of the annual produce is sold by public auction, and no other mode of sale is permitted. There are three principal systems of disposal, viz.:—First, sale of standing trees ; second, sale at a rate per cubic metre, or other unit of the produce, cut, converted, and taken out by the purchaser ; and third, sale of produce cut and converted by departmental agency. The first of these systems necessitates a previous marking, either of the trees which are to be removed, or of those which are to be reserved. There is no guarantee given either as to the number of trees, or as to their species, size, age or condition : but they are bought and sold on the best estimate that either party can make of their value as they stand. The purchaser, as a matter of course, cuts up and exports the wood at his own cost, and in the form which best suits him, being bound under severe penalties to carry out this work in the manner prescribed by the conditions of sale. It has been urged that this system needlessly introduces a middle man between the producer and the consumer, and that thus the profits of the former are reduced, while the regeneration of the forest may be compromised by felling and exporting the trees in a careless or ignorant manner ; but in reply to this it may be said that the wood merchant must always exist, as it is but rarely that the actual consumer can himself go to the forest to get what he wants, and that by strictly enforcing the conditions of sale, which are framed with special regard to this object, interference with the regeneration of the forest is practically avoided.

The second method differs from the first only in that the auction sale determines the rate at which each of the various classes of produce is to be paid for ; but it is open to the objection that the classification of the produce is difficult, and it thus leads to frequent disputes, in the settlement of which the interests of the proprietor (State or Commune) may be allowed to suffer. This method is rarely adopted, except in the case of thinnings, when the quantity of wood cannot well be accurately estimated beforehand.

The sale of timber, cut and fashioned by departmental agency, is rarely resorted to. It has certainly the advantage that the work is better done, and that more complete precautions can be taken to secure the regeneration of the forest ; but on the other hand, the State or the commune, as the case may be, must advance all the money for the work, and the forest officers become charged with a large amount of supervision and accounts, while a number of purchasers are admitted to the forest, and offences of various kinds are from time to time committed by them. But the chief objection to the system is that the wood is not always cut up in the manner which best suits the requirements of the market at the moment, a matter with which the forest officer can never be so well acquainted as the professional timber merchant, and thus not only do the general interests of the country suffer by failure to supply wood in the form in which it is most required by the consumers, but the prices realized are not always so good as those which the produce might have been made to fetch, had it been cut up in some other manner.

Timber sold standing, usually commands a higher rate than it does when disposed of in any other manner, and for this and the other reasons that have been given, the first of the three systems is the one generally adopted in both the State and the communal forests. This method of sale is not generally followed in other European countries; but the French system has stood test of experience; and it is greatly facilitated by the honesty which, as a general rule, prevails in the trade to which it has given rise. In consequence of the absence or insufficiency of export roads in Corsica, and of the difficulty experienced in getting purchasers who are willing to take the produce for a single year only, a law was passed in 1840, which enacted that the timber to be cut in any part of that island during a series of years not exceeding twenty, might be sold at one time to a single purchaser, the State, at the expiry of the term, becoming possessed of all works erected by him without liability to the payment of compensation for them. A few of such contracts exist to the present day; but both the system of roads and the timber trade having largely developed during the last forty-five years, the practice of entering upon such engagements is gradually dying out.

Minor produce.—Receipts on account of minor produce form an insignificant portion of the gross revenue derived from the French forests, the most important item being that which is due to the sale of hunting and shooting permits. Produce of this class is not sold so much as a source of revenue, as to enable the agricultural population to make use of it, without giving rise to the idea that they are entitled to it by right. It is sold by private contract, the price being fixed by the conservator, or by the prefect, or the mayor, in the case of the State and communal forests respectively. The conditions under which such sales are effected in the State forests are determined by each conservator, with reference to local circumstances, and he retains the power to forbid the sale from the communal forests of any classes of produce, the removal of which would, in his opinion, be detrimental from a cultural point of view. Payment for minor produce is often accepted, especially by the communes, in the form of days' work done in the forest.

Wood supplied to the admiralty.—Every year a notice is sent by the forest department to the admiralty, showing the localities in which trees suitable for naval purposes are to be felled; and the latter department then notifies the number and description of those which it desires to have reserved in each forest. The purchaser of the timber sold from these blocks, fells, barks and conveys the trees marked for the above purpose to an appointed place in the forest, where they are inspected and taken over by the admiralty officials, who cut from them what they want, the rest of the wood being sold by the forest department in the ordinary manner. The forest officer and the marine engineer then agree upon the sum to be paid as the price of the wood removed, and as compensation, to cover losses caused by the depreciation in value of that rejected, and the account is subsequently adjusted in the financial department. Up to the year 1837 the admiralty had the right to select trees everywhere, including the private forests; but the system was not found to answer, and it was abandoned in that year. Even under existing regulations a very small proportion of the wood used by the admiralty is obtained directly from the forests, the greater part of it being bought in the open market.

Wood supplied to the war department.—The requirements of the war department are met, as far as possible, from the State forests, the trees being marked and felled by the forest department, and removed either directly by the military authorities, or by the forest department at their cost. The account is adjusted in the financial department. But the amount of wood so supplied is very small, as, except in cases where the State forests lie near the fortifications or garrison towns, it is found more convenient and cheaper to purchase what is required in the market.

ROADS AND BUILDINGS.

Without roads, which are required in order to render the forest accessible and to facilitate the export of produce, this form of the natural riches of a country cannot be utilized ; the construction of good export roads being one of the most important means than can be adopted for raising the forest revenue., Thus in Corsica, where, before 1850, the State forests did not produce more than £200 a year, the annual revenue derived from them was raised in 1868 to £8,000, the improvement being due almost entirely to the development of the communications. At the end of 1867 there were 2,440 miles of metalled, and 5,380 miles of unmetalled, roads in the State forests, and since that year their length has been at least doubled.

The great importance of accommodating the forest guards in suitable houses within the forests is fully recognized ; and out of 3,200 guards, 1,400 are lodged in 1,213 houses, the remainder of them being granted allowances to lodge themselves in neighboring villages. The proportion of roads and buildings in the communal forests is much less than in the State forests, partly because the communes have to pay for their construction, and funds are not always available, but partly also because the average size of these forests, being smaller, roads and guards' houses within them are not needed to the same extent.

At the end of 1867 there were 126 saw-mills in the State forests, all worked by water-power.

Timber-slides, sledge-roads, wire-rope tramways, and such like means of exporting the wood, are very little used in France. A great deal of timber is required for their construction and maintenance, and, considering the price that wood of all kinds can command, it is found better and cheaper, even in mountainous regions, to make permanent roads suitable for timber carriages and carts. They are to be found only in a few localities where the conditions are exceptional.

Portable iron tramways have not yet come into general use as a means of exporting timber from the forests, and it is believed that there is only one in use in France at the present time, viz., that at Baccarat at the base of the Vosges ; but the advantages which the employment of this means of transport affords will doubtless shortly be better understood than at present, and a development of the system is to be anticipated, at any rate, in the forests of the plains. The floating of large timber is almost unknown ; but firewood for the supply of Paris is still floated from the hills of Morvau down to the railways.

FINANCIAL RESULTS OF WORKING.

The profit derivable from a forest is dependent on a number of causes, among which may be mentioned the species of which the crop is composed, the depth and nature of the soil, the climate, the system of culture, the proximity of great centres of consumption of produce, and the existence of good lines of export.

Taking the average of the last three years for which the accounts have been audited, it is found that the receipts, expenditures, and surplus of the State forests were as follows, viz.:—

Revenue............................... £1,297,748 = 10s. 6d. per acre.
Expenditure .... 571,347 = 4 7 "

Surplus .... £726,401 = 5s. 11d. per acre.

But if the money spent on the afforestation of mountain slopes and dunes, and on the purchase of additional areas, be excluded, the expenditure on the existing forests is reduced to about £480,000, and the surplus is raised to 6s. 8d. per acre. The actual profit is indeed slightly more than this; for the figures include both expenditure by the State on the management of the communal forests, and the contributions paid by the communes on this account. The receipts are supposed to cover the payments, but they rarely do so, and some allowance may be made for this fact when calulating the net profits derived from the State forests, which, during the years referred to, probably fell little short·o 7s. an acre. Recent information relating to the receipts, expenditures and surplus resulting from the working of the communal forests is not available.

The latest year for which full details regarding the gross revenue per acre of the State and communal forests are obtainable is 1876, when the figures were as follows, viz.:—

———	State.	Communal.	Mean.
	s. d.	s. d.	s. d.
Principal produce (wood, bark, resin)	12 6	7 5	10 0
Minor produce............................	0 7	0 3	0 5
Total...............	13 1	7 8	10 5

The revenue from the State forests was then, in 1876, considerably higher than that above given as the average of the last three years; and this was due to two causes, of which the first is the exceptionally large number of windfalls which occurred in that year, and the second the comparatively high rates which timber than realized. All but a small fraction of the revenue on the principal produce was obtained by the sale of wood and tanning bark, cork being produced only in the forests near the Mediterranean and in Corsica and resin almost exclusively on the shores of the south-west. The figures relating to the State forests show the results of actual sales, but this is not so in the case of communal forests, as a large proportion of the produce from them is made over to the inhabitants for their own use, and its value is estimated at a low rate, in order to keep down the amount of their contribution for the services of the State forest department, which is levied in proportion to the sum of their gross revenue and the value of the wood delivered to them. In addition to this it should be said that the revenue on minor produce shows cash receipts only, no credit being taken for the payments made chiefly in the communes by means of days' work done in the forests. These circumstances account to some extent for the smaller revenue obtained from the communal forests; but the true explanation of this result is to be found in the important influence exercised by the system of culture adopted. In 1876 it was observed that the highest rate of gross revenue was obtained from high-forest, and the lowest from simple coppice, while coppice under standards occupied an intermediate place. It was also found

9 (F.)

that in the case of high-forest, the area under coniferous trees yielded a much higher revenue than those under broad leaved species, chiefly on account of the form of their stems, which enables a very large proportion of sawn timber to be obtained from them, but partly also from the greater value of the thinnings taken from them during the early stages of their growth—in the form, for example of telegraph and hop-poles, etc. The revenue from forests composed of coniferous and broad-leaved trees mixed together lay between these two. But, of course, this is not an universal rule; for a high forest of beech might yield a better return than a coppice with oak standards, and a similar comparsion might be made between forests stocked with other trees of different relative values, and managed under various systems. The following figures, showing the results of sales in the Nancy conservatorship, will serve to illustrate what has been said :—

		Per acre.	
Simple coppice	Yielded	4s.	4d.
Coppice under standards.	"	11s.	8d.
High forest of broad-leaved species	"	13s.	1d.
High forest of coniferous and broad-leaved species....	"	23s.	10d.
High forest of coniferous species	"	51s.	6d.

Looking, then, at the large proportion of the communal forests, which is under coppice and at the relatively greater proportion of firewood and timber of small size that they consequently produce, the smaller gross revenue per acre that they were able to yield is no longer surprising. Taking the State and the communal forests together, it was found that their gross revenue was 22 per cent., per acre, higher than that of the private forests, notwithstanding that these latter are as a rule, on better soil and are frequently grown under other more favourable natural conditions.

The average all-round rate actually realized in the State forests per load of wood of all sorts, including tanning bark, was 14s. 5d. ; while that obtained in the communal forests was only 9s. 8d. The corresponding rate for the whole of the French forests, including those belonging to private proprietors, was 10s. 7d., so that the rate of the State forests exceeded the general average by 37 per cent., while that in the communal forests fell to 9 per cent below it. The revenue obtained by the sale of minor produce was derived principally from shooting leases and permits.

It is not an easy matter to determine the capital value of a forest, but in 1873 an estimate was made, which put that of the State forests at nearly fifty and one-half million pounds sterling, which is equivalent to a little over fifty pounds per acre. The gross revenue derived from them in that year represented a return of 3.15 per cent., but the net profit did not much exceed two per cent., on the estimated value. The capital value of the communal forests is certainly less per acre than that of the State forests, on account of the younger age at which the trees are, generally speaking, cut; and notwithstanding that their revenue is smaller, it is probable that they pay a higher rate of interest than the State forests.

It has been estimated that the relative rates of interest on their capital value, paid by forests in which the main crop is removed at various ages, is something like the following, viz. :—

	Per cent.			Per cent.
25 years	4	60 years		2
30 "	3½	100 "		1
40 "	3	200 "		½

These figures are intended to give a general idea of the manner in which, notwithstanding the increased value of the produce the relative rate of interest declines as the age to which the trees are left standing is prolonged. They have no claim to absolute accuracy, even as representing the average of French forests, and still less can they be assumed to apply to the forests of other countries. They serve, however, to explain what has been previously said, viz., that on account of the higher rate of interest which coppice, generally speaking, yields, as well as for other reasons, it is a more suitable system for communes than high forest; and this remark applies with equal and even greater force to private forests.

RIGHTS OF USER.

The principal rights of user are those relating to timber, firewood and grazing; but there is also a small number of others, such as those which permit the cutting of turf, the collection of dead leaves, and the like injurious practices. In the State forests the right-holders are, almost without exception, village communities; the instances in which private persons possess rights in them being extremely rare. The communal forests are, comparatively speaking, free from such burdens.

The law of 1827 provided for the investigation and disposal of all claims to exercise rights in the State forests, and barred the acquisition in them of any fresh ones. Hence those only have now to be dealt with which have been formally admitted and recorded in favour of the communities or persons who possess them.

The aim of the department has always been to free the forests from such claims as far as possible, and the law provides for this being done in the following manner, viz., all rights of wood may be commuted by surrendering a portion of the forest itself in lieu of them, the terms being arranged by mutual consent, or in case of disagreement by the courts; but the State alone can demand such a commutation, the right-holder cannot do so. Other rights, including those of pasture, cannot be got rid of in the above manner, but the State can buy them out by the payment of a sum of money, the amount of which is either settled by mutual agreement or by the courts. The sale of pasture rights cannot, however, be enforced in places where their exercise is absolutely necessary for the inhabitants, the question of such necessity being, in case of dispute, referred to the *conseil de prefecture*,* subject to an appeal to the *conseil d'etat*.† The law also provides that the exercise of all rights which have not been got rid of in either of the above ways, may be reduced by the forest department with reference to the condition of the forests and the mean annual production of the material in respect of which they exist; and none can be exercised otherwise than in accordance with the provisions of the law and the rules based on it.

The principal features of the legislation regarding the exercise of wood-rights are the following, viz. No wood can be taken which has not been formally made over by the forest department; persons who possess a right to dead fallen wood cannot employ hooks or iron instruments of any sort in its collection; when firewood is made over standing in the forest, it is felled, cut up and taken out by a contractor, selected and paid by the right-holders, but previously approved by the forest department; the partition of the wood among the inhabit-

---

* An administrative tribunal, established in each department of France.
† The central administrative tribunal, established at Paris for hearing appeals from the decisions of the *conseils de prefecture.*

ants cannot be made until the work is entirely completed; the contractor is responsible in all respects as if he had been the purchaser of the produce, but he acts under the pecuniary guarantee of the body of right-holders, who cannot barter nor sell the wood made over to them, nor put it to any use other than that for which it is given to them; timber made over in satisfaction of a right, but not used in a period of two years, may be reclaimed by the forest department.

No right can exist to take goats into either the State or the communal forests, as the grazing of these animals is considered incompatible with the maintenance of the ground under wood. The old law suppressed without compensation to the right-holders, the practice of grazing sheep in the forests of the ancient royal domain of France, and the law of 1827 suppressed it also, on payment of compensation, in those State forests which are of more recent origin; but the government has the power to permit sheep grazing in certain localities as an exceptional and temporary measure. No right to pasture any kind of animals can be exercised in any part of the forest not declared out of danger by the forest department, which has also the power to limit the number of animals to be admitted, and the period during which they may graze, with reference to the condition of the forest and the quantity of grass in it. Right-holders can only pasture animals which they keep for their own use, not those which they keep for sale.

On the first of January 1877, about one half of the total area of the State forest was burdened with rights of the estimated annual value of £38,400, while only three per cent of the communal forests were so burdened, the annual value in their case being estimated at £6,700. The commutation and purchase of rights, which was commenced in a systematic manner in 1857, is effected by the officers of the ordinary service, as well as by those who are charged with the framing of the working plans. As a general rule, the arrangement with the right-holders is made by mutual consent, appeals to the courts being of rare occurrence. The State is in no hurry to spend large sums in the purchase of grazing rights which will probably disappear with the progress of agriculture; a result which has alredy been realized in the north of France, where the greater portion of these rights has lapsed through failure to exercise them.

GRAZING.

Goats, sheep and cattle have always been the emenies of forests, and they are indeed the principal agents of their destruction, especially in hot and dry climates, where the vegetation is not sufficiently vigorous to resist the effects of over-grazing.

Animals are admitted to the forests under three different conditions, viz :—

(a) In virtue of a right of user.

(b) As a means of raising revenue and of utilizing the grass.

(c) By tolerance, as a temporary arrangement.

Grazing by right. This has been treated of in the preceding section.

Grazing as a means of revenue and of utilizing the grass. Neither goats nor sheep are admitted into the State or communal forests with this object. In the State forests it is sometimes the custom to allow cottagers living near the forest to graze their cattle in exchange for a number of days' work, but this is not done to any important extent. In these forests in fact very little grazing is sold, for the practice can only be permitted in the unwooded portions, which are rarely available for the purpose, because, although they are of considerable extent (about 450 square miles) they are either required as grazing grounds for the cattle of right-holders, or they are being planted up, and hence the revenue from this source is insignificent. It was only £360 during the last year for which the record is available, but it is otherwise in the case of the communal forests, where local custom often necessitates the maintenance as pasture land of blanks, which could otherwise be most advantageously filled up ; and some communes derive almost their entire revenue from this source. The receipts by them amounted in the same year to nearly £15,000.

Grazing by tolerance—It has been said that no right can exist to graze either goats or sheep in the State or communal forests ; and the inhabitants of the communes are specially prohibited by law from admitting their own goats and sheep into their forests ; but the government has the power to sanction the grazing of sheep (not goats) in certain localities under exceptional circumstances. Permission to drive sheep into the State forests is, however, very rarely accorded, except in seasons of extraordinary drought, when the flocks of the neighborhooding communes are sometimes admitted for a single season. But in the case of the communal forests, such temporary sanction is, of necessity, more freely accorded, for the forests belong to the inhabitants, and even through their true interests might be better served by keeping out their sheep entirely, it is not found possible to change their pastoral habits all at once ; and on this account, permission has frequently to be granted them to graze their sheep in their forests, either for a single year or for periods up to five years. They can, however, graze their own horned cattle, horses, ponies, donkeys and pigs there without special permission ; and they usually do so on payment of a fee into the communal treasury. According to the latest available record, the number of animals of all kinds, thus admitted in a single year, was as follows, viz.:

Horned cattle, horses, ponies and donkeys. 359,164
Pigs 48,388
Sheep (by special sanction)..936,960

The animals can, however, only be grazed in places which have been declared out of danger by the forest officers, and their numbers can be limited with reference to the quantity of grass available; but it is not always possible to enforce these restrictions rigidly ; and the forests in certain regions, have much to contend with from the extent to which grazing is practised. The receipts by the communal treasury on this account have been estimated at 4s. 6d. per head of large cattle, 3s. 11d. per pig, and 1s. for sheep ; but this only represents an average revenue of 10d. per acre of the area grazed over, whereas wood yields, on an average, about 8s. 4d. per acre ; and it seems probable that this consideration may gradually lead, in the agricultural districts, at any rate, to the abandonment of the practice of pasturing cattle, on forest lands. There is no doubt that when the grazing even of large cattle, is permitted, it is carried on at the expense of the crop of wood ; and that where it is practised to any considerable extent the forest, properly so called, tends to disappear ; and this is notably the case where, for the time being, local circumstances, such as the absence of export roads, renders wood a less profitable crop than grass. Here the forests, gradually become almost unproductive, and finally succumb from excessive grazing.

About four-fifths of the total area of the communal forests are still used as grazing grounds, nearly one-half of the latter being open each year ; and the average area provided for each class of animals is about three acres per head of large cattle, two acres per pig, and three-fifths of an acre for sheep. Separate grazing grounds are allotted for each class, and these figures represent the average of all qualities of pasture land; they could not therefore, even supposing that the grazing were not excessive, be taken as a guide to the area which should be provided per head in any particular locality, even in France, and still less so in other countries.

OFFENCES.

Until the year 1859 persons who were charged with offences against the forest laws had always to be tried by the courts; but in that year a law was passed which enabled the forest department to take compensation from offenders instead of bringing them before the tribunals, and this method of dealing with them is now largely practised. The department has always the power to charge the delinquents before the courts, while they, on the other hand, have always the right to refuse payment of the compensation demanded, and thus bring about their formal trial. Officers of lower rank than that of conservator are not, however, authorized to deal with cases in this manner, and the power of the conservator is limited to the acceptance, by way of compensation, of sums not exceeding £40; if it is desired to exact a larger amount, the sanction of the government must be obtained.

This system has many advantages. For while it is necessary in the public interests that infractions of the forest rules should be checked, a large proportion of them are usually of a petty nature, and in many cases the persons who commit them hardly deserve the severer penalties that must be inflicted on their being found guilty by the courts. The system of taking compensation, on the other hand, permits the adoption of a scale of punishment more suited to this class of offenders, while it at the same time enables the means of the delinquents, and the attendant circumstances of each case, to be taken into account. The punishment can also be made to follow promptly the committal of the offence, without the necessity for dragging the accused and the witnesses from their occupations to attend before a tribunal, the time of which is thus not occupied in the trial of these petty cases. The present system is easy and simple for the forest department; and that it acts very leniently on the population living near the forests will be seen, when it is stated that the amount of compensation exacted during the last year for which the record has been prepared, amounted to only one-fifth of the sum which the courts must have awarded had the offenders been proved guilty before them. Occasionally the compensation is allowed to be paid in the form of a number of days' work, done in the forest.

With the advancing prosperity of the country, forest offences become less frequent, and the number committed annually is very much smaller now than is used to be a few years ago. It is worthy of remark that they are more than twice as numerous in the communal as in the State forests, probably because individual inhabitants of the communes think that there is not much harm in committing minor depredations on property which they doubtless regard as their own. During the year 1876, the number of offences was 26,377, there being three per thousand acres in the State forest, and seven per one thousand acres in those belonging to the communes. More than half the offences were connected with the theft of wood or injury to trees, and nearly a quarter related to pasture and cattle trespass, 31,231 persons being involved in the charges. As might be expected, wood stealing is more prevalent in winter than in summer, while the reverse is the case with regard to breaches of the grazing laws. Of the total number of charges made in 1876, 7 per cent. were abandoned, either owing to the trivial nature of the offences, or owing to want of sufficient evidence; 70 per cent. were dealt with under the compensation law, and the remaining 23 per cent. were taken into court, convictions being obtained in 99 per cent of these cases.

In addition to clauses dealing directly with wood thefts, illicit grazing, and other fraudulent practices, the forest law provides that no person having cutting instruments in his hand can leave the ordinary roads which pass through the forests, and that no fire can be either lighted or carried within, or at a less distance than two hundred yards from any forest boundaries. A regular tariff exists, which fixes the penalties for damaging trees of various ages and species. The law also prohibits the erection, without permission, of brick-works, or lime-kilns, carpenters'-shops, timber-yards, or saw-mills, within certain distances of the forest. At the time that the law was passed, it was much more necessary than it is at present to check the erection of such buildings, and applications for permission to construct them are now usually accorded on suitable conditions.

Injuries Caused by Wild Animals and Insects, Storms and Fires.

Wild animals and insects.—The principal wild animals which cause injury to forests, either by devouring the seed or the young seedlings, or by peeling the bark off the young plants, are deer, pigs, hares and rabbits. The insects which attack the leaves, the bark, and even the wood of trees, belong chiefly to the families *Coleoptera, Lepidoptera,* and *Hymenoptera.* But the damage done is not excessive, and it is, in fact, far less than that produced by the same causes in many other countries. It is of course exceedingly difficult to put a money value upon injuries of this sort, which include not only the actual death of a certain number of old and young trees, but also a reduction in the growth of others. An estimate was, however, made regarding the damage done in 1876, and it is said to have amounted to about 4s. per 100 acres, taken on the entire area of the State and communal forests. The coniferous trees generally suffer more than the broad-leaved species, as they are more exposed to the attacks of insects which do not infrequently kill them outright, whereas the latter species more often suffer merely a diminution in their rate of increase.

Storms.—The damage done by storms of wind is a much more serious matter. Injuries are caused to the forest by them which it is not always possible either to prevent or even to modify. In the first place, the windfalls interfere with the arrangements laid down in the working plan, and the considerations which guide the execution of felling are thus thrown out; they remove too large a proportion of the seed-bearing trees, and consequently it is sometimes necessary to substitute a difficult and artificial process for the natural regeneration which would otherwise have been effected; while in addition to this they break or otherwise damage neighboring trees by their fall. In the second place, the value of the windfalls themselves is, speaking generally, small, as they are frequently broken or otherwise injured, while most of them have probably not attained the age or dimensions at which it was intended that they should be felled. They are also specially liable to attacks by insects, which often appear in large numbers in forests where many trees were blown down, particularly in case of the coniferous species. Even uninjured windfalls fetch a lower price than trees felled in a regular manner, because they are usually found scattered here and there, instead of being concentrated in one part of the forest.

The year 1876, which is the last for which figures can be obtained, was a disastrous one, the amount of windfalls being exceptionally large, probably double of that which occurs during an average year. The number was put at 1,145,708 trees, and the damage caused was estimated at £10,300, or about £3 4s. per 100 acres in the State forests, and 12s. per 100 acres in those belonging to village communities.

The latter being, for the most part, coppice under standards, suffered less than the former, while the proportion of windfalls in the coniferous forests was greater than that in those composed of broad-leaved species. The windfalls were sold for nearly £621,000.

The forest officers, when arranging the annual felling, are careful to provide, as far as possible, against the effect of storms, by leaving a protecting belt of trees standing on the side of the forest from which the dangerous winds blow, and in other ways; but much depends on natural conditions which are beyond their control, such as the configuration of the ground, the shelter afforded by neighboring hills, the nature of the soil and its physical condition, the kinds of trees and their root development, as well as their size, age, and the system of treatment to which they have been subjected.

It may be added that hailstorms often do great damage by stripping the trees of their foliage, and by breaking or otherwise injuring the young plants.

Fires.—The penal code provides for the punishment of persons who cause forest fires, either intentionally or through carelessness; and the forest law prohibits the lighting or carrying of fire either inside the forest or within 200 yards of their boundaries, but the ordinary laws do not prevent proprietors from lighting fires in their own forests to the danger of their neighbor's property. This an important question in the Maures and Esterel,* where the bad practice is followed of systematically lighting fires in the forests, in order to burn up the heather and other shrubs which interfere with the regeneration of the crop of trees; and in 1870 a special law was passed prohibiting the proprietors of those districts from lighting fires in their forests except at seasons fixed by the prefect; and also compelling them to clear fire-lines around all woods and forests which have not been completely freed from all inflammable shrubs.

In 1876 there were 290 fires in the area managed by the forest department, nearly all of them being the result of accident. The surface burnt over measured 2,350 acres, or about $\frac{1}{3000}$ part of the entire area, and the damage was estimated at £3,280 or 28s. per acre of forest burnt. The proportion of fires was greater in the broad-leaved than in the coniferous forests, but on the other hand, the amount of damage done per acre in the latter was three times as great as in the former, the resin in the trees themselves and in the dead needles on the ground rendering the fir and pine forests excessively inflammable. It is also worthy of remark that, although as a general rule, fires were of more frequent occurrence in the spring than at any other season of the year, the autumn fires were, on account of the recently fallen leaves, by far the most destructive. But this is by no means true of all regions and the general result may be mainly ascribed to the great damage done by fires occurring during the autumn in the south of France. In the north, forest fires are of small importance, and occasion little damage.

HUNTING AND SHOOTING.

The right to hunt and shoot in the State forests is, generally speaking, let out on nine years' leases, which are sold by public auction under the rules for the sale of timber and other forest produce; but when this is not possible, it is sold by means of annual permits issued under the direct authority of the Minister

---

* Low mountain ranges in the south of France.

of Agriculture, the sport being always carried on under the surveillance of the officers of the forest department. No forest officer can become a lessee of the shooting within the limits of his own charge, and the forest guards are never permitted to shoot in the forests under any circumstances.

The municipal councils are, subject to the approval of the prefect, free to dispose of the right to hunt or shoot in their forests in any manner that they wish.

DESTRUCTION OF WOLVES.

The destruction of wolves, boars, and other animals which are considered dangerous or harmful, is entrusted to a corps of 410 *lieutenants de louveterie* (wolf-hunters). These officers, who are unpaid, but have the right to wear a handsome uniform, are under the control of the conservator of forests, and are appointed by the prefect on his recommendation. They are as a rule landed proprietors, who accept their appointment for the sake of the sport it affords them. They are obliged to keep bloodhounds and packs of dogs, and are charged to organize and direct, in communication with the local forest officers, the *battues* which are, from time to time, ordered to take place in the forests. But as this system has not been found a very efficient one, a law has recently been passed, under which a reward, varying from £1 12s. to £7, is payable to anyone who kills a wolf ; and the mayors are authorized, when the snow is on the ground, to organize *battues* for the destruction of wolves, boars, and other animals, anywhere within the limits of their respective communes, on condition only that they give due notice to the proprietors of the lands on which the beat is to take place.

The rewards paid for killing wolves amount to about £4,000 a year.

AFFORESTATION WORKS—WORKS UNDERTAKEN FOR THE CONSOLIDATION AND PROTECTION OF UNSTABLE MOUNTAIN SLOPES.

Excessive grazing, both by local herds and flocks principally of sheep and goats, as well as by vast numbers of these animals which are annually driven up from the plains to the hill pastures, have produced complete denudation over very large areas, and have thus caused incalculable damage in the great mountain regions of France, principally in the southern Alps, and in the level country below them. They eat down the grass to the level of the ground, and then tear out the very roots, breaking up the surface of the soil, and rendering it liable to be washed down by the rain. These hills are of a loose formation, the strata being contorted and dislocated to a remarkable degree, and as soon as the soil is deprived of its protective covering of trees, shrubs and herbs, whose roots hold it together, the slipping and falling of the mountain sides are produced with a constantly increasing intensity. The rain-water, no longer interrupted in its fall, retained by the spongy, vegetable mould, nor hindered in its downward flow by the thousands of obstacles which a living covering would oppose to its progress, flows off the surface of the ground with extraordinary rapidity, and carrying with it large quantities of loose soil, suddenly fills up the torrent beds. These latter, scoured out by the rush of water, charged with mud, stones and rocks, cut their way deeper and deeper into the mountains, and their banks, deprived of their support at the base, fall inward, the *debris* being borne onward to level ground below. The cracks and slips occasioned in this manner extend to a great distance on either side of the torrent, especially on the side on which the strata slopes toward it, and the effect is much increased when the upper layer of rock is loose, and lies upon an impermeable bed ; the water then saturates the loose rock, and penetrating through it, and through the cracks and fissures, flows over the hard surface, the superincumbent mass being precipitated, either suddenly or by slow degrees into the valley below. The same effect is produced in the whole net-work of water-courses, both principal and tributary, which traverse the mountain side; the upper strata over enormous areas, with fields, houses, and even entire villages, being carried down into the valleys, and the whole region, which presents litttle to the eye but a series of unstable slopes of black marl, has an indescribably desolate appearance. It may be added that when the hillsides are covered with trees, the snow, which has accumulated during the winter months, disappears gradually under the influence of the milder temperature which accompanies the advancing spring; but when the trees have been removed, and the masses of snow are consequently exposed to the full force of the sun's rays, they melt rapidly and produce results on the mountain sides similar to those which follow the occurrence of heavy storms of rain.

But the damage does not stop here; for on reaching the comparatively level valleys which form the main lines of drainage of the mountain range, the stones, gravel and sand transported by the numerous torrents are deposited. These valleys being usually very fertile, are occupied by fields, villages, and towns, which are connected by roads and sometimes by railways, constructed with many bridges, retaining walls and other masonry-works, and as, by degrees, enormous areas become covered with *debris*—sometimes this result is produced suddenly without warning—the buildings are either thrown down or overwhelmed, the railways and roads are blocked, and the bridges are overthrown, while the fields are completely and irretrievably destroyed. The damage thus caused is most serious, both in its nature and extent, and to it must be added the great inconvenience and loss occasioned by the interruption of traffic on the roads and railways. But this is not all. If the *debris* transported by the torrent is carried into the river before it can be deposited, it is either borne on at once and thrown on to the level country lower down, or it remains and turns the course of the stream over the fields and buildings on its opposite bank. Occasionally the deposit temporarily blocks up the valley and causes the inundation of villages and fields on the upper side of the barrier, and when this latter ultimately gives way, the most disastrous results ensue, both in the lower part of the valley and in the open country at the foot of the mountain range. It is to mitigate these terrible evils that the vast enterprise of afforesting the mountains has been undertaken as the only means of dealing with them. But, owing to the enormous cost of the work, it cannot be hoped that the forest thus raised will ever prove directly remunerative, and their creation, with a view to their ever becoming so, could not for a a moment be justified.

The works are of two classes, viz.: Firstly, the treatment of the torrent beds by a series of weirs and other structures, destined to bring them gradually and by successive stages to a normal slope, and thus not only prevent "scour," but by the filling up and widening of the beds behind the weirs to afford support to the unstable sloping sides, and thus gradually to consolidate them with a view to their being ultimately planted up. Secondly, the immediate planting up of all areas, the surface of which does not seem likely to be washed down within the period occupied by the construction in that locality of the first class of works. A commencement was made in 1860; but the law passed in that year not having been found sufficient, a new law came into force in 1882 which provides both for the works to be undertaken directly by the State, and for those to be executed by the proprietors of the ground, with or without State aid, as well as for simple measures of prevention.

Works undertaken by the State.—The proposal to take up ground for this purpose emanates from the forest department, and is followed by a formal inquiry, under the direction of the prefect, into the circumstances of the case, regarding which a special commission, with a forest officer as one of its members, makes a report. If the proposal is approved, a law is passed declaring the work to be one of public utility, and under it the ground with all existing rights, either of the proprietor or other persons in it, is bought by the State, either by mutual agreement or by expropriation. The area is then under the forest law, and the works are undertaken at the public cost.

Works undertaken by the proprietors.—If, however, the proprietors, who are for the most part village communities, do not desire to part with the land they must, before the expropriation has been ordered, agree to execute the specified works themselves, within a fixed time, and to maintain them, under the control of the forest department. In some cases, but not always, pecuniary aid is then afforded to them. If the proprietors of land outside the areas which are taken up for treatment as works of public utility, desire to undertake measures for the consolidation of the soil, or for the improvement of their pastures, they can obtain assistance from the State in the way of money, seeds, plants, or of work done for them; but when any such aid is afforded, the operations are under the surveillance of the forest department, and in certain cases the money so advanced has to be refunded.

Preventive measures.—When the condition of the ground is not such as to warrant its being dealt with in the above manner, it may, after the same preliminary formalities as before, be closed against grazing for any period not exceeding ten years, in which case compensation is paid annually to the proprietors for their loss of the use of it. During this interval the State has the power to execute works, in order to promote the more rapid consolidation of the soil, but the nature of the property cannot be changed thereby, and the proprietor cannot be called upon to pay anything for the improvements thus effected; while if, after the lapse of ten years, it is found necessary to continue the exclusion of cattle, the State must buy the land either by mutual agreement or by expropriation.

But none of the measures above described would deal effectually with the situation unless the source of the evil were at the same time attacked, by bringing the pastoral arrangements on the neighboring hills under control, so as to avoid over-grazing; and the law therefore provides that in 313 village communities, all those in which works are undertaken being included, as well as many others, the grazing must be carried out in the manner approved by the forest department. The communes are therefore obliged to submit to the prefect annual proposals on this subject, showing the nature and extent of their pasture lands, the portions that they propose to use during the year, the number of animals of each kind that are to graze, the roads by which they are to reach and return from the pastures and other matters. These proposals are considered by the forest department, and modified if necessary. In addition to this, with a view to encourage the pastoral population of the mountains to take care of their grazing grounds, and to put a stop to abuses resulting from ignorance and from the continuance of injurious customs, the forest department is empowered to grant money rewards to *fruitéres* (associations of cattle-owners for the manufacture of cheese) for improvement made by them to their pastures. It is also desired to encourage, as far as possible, the substitution of cows for sheep; but the population of the mountains does not like the afforestation of their grazing grounds, and the principal reason for the offer of rewards by the State is that it is con-

sidered politic to do something to aid them in their industry, as some set-off
against the inconvenience to which individual communities are sometimes put
by these operations.

Scope and progress of the entire work.—The total surface to be treated as
a work of public utility in the Alps, Pyrenees, and Cevennes, is estimated to
amount to 1,035 square miles, in addition to about 1,900 linear miles of torrent
beds. Up to the end of 1885, 152 square miles of this surface and 373 miles of
torrent beds had been completed, the expenditures having amounted to £819,320,
and the rates having varied from £3 2s. 6d. per acre, and from 2s. to 7s. 6d. per
linear yard of torrent bed. There remain to be treated, therefore, about 883
square miles of surface, and 1,500 miles of torrent beds. In addition to the
above, the State has paid £138,000, or half the cost of treating 212 square miles,
as "permissive works," under the old law, and £12,000 toward pastoral improvements.

DRAINING AND PLANTING OF SWAMPS AND WASTE LANDS.

Measures of the nature above described for the consolidation and protection
of mountain slopes are undertaken in the interest of the population generally.
In the case of sterile unproductive wastes or swamps, not requiring to be dealt
with on these grounds, the government has thought it better, as a general rule,
to leave each proprietor free to do what he considers most to his own advantage,
confining it to the exemption from taxes for thirty years of all lands planted up.
But the State has the right to force the communes to drain their swamps and
wastes, with a view to rendering them suitable either for cultivation or for the
growth of trees; and when this is done advances of funds may be made under
certain conditions, one of which is that the commune has the right to surrender
to the State, in satisfaction of all claims, a portion of the area not exceeding one-
half.

THE DUNES OF THE WEST COAST.

The winds that blow continually from the ocean on to the west coast, carry
with them enormous quantities of sand, which, advancing steadily over the
country at the rate of some fourteen feet per annum, in the form of moving
hills called dunes, bury under them the fields and villages they reach. It has
been calculated that nearly ninety cubic yards of sand per yard of coast line are
thus annually transported inland. Works to arrest the destructive effects of this
invasion of sand have been in progress since 1789; they were originally carried
out under the Department of Public Works, but since 1862 they have been placed
under the forest department. The total area of the dunes is said to be 224,154
acres, a part of which belongs to the State, and a part to private owners, while a
much smaller portion is communal property.

In exposed situations the protective works consist of a wooden palisade,
erected at a short distance above high-water mark, and destined to promote the
formation of an artificial dune, with a view to prevent fresh arrivals of sand from
being blown over the country. Under its shelter, seeds of various kinds, principally those of the maritime pine (*Pinus maritima*), broom, gorse, and gourbet
(*Arunde arenaria*), are sown; and the seeds being covered with brush-wood to
prevent the sand in which they are sown from moving, and the sowing is thus
continued inland, in successive belts, until a crop of trees is raised on the entire
area. In less exposed situations a wattled fence is substituted for the wooden
palisades. In the Departments of Gironde and Landes, forests of the maritime
pine have been most successfully raised in this manner, the trees being tapped for

resin, and the wood of those which have been exhausted being sold for railway sleepers and other purposes. But north of the Loire the maritime pine is not sown, as in that region it does not yield a sufficient quantity of resin to repay the cost of its introduction, and here it is sought merely to establish a crop of grass on the ground.

The law of 1810 relative to the treatment of the dunes, which is still in force, provides that the government can order the planting up of any area which in the public interest requires to be so dealt with. When the land or any part of it belongs to communes or private proprietors, who cannot or do not wish to undertake the work, the State can execute it, reimbursing itself with interest from the subsequent yield of the forests. As soon as the money so advanced has been recovered, the land is restored to the proprietors, who are bound to maintain the works in good condition, and not to fell any trees without sanction of the forest department. This system of raising forests on private lands would not be likely to succeed elsewhere; but here the extremely profitable cultivation of the maritime pine, due to the large quantity of valuable resin that it yields in the hot and moist climate of the south-west littoral coast, renders it a safe transaction for the State to engage in.

Before the forest department took over the work in 1862, 111,787 acres had been dealt with; and the entire area has now been completed. The works have to be most scrupulously maintained, in order to prevent a recurrence of the evil.

ADMINISTRATIVE ORGANIZATION AND DEPARTMENT STAFF.

Administrative Organization.—In order to carry out the work which has been briefly described in the preceding chapters, a corps of professional foresters, composed as follows, is maintained, viz. :—

1	Director of the forest department.	} superior staff.
9	Inspectors-general.	
39	Conservators.	
245	Inspectors.	
234	Assistant-inspectors.	
308	Sub-assistant-inspectors (*gardes generaux*)	
3532	Brigadiers (head guards) and guards,	subordinate staff.

This body of officials is employed, partly in the ordinary duties of the department, as being in administrative, executive, or protective charge of the units into which the forests (including those of Algeria) are grouped, for their more effective and covenient control; partly in special branches, such as those which are charged with the preparation of working plans with the treatment of unstable mountains, and with the communal grazing arrangements; and partly also in the central offices at Paris. The following statement shows the number of officers of the superior staff employed on each kind of duty :

	Director.	Inspectors-general.	Conservators.	Inspectors.	Assistant-inspectors.	Sub-assistant-inspectors.	Total.
Central offices........	1	8	...	10	12	2	33
Ordinary duties	.....	.....	35	180	177	209	601
Working-plans branch	.....	... ..		15	14	6	35
Consolidation of mountain slopes.........	...	....	..	15	12	49	76
Communal grazing	....	.....	......	2	2	1	5
Schools	.....	1	1	3	6	....	11*
Algeria	.....	......	3	17	10	37	67
Detached duty	....	......	......	3	1	4	8
Total on active list	1	9	39	245	231	308	836

* Exclusive of two forest officers who have been removed from the active list as professors and three professors who are not forest officers.

THE CENTRAL OFFICES AT PARIS.

Since 1877 the forest department has been under the Minister of Agriculture instead of, as formerly, under the Minister of Finance. And the change has proved a most beneficial one ; for the forests are now regarded more from the point of view of their utility in augmenting the general prosperity of the country, than from that of the money revenue they can be made to yield ; and they are no longer looked upon as available for sale whenever the low state of the exchequer may seem to suggest this course, which was not seldom in olden days. The Minister of Agriculture is the president, and the director of the forest department is the vice-president, of a council of administration formed by the eight inspectors-general, which considers all questions submitted for the orders of government. The central office is divided into seven sections, each of which deals with certain branches of the work, and is presided over by an inspector-in-charge, who is assisted by two or three other forest officers and a number of clerks.

Ordinary duties in the forests.—The unit of administrative charge is the division (*inspection*) which is held by an inspector ; but for purposes of executive management this charge is split up into sub-divisions (cantonments), under assistant or sub-assistant-inspectors, who are also at the disposal of the inspector for any special work that he may require of them. Occasionally, when the division is a small one, the inspector himself holds charge of a sub-division. The divisions are grouped into conservatorships, and these again into six circles (*regions*) each of the latter being assigned to an inspector-general. The forests, State and communal, managed by the forest department are 11,508 square miles in extent, and they are divided into 414 sub-divisions, 192 divisions, and thirty-five conservatorships ; consequently, the average area of each of these charges is as follows, viz., sub-division, twenty-eight square miles ; division, sixty square miles ; conservatorship, 329 square miles. The average area of an inspector-general's circle extends over 1,918 square miles.

The sub-divisional officer is essentially an out-of-doors man, who personally directs all work going on within the limits of his charge, in accordance with the instructions given to him by the inspector, whose assistant he is, and who can at his discretion employ him on special duties outside his sub-division. The divisional officer is the manager of the forest estates. He prepares projects for the various works that are to be undertaken, and directs the subordinate officers in their execution; he is also the prosecutor in all cases taken into court for the suppression of forest offences. The conservator exercises a general control over the divisional officers employed under him; and it is his duty to see that all work is directed in accordance with the views of the government, as they are from time to time communnicated to him from the central office. He alone has control of the expenditure, and has power to issue orders on the public treasury. As regards his circle, the inspector-general is not an administrative officer, but he makes an annual tour and is required to become personally acquainted with all the work going on, and with the qualifications of all ranks of officers employed within it, seeing that each fulfills his duties properly. During the remainder of the year he is at head-quarters, where he is able to make use at the council-board of the information collected during his tour, by advising the government both in the issue of orders for works and in the selection of officers and subordinates for promotion to fill the vacancies that may occur.

It may here be mentioned that in addition to the charge of the State and communal forests, the officers of the department are called upon to exercise certain functions in the private forests, which will be explained hereafter.

Working plans.—A separate branch of the department is charged with the framing of working plans for the most important forests, those for the smaller ones being prepared by the local officers. The thirty-five inspectors, assistant and sub-assistant-inspectors, who are thus employed, are divided into nineteen sections, which are at present working in twenty-four conservatorships. As the operations are concluded in one locality, the sections are moved to another. The officers are under the orders of the local conservator, who transmits their proposals to head-quarters with his own opinions and recommendations.

Consolidation of mountain slopes.—The branch of the department to which this vast undertaking is intrusted, is presided over by an inspector-general and is composed of seventy-six officers of the superior staff, working in eighteen centres. These officers are under the orders of the conservator within whose charge they are employed; and he transmits their projects and proposals to the inspector-general, who is thus enabled, by the exercise of his supervision, to utilize the experiences gained in the various localities for the benefit of the entire work. The inspector-general reports to the director of the department all matters relating to this undertaking which are to be laid before the council of administration.

Communal grazing arrangements.—The five officers who are employed in the three great mountain regions to prepare projects for the control of the communal grazing arrangements, and the issue of rewards for improvements to the pastures effected by the *fruitieres* (associations for cheese-making), are placed in the same relation to the conservators as are the officers employed on the consolidation of mountain slopes.

Accounts.—It is a fundamental principle of the French system of forest administration that the forest officers have nothing to do with either the receipt or the payment of money. They sell the produce by auction, or by the granting of permits, as the case may be; but the sums realized on account of such sales are paid by the purchasers directly into the public or communal treasury. The inspector prepares a budget estimate for his proposed expenditure on works, and when this has been sanctioned the various undertakings are commenced. Towards the end of each month he submits to the conservator an estimate of his proposed expenditure for the following month, during the last days of which that sum is paid to him, and he disburses it at once, transmitting the vouchers, together with the unexpended balance, should there be any, to the treasurer-general; he keeps no money in his hands. In exceptional cases, however, the conservator can grant orders for advances to the officers employed under him; but in this case they must, at the end of each month, adjust the advance by vouchers handed in to the treasurer-general, along with any balance of cash that may remain unexpended in their hands. The treasurer-general thus keeps all accounts, both of receipts and expenditure of the department.

Departmental staff.—Members of the forest department are ineligible for any other office, either administrative or judicial; they are prohibited from engaging in trade, or in any industry connected with wood, and they must be regularly sworn in before they can enter upon the exercise of their functions.

They have as regards forest offences, the powers of police, including the right to make domiciliary visits for purposes of investigation and to arrest suspected persons; but these powers are exercised chiefly by the members of the surbordinate staff. Officers of the superior staff act as public prosecutors in forest cases.

Superior staff.—Candidates for the superior staff are, as a rule, trained at the national forest school at Nancy; but one-third of the appointments to the lowest grade (*Garde general*) are reserved for the promotion of deserving subordinates. A young forest officer on leaving the school is employed for a time, usually about a year, in learning his duties under an inspector; and his advancement from this probationary stage, as well as his further promotion through the higher grades depends on his own qualifications and exertions, as reported by his immediate superiors.

A promotion list is drawn out every year by the council of administration and published for general information. On it are inscribed the names of those officers of each grade who are considered to be the most deserving of immediate promotion, the number of names on the list being limited to three times the number of the anticipated vacancies.

The minister of agriculture makes all promotions up to and including the grade of inspector, but the conservators, the inspectors-general, and the director of the department are nominated by the president of the republic. No officer can, however, be selected for promotion whose name is not found on the list and who has not served at least two years in the lower grade.

The yearly pay of the various grades is as follows:

	£	£
Director of the forest department		800
Inspectors-general, three classes	480 to	600
Conservators, four classes	320 to	480
Inspectors, four classes	160 to	240
Assistant-inspectors, three classes	120 to	152
Sub-assistant-inspectors, three classes	80 to	104
Sub-assistant-inspectors, on probation		60

In addition to their salaries, the officers receive travelling allowances, usually a fixed sum per annum, at various rates according to local circumstances.

A pension, at a rate which varies according to the grade of the retiring officer, is obtainable after the age of sixty years; but no inspector can become a conservator after he has passed the age of fifty-five years. Conservators are usually pensioned at the age of sixty-two and inspectors-general at sixty-five.

Subordinate staff.—All members of the subordinate staff must have served in the army, and as a general rule, they must have attained the rank of non-commissioned officers; they cannot be less than twenty-five or more than thirty-five years of age at the time of their appointment. They receive their first nomination from the minister of agriculture, who promotes them from a list similar to that which is annually prepared for the superior staff. The scale of annual salaries is as follows, viz., head guard, three classes, £36 to £44; guard, two classes, £28 and £30 with an additional £2 after fifteen years service.

They must live in or near the forests, where they are provided, as far as possible, with accommodations for themselves and their families in houses specially built for them; but if such houses are not available, they receive a lodging allowance. In addition to their pay, they are given a fixed quantity of firewood per annum and they are allowed to cultivate a plot of ground not exceeding two-and-one-half acres, and to graze two cows in the forest.

Each guard has a beat which he is bound to visit daily, the average size of such charge being about 1,200 to 1,300 acres, or say two square miles. The head guard has four or five guards under his orders: he superintends their work, and communicates to them the instructions received by him from the sub-divisional officer.

The duties of the subordinate staff are chiefly those of protection; they act as forest police, and have the power to serve summonses as well as to arrest delinquents. They are bound to report all offences committed within their beat: and should they fail to do so, they become responsible for the payment of any fines or compensation money which might be levied from the offenders. Acting under the orders of the sub-divisional officer, they superintend all work going on within the limits of their charge; and in addition to this they, under his direction, tend the young plants, prune the stems of the reserve trees, fill up small blanks in the forest, and perform such like minor operations with their own hands. Rewards are given annually to men who have specially exerted themselves in this manner; but they are forbidden to accept without special sanction, any gratuity from "communes" or private proprietors for services rendered by them in the execution of their duties. They are entitled to a pension when they have attained the age of fifty-five years, and have completed twenty-five years' service, including the time spent in the army.

As above stated, one-third of the appointments to the grade of sub-assistant-inspector are reserved for the promotion of deserving members of the subordinate staff. Ordinarily men so promoted must have at least fifteen years' service,·and be less than fifty years of age; but they can be promoted after four years' service, if they have passed successfully through the secondary school at Barres.

Military Organization.—Under the law which provides that all men belonging, in time of peace, to regularly organized public services can in time of war, be formed into special corps, destined to serve with the active or with the territorial army, the members of the forest department form a part of the military forces of the country; and the officers of the superior and the subordinate staff are organized by conservatorships into companies or sections, according to their numerical strength. In case of the mobilization of the army, the forest corps is at the disposal of the war minister, and its various units assemble at previously determined points. The students of the forest school at Nancy receive military instruction and are drilled, the time passed at the school counting as service with the colors. The officers of the superior staff hold rank as officers of the reserve. or of the territorial army, and in time of war may be employed either in command of the companies and sections of the forest corps, or otherwise as may be ordered. From the day that they are called out, the companies form an integral part of the army and enjoy the same rights, honors, and rewards as the other troops which compose it. They are inspected by their own officers annually in time of peace, and the head guards, and guards who form the the non-commissioned officers and rank and file of the companies enjoy at all times certain privileges as soldiers.

In virtue of this service, a military uniform is prescribed for all grades, including the students at the schools. The subordinates wear it always; and the officers do so on all ceremonial occasions, including official inspections of the forests by their superiors.

FOREST SCHOOLS.—THE HIGHER SCHOOL AT NANCY.

The forest school at Nancy is the only one existing in France for the training of officers of the superior staff. It was founded in 1824, before which year the department was recruited either by means of young men, often of good families, who worked gratuitously in the inspectors' offices in the hope of ultimately obtaining an appointment, or by means of retired officers of the army. Very few forest officers received under the old system, a professional training sufficient to enable them to discharge their duties satisfactorily; and it was to remedy this state of things that the school was established. The arrangements were modest at first; but a great development has taken place during the sixty-two years that have elapsed since 1824. The present organization of the school will now be briefly described.

The controlling and teaching staff is composed as follows, viz.:—

1 director, with the rank of inspector-general (professor of political economy and forest statistics).

1 deputy-director (professor of forestry).

1 assistant professor of forestry.

1 inspector of studies (professor of law).

1 assistant-professor of law.

1 professor of natural history.

1 assistant-professor or natural history.

1 professor of applied mathematics.

1 assistant-professor of applied mathematics.

·1 professor of agriculture.

1 professor of German.

1 professor of military science.

1 assistant-inspector for experiments.

All these are forest officers except the professors of agriculture, German, and military science; and none of them, except the professor of agriculture, who is dean of the faculty of science at Nancy, have any other duties.

The salary of the director rises from £360 to £480, with £80 a year sumptuary allowance. The professors of forestry, natural history, law, and applied mathematics, receive, on first appointment £80 a year in addition to the pay of their grade, whatever it may be; but if, after some years, they desire to be permanently attached to the school, they may be removed from the active list, on a salary rising from £280 to £360 a year, when they are entitled to a higher rate of pension than they would otherwise receive. The assistants take part in the instruction under the control and guidance of professors, whom they are in training to succeed; they receive £40 a year in addition to the pay of their grade. The salaries of the professors of agriculture, German, and military science are fixed from time to time, the maximum rate being £240. The appointment of deputy-director and inspector of studies do not entitle their holders to any extra pay; but these officers, as well as the director, have free quarters at the school. The staff is completed with an accountant, two adjutants (corresponding to sergeant-majors), a librarian, a gate-keeper, and other subordinates.

The director of the school is the president, and the professors and assistants are the members of a council of instruction, which assembles at the school from time to time to consider any matter which may be brought before it by the director.

A council sits at Paris at least once a year for the consideration of such general questions as may be brought before it, relative both to the instruction given at the forest schools at Nancy and Barres, and the conditions of admission to, and the regulations in force at, those institutions. President; the minister of agriculture : members ; a senator, a member of the *conseil d'etat*, the director of the forest department, the director of agriculture, the agricultural hydraulics, and inspector-general of forests, the directors of the forest schools at Nancy and Barres, a conservator of forests, a retired forest officer, the director of the agronomic institute, a member of the national agricultural society, an inspector-general of mines, a chief engineer of naval construction, the professor of surveying from the military school, and an officer of the army.

Admission to the school is obtained by public competition. Candidates must be between the ages of eighteen and twenty-two years ; they must be in sound health, and hold a certificate showing that they have completed their course of general studies at the *lycée* (high school). The subjects in which they are required to pass the entrance examination are as follows, viz. :

Arithmetic, elementary geometry, algebra, trigonometry, analytical geometry, descriptive geometry, natural philosophy, organic and inorganic chemistry, cosmography, mechanics, the German language, history, physical and political geography, and plan-drawing. Two passed students from the agronomic institute and two from the polytechnic school, can if otherwise qualified be admitted every year without further examination. The number of candidates admitted annually is, as a general rule, from fifteen to eighteen, and the course of study extends over two years, so that there are from about thirty to thirty-six regular students at the school at one time. The young men while at Nancy are housed in the school building, but take their meals in the town. Their parents deposit £60 a year for their maintenance including the purchase of books and instruments, but they do not pay anything for their instruction, or toward the annual expenses of the school, which may be estimated as follows, viz.:—

Salaries, scholarships, tours and examinations	£4,170
Maintenance of the buildings, library, museum. etc	742
Total annual payments by government	£4,912

If the number of students passed annually through the school be taken as sixteen and a-half, the actual expenditure per head, for the entire period of two years' residence is £298 ; but if interest at four per cent. on the estimated capital value of the buildings and collections (£22,000) be added, the annual expenditure becomes £5,702, and the amount spent by the State on each student during the period of his training is raised to about £350.

Each year of study at the school comprises six and a-half months of theoretical and two and a-half months of practical instruction ; one month being devoted to examinations, and there being two months of vacation. During the period devoted to theoretical instruction, the following subjects are taught, viz. : First year, sylviculture in all its branches ; botany, including vegetable anatomy and physiology, as well as the classification of plants and their geographical distribution, special attention being paid to forest trees and shrubs ; political economy with special reference to forests ; forest statistics ; law, including forest laws and rules ; together with such general knowledge of the common law of the country as is judged necessary; surveying and the construction of roads; the German language; military science; riding. Second year, working-plans or schemes of forest management; mineralogy and geology, with special reference to the chemical and physical properties of forest soils; zoology, especially the branch relating to the insects which attack trees; agriculture; buildings, including houses, saw-mills and bridges; the treatment of torrent beds, including the construction of masonry and other weirs. The teaching of surveying, law, the German language, military science and riding is continued. During the last month of each theoretical course weekly excursions are made into the forest, but with the exception of this and the riding drill the whole of the instruction is given in the class rooms.

The practical course which occupies two-and-a-half months of each year, or five months in all, consists of tours made into the forests in the neighborhood of Nancy as well as into those of the Vosges and Jura, and occasionally to other localities for the purpose of studying forestry, natural history and surveying, a part of the time being devoted to military exercises. An area of 7,500 acres of forest situated near Nancy and placed under the director of the school is used as a field of practical instruction as well as for various experiments and researches, to carry out which an assistant-inspector is attached to the staff. The subjects dealt with by him are principally meterology, the growing of plants in nurseries, various methods of pruning, the effects of different systems of thinning, the rate of growth of various kinds of trees living under different conditions and many other things.

The school is well equiped in every way. Besides commodious buildings to accomodate the director, the deputy-director, the inspector of studies, the students, the adjutants, and other subordinates, there is a spacious amphitheatre with halls of study; a recreation room, and an infirmary are also provided. The museum contains very complete collections, illustrating the courses of mineralogy, geology, palæontology and botany, with woods, fruits, seeds and carefully arranged dried specimens of the foliage and flowers of trees and other plants, as well as raw forest products. There are also stuffed mammals, birds, reptiles and fish, and a collection of insects, with sections of wood showing the damage done by them to the trees. The school possesses an excellent professional library, comprising about 3,350 volumes and a number of maps, It has also a chemical laboratory, in which many interesting researches are made either at the instance of the professors or of forest officers of the ordinary service who may desire the investigation of questions which have arisen in the course of their work. There is a collection of models of saw-mills, of torrent beds treated with weirs, and of sand dunes, etc., as well as a fencing hall and a botanical garden. It is estimated that the buildings are worth about £12,000, and that the library and other collections are worth £10,000; total £22,000.

The students having passed out of the school at the end of their course of instruction are appointed to the forest department as *Gardes generaux* (sub-assistant-inspectors), and are employed on special duty for a time before being instructed with the charge of a sub-division.

Both Frenchmen and foreigners can obtain permission to follow the course of the school as "free students" without the payment of any fees. Since the foundation of the school in 1824, 1,334 regular students, candidates for the French forest service, have been received; and complete or partial training has been afforded to 239 free students of whom 30 were Frenchmen, 73 Englishmen and the remainder were foreigners of other countries.

The Englishmen are sent by the Secretary of State for India to be trained for the Indian service, under a special arrangement made with the French

Government. Ordinarily the free students merely attend the lectures, and as a matter of course, are not examined; but the English students have to pass all the school examinations.

THE SECONDARY AND PRIMARY SCHOOLS AT BARRES.

The secondary school was established in 1883, in order to train a class of men who should occupy an intermediate position between the officers of the superior and those of the subordinate staff. Of the students who entered in that year, seventeen passed out as head guards, and one of these has been promoted to the superior staff as a sub-assistant-inspector. But the school was reorganized in 1884, and it is now maintained in order to facilitate the entrance of subordinates into the superior staff, by completing the education of such of them as may be deemed otherwise fitted for advancement. Candidates for admission to the school are selected by the conservators from among those of their head guards and guards who are thought to possess the needful qualifications, and to be capable of passing the required educational tests: ordinarily, they must have completed four years' service in the forests, and be under thirty-five years of age, but passed students of the primary school can be admitted after two years' service in the forests. They are subjected to an entrance examination in the following subjects, viz., Dictation, elementary geometry, French history, French geography, timber measurement, the selection and marking of trees to be felled or reserved, and the duties of forest subordinates generally.

The director of the school is a conservator of forests, who receives the pay of his grade and free quarters; he is aided in the administration and teaching by two assistant-inspectors, each of whom receives an allowance of £40 a year in addition to his pay. Teachers who are not forest officers can be employed when their services are required. As is the case at Nancy, the director and the professors form a council of instruction and discipline. The students all hold the rank and wear the uniform of a head guard. They are lodged at the school, and receive an allowance of £2 a month to provide themselves with food and clothing.

The instruction, which extends over two years, is both general and special or technical; the object being to improve the general education of the students, and also to give them such a professional training, theoretical and practical, as may fit them for the position they are to occupy. The course is arranged as follows, viz.:—

First year,—Sylviculture, the cutting up and export of wood, estimates of quantity and value of timber, sales of forest produce, arithmetic and geometry, the elements of algebra and trigonometry, surveying and map drawing, levelling, forest law, the elements of forest botany, (including vegetable anatomy and physiology, and the classification of the principal forest trees), planting and sowing, and geography.

Second year,—Working plans, buildings and roads, the elements of mineralogy, geology, and zoology, the treatment of torrents and dunes, forest law and administration, the elements of inorganic chemistry, agriculture and agricultural chemistry, literature and the geography of France. Most of the above subjects are taught not only in the class room, but also practically in the forest. The school is established on a property purchased before 1873 for the primary school from M. Vilmorin, who had raised on it a large number of exotic trees of many kinds. There is also on the estate, a small forest treated as a coppice under standards, which, with the State forest of Montargis, situated at a short distance from the school, is used for the practical instruction of the students. The buildings comprise the residence of the director, the class-rooms, and students' quarters, as well as a museum, containing collections to illustrate the various courses of study.

The examinations are conducted before the director of the forest department, or an inspector-general deputed by him for this duty, and the students who pass will, under the new organization, be appointed to the superior staff as sub-assistant-inspectors. Like the officers trained at Nancy, they will be employed for about a year in learning their duties under an inspector, after which they will become eligible for further promotion on their merits, as are the other officers of the department. Subordinates from the communal forests are permitted to pass into the superior grades of the government service through this school. Nine students entered it during 1884 and 1885, and are still under instruction, eight of them having previously passed through the primary school. One free student followed the courses for a short time in 1883.

The primary school is a branch of the establishment at Barres, the instruction being given by the director and professors of the secondary school. It was established in 1878 for the training of young men who desired to enter the service of government as forest guards, or that of private proprietors as guards or wood managers, there being no restriction as regards their parentage. Up to the year 1883, 148 students had passed through it into the government service, and eight of these have since entered the secondary school. But in 1884 the primary school was reorganized, and it is now reserved solely for the education of the sons of forest officers and subordinates who may desire to enter the government service as forest guards, with a view in most cases, of their ultimately gaining the ranks of the superior staff through the secondary school.

Candidates must be between twenty-four and twenty-seven years of age; they must have completed their military service and be of good character, with a sound constitution. They are obliged to pass an entrance examination in dictation, French composition, arithmetic, elementary geometry, and French history and geography. While at school they are styled "Student-Guards"; quarters are provided for them, and they receive from government a part of their uniform, and an allowance of £1 16s. a month, to provide themselves with food and clothes.

The course occupies eleven months, and embraces the following subjects, viz.: Arithmetic, plane geometry, algebraical signs, surveying and levelling, the French language, French history and geography, the elements of sylviculture, the elements of forest botany (including vegetable anatomy, physiology, and the classification of the principal forest trees), and the elements of forest law, and adminstration. The instruction is given partly in the class rooms and partly in the form of practical work done in the forests.

Passed students are, as vacancies occur, admitted to the government service as forests guards of the second class; and after two years passed in the forests in that capacity they are eligible for entrance into the secondary school. During 1884 and 1885, however, only three students entered the primary school, two of whom are still there and one has received his appointment.

Free students can be admitted with the sanction in each case of the director of the forest department, but as yet none have entered the school.

THE PRIVATE WOODS AND FORESTS OF FRANCE.

Those woods and forests which are neither State nor communal property belong principally to private proprietors, of whom the number is very great, but also partly to civil, religious, commercial and other societies. Their extent varies of course from year to year, according as clearances are made for cultivation or planting work is undertaken. No very exact record of the area is available, but the latest figures show it to be 23,657 square miles, or about two-thirds of the total wooded surface of France. It is probable that at the present time the private woodlands are being somewhat added to, rather than reduced, for it is believed that the areas annually planted up or sown exceed in extent those which are cleared. The private forests are not entirely free from State control ; while at the same time they are protected by the legislation almost in the same manner and to the same extent as are the State and communal forests. For instance, private owners, in common with the government and the communes, enjoy the power to free their forests from wood-rights by making over a portion of the ground to the right-holders in lieu thereof. Grazing rights can only be exercised in those parts of them which are declared by the forest department to be out of danger from the entrance of cattle, and the number of animals can be limited with reference to the supply of grass, while no right can exist to graze sheep or goats in them. Owners have also the power to free their forests of all rights, except those of wood, by the payment of compensation ; and speaking generally, it may be said that they have the same protection against injury to their property by right-holders as is enjoyed by the State and the communes. The law also places them in the same position as regards the punishment of forest offences, including trespass by persons carrying cutting tools, cattle tres-pass, and the lighting or carrying of fire in or near the forests. with a claim to damages for injury caused. Proprietors can obtain for their forest guards, if they have them regularly sworn in, the same powers for the protection of their property as are exercised by the State and communal guards.

On the other hand, private owners cannot cut down and clear their forests without notifying their intention to do so at least four months beforehand, and the forest department can, with certain exceptions, successfully oppose the clearance if the maintenance of the woods is desirable on any of the following grounds, viz. :—

1. To protect mountain slopes.

2. To protect the soil from erosion, and to prevent encroachments by rivers, streams or torrents.

3. To preserve springs and water-courses.

4. To protect coasts against the erosion by the sea and the encroachments of moving sand.

5. For the defence of the national frontier.

6. For sanitary reasons.

The minister of agriculture decides whether the clearance may be made or not. Between the years of 1828 and 1884 sanction has been accorded to the clearing of 1,795 square miles of private woodlands, but there is no record showing what proportion of this area has actually been cleared ; and it is known that sanction is sometimes obtained merely to give an enchanced value to the property by the removal of restrictions on it. It is worthy of remark, however, that while the average area of which the clearance was annually authorized during the whole period above mentioned, amounted to 20,160 acres, the average during the last ten years was 5,404 acres, and during the last five years it was only 3,731 acres. These figures seem to show that woods are acquiring an increased value in France, and that they are cleared for cultivation to a less extent than formerly.

It has already been said that there is a special law relating to the forests of the Maures and Esterel, where fires are systematically lighted in order to get rid of the injurious undergrowth, and that under it private proprietors in those regions are only permitted to light forest fires at certain seasons, while they are compelled to cut fire-lines round all woods which are not completely cleared of inflammable shrubs.

The manner in which the laws relating to the consolidation of mountain slopes and the planting of the dunes affect private owners has also been briefly explained in a previous chapter.

What has already been said regarding the systems of culture generally adopted for the State and communal forests respectively, will lead to the correct conclusion that those belonging to private owners are, as a rule, treated as simple coppice, or coppice under standards, private high forests being usually composed of coniferous trees, and situated in mountainous regions. But many of the forests that have been planted in the plains of the Landes, Salogne and Champagne are stocked with coniferous species, which are frequently more suited to the local conditions, under which they yield a better revenue than could be derived from other kinds of trees. Notwithstanding that the private forests are, as a rule, more favourably situated than those owned by the state or by communes, the gross revenue per acre derived from them is considerably less; because the trees, being cut down at a young age, yield a large proportion of timber of a small size and firewood. On the other hand, their capital value is less, and when they are properly managed they should give a higher rate of interest.

But, unfortunately, although there are exceptions to the general rule, and some of the private forests are maintained in an excellent condition, it cannot be said that, generally speaking, they are so. For while coppice, and particularly simple coppice, is exhausting to the soil, from the young age at which the crop is cut and removed, and, in consequence of the comparative frequency with which the ground is denuded, tends to its physical deterioration, working plans are rarely prepared, and there is consequently no guarantee that the cuttings are confined within proper limits. The fellings are, in fact, too frequently regulated according to the financial requirements of the owner, rather than by the considerations which ought to govern such operations; and hence it follows that the condition of the private forests is not always such as could be desired. This is found to be the case in all countries; but it is probably especially so in France, where the laws relating to the division of the land on the death of its owner, and the custom of the country tend constantly to diminish the number of large properties, and to leave in the hands of each proprietor an area of woodland too small to admit of its management on a regular system.

The produce derived from the private forests is, however, large in amount, and of very great value. Exact figures are not obtainable; but it is probable that the 26,657 square miles yield annually over 12,000,000 loads (of 50 cubic feet) of wood, with about 270,000 tons of tanning bark, 2,250 tons of cork bark, and 30,000 tons of resin—worth altogether, more than £6,000,000; while the isolated trees and vines yield another three and a half million loads of wood, valued at £1,000,000. The number of foresters and guards employed in these forests, is, however, comparatively speaking, very limited; this being due, in a great measure, to the small size of the individual properties, which are consequently in a very large number of cases, managed directly by their owners. There are no private institutions for the training of foresters and woodmen; and although the State forest schools are open to receive "free students" very little advantage is taken of this privilege. The Nancy school has only trained thirty such students since it was established in 1824, and the secondary and primary schools have only received one student between them. Neither the owners, nor their managers or guards, have then, as a rule, had any professional education, notwithstanding that the means of obtaining it is open to them; and it is not to be wondered at if grave mistakes in the management of their forests are of frequent occurrence. In some places they have the means of getting a certain amount of advice from the State forest officials, who are occasionally permitted to render assistance in this manner; but they frequently attempt to imitate what is being done in the State forests, without knowing the reasons for what they see; and they are thus led to commit serious mistakes, as, for example, when, in treating a forest which is to be permanently maintained as coppice under standards, they follow the procedure adopted in a neighboring State forest which is undergoing conversion into high forest. In many cases, of course, the private woods are too distant from the State or communal forests to permit of their owners obtaining any advice or assistance from the officials of the forest department, and they are then thrown entirely on their own resources.

THE TIMBER RESOURCES AND TIMBER TRADE OF CANADA.*

From the geographical position of Canada and the United States, and the natural and artificial routes of transportation that exist along the line, and across the boundary between them, it is reasonable to expect that the interests of trade will, in the future as in the past, draw from the timber resources of both countries for their respective wants, so long as either of them has these commodities to supply Besides this common interest in the forests, for meeting the demand for consumption, both countries have, for a long period, been competitors in the foreign lumber and timber trade, and have shared alike in the vicissitudes that have attended it.

It therefore appears proper to present in connection with the statistics already given concerning our foreign commerce in forest products, as full information as can be derived from official and trustworthy sources, as to the nature and extent of this business in Canada, extending back to the date of the present Dominion Government, and in some instances to an earlier period. The series of tables that we present will sufficiently represent the tendencies of the trade during the years they embrace, and its extent as compared one Province with another, and in different years.

The great prominence of the timber interest of Canada has in recent years led to thoughtful inquiries into the extent of these resources, a synopsis of which will be first presented.

INQUIRIES CONCERNING THE TIMBER INTERESTS OF CANADA.

A Select Standing Committee on Immigration and Colonization, appointed by the Dominion Parliament, has at recent sessions thought proper to institute inquiries having reference to the condition of the forests of the country, and the extent, value, and prospects of the lumber trade. The chairman in the session of 1878, (Mr. James Trow,) in introducing the subject remarked :

That the actual condition of the timber supply of the Dominion was a subject of the utmost importance, and one that deserved the special attention of the committee. It involved not merely the prosperity of the greatest of the manufacturing industries of the country, and the main staple of its foreign commerce, but exercised also a controlling influence in regulating the extent of future settlements, in as much as the forests tempered the climate by rendering it more equable—maintained the regular flow in rivers by preventing inunda-tions, and furnished new settlements with the cheapest building material and fuel.

Mr. Stewart Thayne, an English journalist of some years' experience, who had for the last five years been engaged in researches having reference to the lumber interest, and who had been for two and a half years exclusively engaged in studying this subject of timber resources in Canada, appeared before the committee, and gave in substance the following information :—

" The advantages which Great Britain derives from the Canadian supply of timber are numerous, the principal being :—

1. The best quality of Canadian pine is the most valued of the soft woods. used in the United Kingdom.

2. The dimensions of the soft woods shipped from Canada are larger than can be procured from the timber-producing countries of Europe

*F. B. Hough : Report upon Forestry (U. S.) 1878-79.

3. The colonial supply maintains a healthy competition in the trade, decidedly favorable to the interests of the British consumers.

This trade affords employment for a large amount of British and colonial tonnage.

The kinds of wood exported are, among the hard woods, oak, elm, ash, birch, etc., and of soft woods, the white and red pine, and spruce.

" The dimensions now exported are less than formerly. It was quite usual for the square-timber shipped from the St. Lawrence about thirty years ago to average from 70 to 75 cubic feet per log, whereas, at the present day, the average of the season's log crop does not range beyond 55 cubic feet. Then, in regard to the quality, it was no unusual thing at the period just referred to, for the pine rafts to yield from 70 to 80 per cent., of first quality of wood. I think it would be within the mark to state that the pine at present sent to the Quebec market does not furnish 20 per cent, of first quality. About two years ago I took the trouble to ascertain the qualities of the stock wintering at Quebec, and the estimate I then found was lower than the one just quoted ; indeed, the deals, in my opinion, did not show 15 per cent. Perhaps however, some allowance should be made for the fact that this stock was that which was left after the season's shipments.

" The quantity of lumber that passes through the lakes and down the St. Lawrence is comparatively small, and I am not of opinion that it is all of the first quality. The British Board of Trade returns estimate the value of the Canadian wood imported during the year 1877 as something like $26,000,000. The total imports of hewn timber, during the year, amounted to 103,980,650 cubic feet, of which quantity British North America furnishes 24,286,000 or a little less than one-fourth. This included every description of wood not sawn or split. Of sawn wood there was imported during the same period 228,637,400 feet, of which the Dominion supplied 62,810,600 cubic feet. So that in rough numbers it may be said that Canada supplied the United Kingdom with one-fourth of its timber imports. The total estimated value of these imports, exclusive of furniture wood, is set down at £19,705,447, and the value of the Canadian goods at £5,500,000 sterling. It may be gathered from these figures that a higher value is given to the Canadian produce than to that received from other countries.

" In respect to the present timber trade of Canada, as compared with that of thirty or forty years ago, there is a very great difference in the proportion. For instance, in the year 1831 the total importation of hewn wood into Great Britain amounted to 28,109,950 cubic feet, and of this quantity 20,943,950 cubic feet were sent from British North America.

" In 1832, 1833, 1834 and indeed up to 1840, Canadian shipments held their position ; the total quantity imported by Great Britain is gradually increasing, but the exports from this country do not bear the same ratio to the general trade. Thus in the latter year, the total importation of hewn wood reached 40,858,150 cubic feet of which Canada contributed 32,497,650.

" The square timber trade of Canada held its position in the English market up to the change in the tariff, during Sir Robert Peel's administration. The immediate result of the reduction of the duty on foreign wood was to increase the importation of the latter very considerably, during the years 1845, 1846, 1847 and 1848. During these years the exports from Canada increased also, but not in the same ratio as the foreign. In 1850, the figures representing the then volume of trade are as follows : Total imports of hewn wood, 43,408,950 cubic feet ; from Canada, 30,901,950 cubic feet ; sawn wood, total, 39,708,900 cubic feet ; from Canada, 21,740,900 cubic feet.

" The following table shows the expansion of the trade in recent years, the quantities being cubic feet:

	Hewn wood.	Sawn wood.
1872 Total imports.....................................	89,131,650	154,167,450
From British North America..........................	22,174,200	39,414,400
(Percentage from British North America).............	24.7	25.6
1873 Total imports...............	103,569,500	170,786,150
From British North America).........................	18,293,750	47,717,800
(Percentage from British North America).............	17.6	27.9
1874 Total imports......................	122,369,700	190,262,350
From British North America.......................	23,818,750	53,809,400
(Percentage from British North America).............	19.5	28.3
1875 Total imports........................	84,396,950	164,891,500
From British North America..........................	16,843,350	47,661,400
(Percentage from British North America).............	19.9	28.9
1876 Total imports........................	107,914,750	205,130,900
From British North America.................	23,527,450	55,367,850
(Percentage from British North America).............	21.8	27.0
1877 Total imports...........	103,980,650	228,637,400
From British North America....	24,286,000	62,810,600
(Percentage From British North America)....	23.3	27.4

" All the timber-producing countries of Europe have participated in furnishing these immense supplies of wood; but the most notable increase apparent, during the past few years, has taken place in the quantities of pitch-pine imported from the Southern States. A few years back the demand for this wood in England was limited, being used only for a few special purposes. Immense quantities have been shipped to Europe during the last few years, and, having been sent on speculation, it was sold frequently at very low prices, in some cases at rates that did not cover the freight and expenses, hence it has been introduced into many districts where it was formerly unknown, and competes with the lower grades of Canadian pine, but more particularly with the red pine.

" Sweden and Norway supplied the United Kingdom with from 4,000,000 to 6,000,000 cubic feet of hewn wood during the last few years more than Canada.

" But a very large proportion of the goods under this heading consists of pit-props, and spars, and other small wood of little value. In the matter of sawn wood, these countries furnish Great Britain with some twenty or twenty-five million cubic feet more than the Dominion. The best Swedish deals do not compete with the best quality of Canadian pine, but find a readier sale than the second and third qualities of the latter.

" This trade must be of great value to the shipping interests of Canada and Great Britain, but I have no means of ascertaining the exact number and tonnage of the vessels engaged in this trade during the last few years. The quantity of timber shipped to the British Islands alone must require a carrying capacity of something like 1,500,000 tons. The timber carrying of Europe is confined almost exclusively to foreign bottoms, and though these latter figure largely also in the colonial trade, still British shipping finds in it a source of profit, particularly since the construction of so many new iron vessels has deprived the wooden ones of the carriage of much valuable freight over long sea voyages. Another advantage the shipping interest derives from this trade is the fact that the vessels can be employed in it for a longer period than in almost any other.

" As a matter of fact, there is no other soft wood imported into Great Britain that finds more favor, or that can command a higher price than the first quality of Canadian pine. The consumption is increasing (as shown by the figures above quoted,) at a rapid rate. In 1831 the import of hewn timber amounted to 28,000,-000 cubic feet, while in 1877 it exceeded 100,000,000 cubic feet. The increase in the import of sawn wood is still more extraordinary. The trade has never ceased to expand. No doubt the annual returns show occasionally very serious reductions in the quanties imported. The timber trade has experienced seasons of depression, but they have always followed periods of inflation. Such vicissitudes are inevitable in any branch of commerce where the speculative element has full play. The averages for any given series of years prove, however, that the consumption has advanced with remarkable regularity.

" The common pine and spruce from Canada are used in England for general purposes, but the best quality of pine is now extensively employed in the finishing work of the higher class of dwelling-houses. This wood, when very clear and soft, commands a high price among engineers, metal founders, etc. Its advantages are: that it is easy to obtain a remarkably smooth surface, and the wood is susceptible of being worked to the highest degree of finish, and to the finest edge, without the risk of chipping or breaking like other woods, rendering it very useful to moulders, and I understand that the quantity purchased by them for this purpose is very considerable.

"As to its preference over other woods for finishing purposes I should consider, judging from its frequent appearance in architects' specifications—that it is a favorite wood with the profession, but its merits are so transparent that I do not consider this surprising. No doubt very strong prejudices existed against Canadian wood in England at one time. A constructor of the royal navy stated before a parliamentary committee that a ship constructed of colonial timber could not be depended on for more than twelve months on accounts of its partiality to the dry-rot. Builders came forward on the same occasion to allege that a house having a covered beam of Canadian pine was dangerous to human life, because it might cave in at any moment, while there were some who did not hesitate to maintain that a building containing any portion of this despised wood would speedily become uninhabitable owing to its tendency to breed bugs. One gentleman who boasted of his experience, said that the pine in its native woods harbored myriads of these insects; that they might be seen swarming the logs at Quebec; that they infested the ships that brought this kind of timber to Europe, and finally thronged the woodyards of Liverpool."

To an inquiry as to the quantity of first quality of pine now at Quebec, as compared with that of former years, Mr. Thayne replied: " I saw only a small proportion of the stock that could be considered first quality, and should imagine , therefore, that it must be much less than in former years. By quality, I mean not the size but the texture of the wood."*

In answer to the question as to why the importation of timber into England from Canada had fallen off, it was replied : "I imagine that the reason why the export of square timber from Canada has not kept pace with the home demand, is your inability to supply the description of it that is most particularly wanted. I think also that your profits have diminished because so much of your timber is of poor quality. I think it is safe to contend that the reason why more of your best pine is not purchased is that it cannot be had, and I fear that your power of producing it is not likely to flood the home markets.†

"No doubt there is still some excellent timber in Canada. What I have been attempting to explain is, that however good the produce of certain sections may be, or however well some portion of the present supply may compare with that of former years, still the total quantity of such wood brought to market is small when compared with that of former years, perhaps not one-fifth of a season's manufacture."

With respect to the probable duration of timber supply, at the present rate of consumption, exportation, and waste, Mr. Thayne did not like to give a definite opinion for the following reasons :—

1. Because he could not find data sufficiently reliable to guide him to a safe conclusion ;

2. Any calculation that would ignore the quantity of young timber standing in the woods, but which may become available in the course of twenty or thirty years, would rest on an unsound basis ; and,

3. Because there are so many sections of timber-producing land in these Provinces which, though not extensive when considered separately, still form, in the aggregate, no mean source of supply, and which, though now lost sight of, would soon be opened up, provided a profitable demand should spring up. Having made this statement, he added : "I feel bound to say that every test I have applied to ascertain the quantity of merchantable timber actually standing in any section of the country has convinced me that the resources available are much smaller than public opinion supposes them, particularly those woods adapted to the export trade."

In reply to a member, the witness said :—"No doubt the duration of the timber supply of the United States is a point of much interest to this country. Any interruption of the supplies now drawn by the eastern States from the west would at once compel the former to resort to your markets. Under such circumstances it is easy to foresee that Canadian lumbermen would seek an outlet nearer home for their produce. It would, moreover, be easy for the New England dealer to compete with the English buyer, burdened, as the latter will always be, by a heavy ocean freight."

* To this statement Mr. Cockburn, a member of the committee said : "I must join issue with you on this point, as the quality we are getting now is very fine. In fact, I believe that the soft pine now is of better quality than that formerly dealt with. The pine growing in the free grant lands, or in Northern Ontario, meets with a very ready sale. The quality is found by experience to be very fine. At one time it was supposed to be very inferior, but happily, experience has shown it is of a very superior quality, althought not so large. Though smaller it can take its place beside the larger Michigan pine." Another member remarked that formerly the difference in price between first quality and fair average was less than now, but that at present regard is had, not so much to size as to quality, a small log being sometimes worth more than a large one.

† It was here remarked, by a member, that large pines came from Michigan, up to 22 inches. Good pines were obtained from Laurentian range region of Ontario, of a size that only goes to 18 or 19 inches, strong and clear, which sells as fast or faster than the Michigan, though smaller. Another member remarked that the texture of Canadian wood is not so open as the American ; it is closer in grain. But we should bear in mind that these woods, although competing favorably with Michigan timber of the present day, do not compare with the larger timber produced in Canada some years ago. We produced as good a quality of a larger size, fifteen years ago.

With respect to principal lumbering districts, the Ottawa valley, so far as the export trade is concerned, was by far the most productive, the area drained by the Ottawa and its branches being about 8,000 square miles. Over four-fifths of the square white pine shipped to the United Kingdom is manufactured in this valley.

The chairman remarked that altogether the average area of timber lands in the Dominion is about 280,000 square miles, The river Saguenay and St. Maurice drain large regions extensively timbered.

Great Britain imports masts and spars from Puget Sound, and some splendid pine boards from British Columbia find their way to the workshops of furniture manufactories in London ; but the cost of freight is so great that it will effectually preclude importation from that quarter on a large scale. There is not sufficient freight outward to occupy a small fleet, and the journey is too long and costly to entice vessels merely for a return cargo.

Of the $30,000,000 to $35,000,000 worth of soft woods imported annually by France during the five years preceeding the late war, only a very small proportion was obtained from Canada—a few cargoes of spruce and red pine. The French do not seem to value the white pine. This may arise from the fact that the native hard woods are used very extensively in household construction. Of birch, a very fair quantity enters into consumption in England, large shipments being made from the Maritime Provinces.

It has been computed that the lumber trade of the Ottawa valley alone affords employment to upwards of 25,000 men.

In regard to the duration of timber supply in the north of Europe, a definite answer could not be given. Russia is credited with a large forest area, that might be made available by railroads. Austria likewise possesses some magnificent forests in the centre of Europe, which can only be reached by similar means. Whether so bulky an article as timber can bear the expense that such transportation in Europe would involve, can only be decided by experience. It is true that the European governments are beginning to show a great deal of interest in protecting the forests ; but this newly awakened feeling does not owe its existence entirely to any desire to promote the exportation of forest products, but rather to the fact that they are alarmed at the injuries sustained by the arable land consequent on stripping the hills and river banks of their wood.

With respect to the waste attending the system of leasing limits, by selecting the best logs and allowing a large portion of the trees to rot in the woods, it was deemed to have been greater formerly than at present, the present tenure of these leases being looked upon as so secure that no apprehension of arbitrary interference on the part of the government is now entertained. There is, of course, great waste in the manufacture of square timber, as one-fourth of the best part of the tree was left in the woods in the form of chips. The present system of imposing dues does not present an inducement to waste, but there was a time when a sort of premium was paid for cutting only the largest size of timber because the dues were the same on the smallest sticks as on the largest. At the time referred to, the dues were computed by the piece, red pine at thirty feet average, and white at seventy feet, and the object of the lumberman was, consequently, to cut sticks as large as he could.

With respect to the replanting of denuded timber lands, the witness replied to an inquiry touching the feasibility of the measure, as follows :—

" It is difficult to understand how some steps in this direction have not been taken. In the Provinces of Ontario and Quebec, the local governments derive a handsome revenue from the timber-lands, and yet they seem to regard their disappearance with perfect indifference. Every tree that is felled contributes to their exchequers, still millions have been destroyed by fire without exciting the least effort to prevent such wholesale destruction. These Provinces are spoken of as the future home of millions of people, and yet there is no foresight displayed in reserving, for their use, such indispensable necessaries as cheap building material and fuel. In these two Provinces there exists an immense area that will never be fit for settlement, but which, if judiciously managed, would place Canada in the front rank as a timber-producing country, thereby affording constant employment to a large section of the population, and supporting both commercial and shipping interests. To attain these results, it is neither necessary to injure or disturb such vested rights as have been acquired, nor to adopt very extraordinary or costly expedients.

" Either to lease such lands for long terms, on condition of keeping up the supply, and restricting the cut according to the growth and species of timber on the limit, or by resuming possession of those lands which have been cleared of their pine, and placing them under the charge of practical foresters, replace the pines by varieties that would repay the cost of culture. I am aware that the mere mention of forest-culture seems something far fetched and impracticable to, Canadian ears, but that does not alter the fact that, of all descriptions of cultivation, it is the most profitable. When, further, in a country like this, it becomes a question of utilizing a territory not adapted to any other purposes, and which otherwise must remain barren and unproductive, there should be no hesitation respecting the course to be pursued.

" It is, no doubt, very unfortunate that a line of policy, which is calculated to stir up some grumbling and opposition, and of which the advantages can scarcely be fully appreciated for one or two generations, is not likely to enlist the sympathy of politicians, but this very reason should decide a patriotic statesman to undertake it with determination."

The opinion was expressed that white pine, valuable as it is, would scarcely pay to cultivate. By preserving the young trees, it may still last for a number of years, particularly as there is not much likelihood that the soil which it occupies will be required for other puposes. It requires something near a hundred and fifty years to attain maturity. It was remarked that of late years experiments made in various countries having widely different climates have established the fact that trees may be successfully grown in regions far removed from their original *habitat*, and can already compare favorably with those of mature growth.

There is, therefore, no reason why similar results should not be attained in this country.

The Eucalyptus, an Australian gum-tree, was mentioned as an illustration of this fact, it having been found to thrive remarkably well in the south of France, in Algeria, Hindostan, and California, but it would not survive the winters of Canada.

As to the appointment of inspectors of forests, to report on the timber, and enforce the laws for prevention of forest fires, it is said :—

" The appointment of such a staff would supply one of the most urgent needs of the country—the prevention of forest fires. If it were generally understood that the lowest estimate of the average annual loss through the forest fires, places it at $5,000,000 in the Ottawa valley alone, it appears to me that public opinion would soon interfere to prevent such a fearful waste of the national wealth, for it should be remembered that in the great majority of cases these fires originate in causes that could be easily controlled. But that the country should derive the fullest benefit from the services of such a corps, it is necessary that these inspectors should be practical foresters, of high education and ample experience in the best training schools of Europe. It would be comparatively easy to secure the services of such a class, who, when once established in the country, could train their assistants. Officers of this stamp would, in the course of a few years, be in a position to furnish the government employing them with such information as would render the inauguration of a sound forest policy comparatively easy. It may be objected that this plan would involve considerable expense, but what would the heaviest outlay under this head amount to after all, but an infinitesimal premium of insurance against the average annual loss sustained through these fires, leaving all other considerations out of sight."

To the question as to whether it would be deemed arbitrary on the part of government to make it imperative upon the settlers to plant a certain number of trees on their homesteads, it was replied : " I would consider such a provision one suggested by ordinary prudence. In the treeless districts these plantations would insure a continual supply of fuel, and afford shelter to the land. And here again the necessity of practical foresters in a district makes itself apparent. In order that the settler may derive the fullest benefit of such woodlands, the trees should be planted in positions where they would be of real service to the arable land. I would go even further in suggesting that where new town lands were laid out for settlement, the position to be occupied by the plantation should be selected in such manner as to afford protection to the more exposed districts. The newcomer should also be advised as to the description of timber best adapted to the soil, etc."

Returning to the subject of the difficulty of raising white pine, the question was raised as to whether it would not be advantageous to replant sections of the country with spruce—a rapid-growing timber—the witness said : " Most decidedly. I imagine however that it would be only in rare instances where it would be necessary to incur the expense of planting ; regulations providing for the proper protection of the young trees would answer the purpose in view. At the same time the government should offer inducements, either to farmers or limit-holders, to devote a small portion of their lands to the cultivation of both native and foreign trees, and ascertain from time to time the rate of growth, etc. The government should provide either the seeds or saplings upon which the experiments were to be made."

A member remarked that in the spruce country, by ten or fifteen years, you would get quite a good crop ; but it would take a long time to grow trees from the seed. When eight or ten inches in diameter let them stand ten or fifteen years, and they will yield good cutting timber.

Spruce is used in England in very large quantities. The Maritime Provinces, the Gulf ports and the lower St. Lawrence ship a very considerable amount. Norway is the principal source of the European supply of this wood, but it is a very small size, battens 6½ inches wide and boards as narrow as 5 inches.

A considerable portion of the trade between the north of Europe and Great Britain is in the shape of manufactured goods—flooring boards, window sashes, doors, mouldings, frames, etc. There are many obstacles to the successful prosecution of this trade with respect to Canada. In the first place the manufactured goods imported from the north of Europe are used principally in the construction of the inferior class of houses, and of factories, warehouses, etc. These manufactures are cheap; orders for them can be speedily executed, and can be forwarded with dispatch at a moderate rate of freight to all the principal ports of Great Britain; such as are consigned for sale are also sold at very low prices, labor being cheap in those countries, and the mills close to the seaboard. On the other hand the builders of first-class houses in which Canadian lumber is probably used, have their orders carried out under their own supervision, and were it otherwise the time necessary to forward orders, the delay that might attend their expedition to

11 (F.)

any but one or two ports, and above all the short season of open navigation, are so many obstacles in the path of the Canadian manufacturer.

It might be added to the foregoing that English dwellings of the best class are not constructed with so much uniformity of style as they are on this side of the Atlantic. An enterprising firm might, no doubt, surmount some of these difficulties by establishing a depot for the sale of its goods, and forwarding a plentiful supply of stock during the summer season ; or, better still, appoint as agents in Europe firms of high standing in the trade, likely to be able to dispose of large consignments. But to succeed it would be necessary to possess enter-prise, capital, and an intimate acquaintance with the details of English building operations.

As to hemlock, it was thought that when pine becomes more scarce and costly it would come into demand. If its peculiar qualities were as well known in Europe as in the United States, it would be generally used there also for the flooring of large warehouses, particularly where grain is stored.

"In respect to fires, forests in Europe are differently situated from those in this country. They are not in such unbroken stretches as they are here. Except in parts of Russia and the north of Sweden, there are numerous villages scattered through them. Most of the inhabitants of those villages are employed in the forests, either as charcoal-burners or otherwise. Every forest of any extent has its regular staff of officers and rangers, whose special duty it is to watch over its safety. Open spaces and broad belts of cleared land are kept up on purpose to prevent fire from spreading. The ground is not encumbered with such quan-tities of *debris* as is usual in this country. There are no inexperienced settlers, no reckless workmen, and no careless hunters at hand to court the ravages of this destructive element. The people employed in the forests are interested in their preservation, and stringent police regulations control all others. Notwith-standing all these precautions, fires do occasionally occur ; but of late years they are becoming rare, and on a smaller scale. Probably very few fires occur from lightning, as it is almost invariably accompanied by heavy rain-storms and if a fire should occur from lightning the rain would almost invariably put it out. In quiries directed to this point had resulted in tracing two great fires in the Ottawa Valley to lightning, but they occurred some time ago."

The question of the influence of forests upon the climate of the country and of the effect of clearings being raised, Mr. Thayne replied : "I have endeavored to obtain information upon this point, but without results that would enable me to form a definite opinion. Unfortunately such meteorological observations as have been registered were made at points too far from the influence of forests to be able to denote any but the most trivial changes during the comparatively short period that the subject has received attention. These observations, to be of real use for the purposes referred to, should be made at many points scattered over a wide area. There can be little doubt that the clearings made by the settlers, and more particularly by the forest fires, must already begin to exercise a certain amount of influence on the climate of this portion of the Ottawa Valley. Still, the total absence of any observations at or above this point renders it impossible to express any opinion on the subject."

The effect of planting upon the prairies being referred to, its importance was urged in the strongest terms : "In the various accounts I have read of the prairie land of the North-west, I find frequent mention of the sudden changes of tem-perature. Severe frosts occur sometimes after the crops have been sown, and again before they have been reaped ; or the temperature of the night is often much lower than that of the day. Then these plains are exposed to violent tempests through the cold currents of the Arctic regions coming into contact with the heated ones of the plains. To ameliorate a climate presenting such contrasts there is only one method—that of planting wherever the nature of soil will permit it, and forming the settlements under the shelter of these plantations. Of so great importance is this to our western country that, in my opinion, upon its solution depends whether that region will realize the sanguine expectations now entertained of its being able to support an immense population ; or whether, after many sore disappointments, perhaps, it will deserve the name of the Lone Land. If some of the most fertile regions of the earth have been reduced to the condition of sterile wastes through the destruction of their wooded lands, I think it not unreasonable to infer that a country exposed to a severe climate cannot continue to be productive when, instead of being vigorously planted its already scanty stock of timber is further encroached upon by the new settlers."

The inquiry being raised as to whether a reduction in the amount of exports would not tend to enhance prices, and thus bring increased profits to the business, the opinion was expressed that any further reduction in the export of the first quality of pine would make it so scarce that its sale would be restricted to a few markets of England, and a substitute would be found for it in many quarters where it is now used. The best means of preventing fluctuations in the market would be to export no more than experience had proved to be a fair average demand. So long as lumbermen manufactured in defiance of every law that should regulate the rate of supply, they must take the chances as to the prices which their goods will fetch in the foreign markets.

With regard to the demand of timber for ship-building, a tendency had been observed towards its decline, sailing vessels being superseded by iron steamships, in the carriage of all the costlier and finer classes of merchandise.

In reference to some remarks on the lumber and ship building trade of Prince Edward Island, a member stated : " We import some of our large beams used in ship-building, for keelsons, etc., from Quebec ; they are of pine and tamarac. We build our vessels of a class just about the same as our juniper vessels formerly. We can class from seven to nine years. We own vessels in Prince Edward Island, and can produce them cheaper than in Quebec. We find that wooden ships are taking the place of iron ships, and derive a great advantage from the fact."

In a report of a similar committee upon immigration and colonization, made in 1879, in considering the capabilities and prospects of the north-western regions of the Dominion of Canada, the following answer was given by Mr. Thayne to the question as to how the present growth of wood might be maintained, so as to prevent its exhaustion :

" By a system very different to the one pursued in the older Provinces of the Dominion, where the forest lands have been treated without due regard either for present purposes or for the future wants of the country. Here, however, the opinion was universal that the timber was inexhaustible, and that its destruction was advantageous to the country. It is only of late years that the fallacy of this belief has been brought home to the minds of those who have examined the matter. In the North-west the case is very different ; no competent authority affects to maintain that the timber supply is equal to the wants of such a population as the fertile lands might be expected to support. The obvious policy of the government would, therefore, be to have the timber-producing regions surveyed at the earliest date, before any vested interests are created, and set apart permanent reserves wherever the adjacent lands require shelter, or where a large population is likely to settle. These reserves should be under the direct control of the government, who might either lease them, subject to the condition that the lessee should maintain a regular supply or, better still, according to the system followed in the state forests of Northern Europe, a certain proportion of full-grown timber should be disposed of by public competition the trees to be removed by the purchasers, the number being regulated by the requirements of the locality and the yield of the reserve."

To the question as to what would be the probable effect on the prairies of the North-west if the settlers were under obligation to plant a certain number of acres of trees about their farms, it was replied that no provision would be likely to be of such general advantage, nor one better calculated to promote the welfare of the inhabitants, but under existing circumstances such a proposal was hardly practicable. It would be unreasonable to expect that an immigrant should know what plants would thrive on the soil he occupied. Very few settlers, indeed, are likely to have any experience in arboriculture, and with the great majority it is only too probable that the struggle for existence during the first years of their occupancy would preclude experiments involving any additional outlay of money or labor. To impose such an obligation on the colonist, it would be incumbent on the government to provide him with the means of fulfilling it, and this could be done only by establishing nurseries in the treeless sections, whence the seedlings adapted to the locality might be distributed, either gratuitously, or at a very low price, with the needful instructions for their cultivation. These nurseries might be owned by the government or their formation encouraged by grants in aid to county or municipal authorities or associations.

In reference to the maintenance of supplies in the north of Europe, Mr. Thayne replied that where the supply is limited, as in Germany, the laws are very stringent in some States, going so far as to prevent lands once under forest being devoted permanently to any other purpose; in others, again, private landholders have been prohibited from felling timber in the vicinity of streams, or wherever the forest inspectors consider the arable land adjacent requires shelter from the wind. Throughout the whole empire the forests are subjected to the watchful supervision of a specially trained corps of officials, and no efforts are spared to render them as productive as is compatible with their preservation, which latter is the first consideration. In Sweden, the large forests owned by the government (over 5,000,000 acres) are strictly preserved, trees of mature growth being sold at so much per stump, standing, the felling and removal being carried out at the purchaser's expense, but under the supervision of forest officials. Quite recently a law has been passed prohibiting the felling of trees under certain dimensions, but it only applies to the northern portion of the kingdom.

It was proposed to apply it throughout the southern portion as well, but the opposition was so strong that the minister who introduced the measure resigned in consequence of its partial defeat. However, the whole tendency of legislation in the timber-producing countries of Europe is towards imposing restrictions upon forest owners, and investing the government with greater control over their lands, and there is little doubt that any marked decrease in the supply would be the signal for measures of a far more stringent character.

Being asked as to whether the supply in Norway and Sweden was diminishing, notwithstanding the precautions that had been taken, the witness replied, that in the former country the decrease had been very considerable, many of the mill-owners being now compelled to purchase logs of large dimensions from Sweden. In the latter country there are many districts denuded of all the best timber, and it may be said that the annual consumption is, throughout, larger than the annual growth. The falling off in the over-worked districts has hitherto had no perceptible effect upon the export trade, owing to the extension of the railway system, which has opened up many sections of forest land previously untouched. It is alleged by some that the extent of forest that may be made

available by railways is very large, while others assert that in these comparatively unknown districts the quantity of purely merchantable timber is very
limited. What may be taken for granted is, that while the area under wood
suffers no perceptible decrease, the requirements of the home and foreign markets
are augmenting in a ratio far beyond the productive power of the soil.

To the inquiry as to how the government could promote forest culture in
the North-west this witness replied:

"In my opinion, their first duty is to ascertain the exact nature and extent
of the timber supply in the wooded region, and this can be effected only by an
exhaustive survey. It would then be possible to determine the area that should
be set apart for the support of a permanent forest growth, due consideration
being given to the nature of the climate, the condition of the river and other
water sources, and the wants, present and prospective, of the population which
the arable soil within access is likely to maintain. It would then be in order to
reserve certain tracts for the growth of timber in the most fertile sections of the
prairie lands. County or municipal authorities should be directed to establish
reserves for the protection of river sources or to act as wind-breaks. Railway
companies should be compelled to plant the waste lands bordering their tracts,
and road-boards or trustees should be under a similar obligation wherever violent
winds or snow-drifts were likely to impede traffic or endanger life. Finally,
settlers should be encouraged to plant trees for shade and shelter. It would be
erroneous to suppose that this system of forest preservation and extension would
entail a burden on the exchequer; the forest lands in the actual possession of the
government may, by judicious management, be made to yield a large revenue
beyond their expenses, and a portion of this income spent in planting the new
reserves would, in course of time, become in such a country the most profitable
of investments."

The possibility of raising a second crop of timber in places where it has been
consumed by fire being a subject of inquiry, no definite opinion could be
expressed. The only experience in Canada was that of nature left to her own
resources. What might be done by systematic culture is doubtful, as no experiments had been made; nor, indeed, could it be said that the consequences arising
from forest fires had been examined and reported upon by persons of sufficient
authority to have any weight attached to their opinions. This much was certain,
that wherever a fire ravaged a pine district, the new growth was of a totally
different species, and in hard woods the result was similar. When fire runs
along the soil it effectually destroys all vegetative power wherever the rock is
thinly covered, but when it is confined to the branches and trunks, no reason
could be seen why the same species might not be regrown. It was feared that
the pine would never pay to cultivate in Canada. At a small outlay the Crown
Lands Department might easily ascertain what species of timber could succeed
in pine-growing lands, and settle this and many other points of no small moment
not only to forest science, but to the whole community.

For the renewal of supplies in Norway and Sweden up to a very recent
date, the natural growth has been depended upon to replace the timber felled for
commercial purposes and that destroyed by fire. Of late years, however, the
growing scarcity of wood has induced many Norwegian mill-owners to purchase
cleared or partially-exhausted woodlands, and attempt planting on a large scale,
and this movement is extending. Something similar has been undertaken in
Sweden by the same class as also by the iron manufacturers, who are at the
same time owners of extensive forests. The impression seems to be gaining
ground in both countries that the present rate of consumption cannot be maintained, unless steps are taken to assist the efforts of nature. Were the govern-

ments of those countries to introduce measures for the promotion of timber culture, they would not be under the disagreeable necessity of imposing restrictions that operate frequently to the disadvantage of trade. There is no fact better proved than the one that capital invested in the cultivation of timber yields immense returns.

It is claimed that the area under forests in Sweden amounts to 150,000 square miles, and in Norway to 50,000 square miles. Competent judges are of opinion that, in the former country less than one-half, or about 40,000,000 acres, represent the total quantity of land bearing merchantable timber ; an area not larger than that which the Province of Quebec might set apart for the production of timber without encroaching on lands adapted to agricultural purposes. In the older Provinces of the Dominion there is an extent of forest territory far greater than there is in the north of Europe, if we except Russia.

OTHER STATEMENTS CONCERNING THE FOREST RESOURCES OF CANADA.

The annual reports of the Montreal Board of Trade and Corn Association, in giving statements of the dealings in forest products from year to year, have repeatedly called attention to the great and needless *waste* that was continually going on, and have suggested the propriety of compulsory regulations to enforce economy in lumbering operations. The custom of levying dues upon logs by number, without reference to quality, naturally leads the lumbermen to select only the best leaving the poorer grades to rot in the woods. But if these dues were imposed on the basis of quality an expensive system of inspection in the woods would be involved, and it has been suggested that the most satisfactory means of collecting the revenue would be by an *ad valorem* rate on the timber sawed and exported, as could be best done by the inland revenue department.

It has been further suggested that rigorous measures should be devised and enforced with the view of preventing the vast damages annually done by forest fires, and that inducement should be offered for information that should lead to the punishment of those originating them, whether wilfully or by accident. " Such a fire may have been set by a stray hunter or fisherman bent on sport, or by the clearing of some pioneer far in advance of the frontier settlement, or, as is often the case, by some of the lumbermen's employees, who, troubled by the flies on the banks of a stream, may have kindled a fire to secure protection from their tormentors which the smoke affords." The remedy against these acts of carelessness or malice must be found in adequate penalties rigidly enforced, and of such degree as to render it certain that due care shall be taken in the handling of fires in the woods.

With respect to the rate of reproduction of woodlands and the measures that deserve attention in securing that end, the reports in former years offer the following statements and opinions :

" To obtain an idea of the regular increase in the value of growing timber it may be supposed that it grows one-quarter of an inch in diameter yearly, which is not over the mark ; and as the trees cut will average about twenty inches in diameter, the increase in size will, therefore, be about seven and a-half feet per log, board measure, or over three and a-half per cent. If to this, three and a-half per cent, be added to the sum lost by over-production, an idea of the foolishness of such a policy may be had. It is quite certain that as timber gets cut away and becomes scarce, prices will rise ; and that the lumberers of the present generation are actually killing the hen that would, if properly treated, continue to lay golden eggs.

"Government would deserve the praise of the future inhabitants of the country if they would originate a scheme for planting with young timber trees the immense wastes of the Province of Quebec. Such an investment would certainly not pay a dividend to this generation, but it would utilize what will only be a wilderness when the present trees are all cut, and would be a mine of wealth to those who possess it when the timber becomes large enough to be merchantable. By maintaining a judiciously-matured system of planting, the supply might be prolonged indefinitely; as it is, the forests are denuded of all their valuable timber and comparatively nothing grows up to supply its place. A very large proportion of the country north of the Ottawa is not fit for farming, and never can be properly made fit for grain-growing or pasturage, but it is admirably suited for the growth of timber; and even a limited experiment would soon convince all as to the good results likely to accrue. The cost would be small, there being many large tracts so cleared by repeated fires that there is nothing left to burn. The expense would only be the cost of the plants and their planting, and that would not be much; for the seed could be sown in a cleared spot, where the plants would be set out. The whole arrangement would, of course, require to be planned by a practical man and properly carried out; and, such being the case, there need be no fear of the result. What is above suggested can be done, and may yet be accomplished; and he who does it will be a greater benefactor to Canada than any of the statesmen of the present day."

An English traveller, after noticing the apparently abundant supply of woodlands observed in passing through the settled parts of Canada—and the same remark would eqally apply to considerable portions of the once heavily-timbered regions of the United States—thus remarks, concerning the actual resources of these forests in meeting the demands of commerce :—

"It must often happen to the traveller who travels only the more frequented routes, when he sees great rafts made out of huge blocks of timber floating down the Canadian rivers, to wonder what part of the country produces trees so much larger than any to be seen along its way. Near the thoroughfares of Canadian travel hardly any trees of great bulk remain. . . . The fact is that, for the very large timber, the lumbermen have now to go deep into the country, and far out of the common way. Along the travelled routes you see woods out of which all the finest trees have been long ago cut; and even where you do see trees of large girth in Canada, they have seldom had such room to spread and such free air around them as would have enabled them to develop into objects magnificent in themselves. On the Ottawa, for instance, you may often observe how some one tree in the thick forest having somehow been endowed with a little more hardy vitality than its young and half-smothered fellows, has forced its way right through their competing branches, got its head well into the clear, open daylight, and so vigorously prospered as to have grown to immense sturdiness of trunk; but even it is pretty sure to bear marks of the hard struggle undergone, and to have had its branches and off-shoots, on some side or other at least, checked and hindered in their development, if not crushed and blackened into utter deadness. Whatever charm Canadian woods may have, Canada is not the place to see the beauty of fine single trees. To go in amongst Canadian woods are poor in comparison with the new forest; but when the eye ranges over a great tract of them, often they are indeed most beautiful; as, for example, where they rise and fall over hillsides or undulating ground, or are interspersed with gray boulders and sharp points of jutting rock, and are set off by contrast with waters brightly glimmering at their foot. Such are the woods along much of the Ottawa's course, picturesque and lovely at any time, magnificent when kindled with the colouring of an American autumn. Scenery like this will not easily pall upon the eye, but to travel miles after miles with your view narrowly closed upon either side by flat, unrelieved, unbroken woods, of ungainly and half-developed trees, is a thing far more wearying to the sight than even a journey over the bare wilderness of the prairie. Then, whenever the continuousness of thick woods is broken, it is apt to give place to something not more cheerful. Here you come into the clearing made by some recent fire, where the crowd of living and struggling trees has been burned into a few bare, blackened poles, standing in their gaunt unsightliness, the ghosts of their former selves, with other blackened logs and branches lying strewn over the scorched ground. Again, you plunge into the forest, and see it as it makes itself without the ordering hand of man—now dense, and now thin—trees of different kinds, not generally blended together in intermixture, but standing apart as nature has sorted them; and as, in the great struggle for existence, every kind ousted from elsewhere has been forced into the station best fitted for its support; trees of all ages fighting together for bare life; some vigorous and freshly green, and feathered down to the very ground; some weakly and faded, and only flinging out here and there ragged and ill-balanced branches; some that are mere dead corpses and have fallen aslant out of their places, bruising and breaking the living; some that, with their lower branches all torn and maimed, have yet stretched up out of the throng, and seem as if straining all the life within them to peer over the heads of their fellows, and catch glimpses of how the fire, their deadliest enemy, is spreading havoc nearer and nearer. Again, you are once more in open grounds, lately cleared by some settler, who has ploughed and sown among the tree stumps, those broken columns of the forest ruin; fenced in his clearing with the rude zig-zag wall of logs, the universal snake-fence of the country; built up his log hut in the midst, and set himself to that task which takes half the lifetime of a man to carry out, the turning of forest land into a farm. After many hours of such a journey, and after many days of similar journeyings, the English traveller will not find himself thinking less fondly of the more smiling landscape at home."*

The Waste in Working Square Timber.—Economies in the Timber Trade.

Concerning the great waste from the preparation of hewn timber, as heretofore practiced, the Commissioner of Crown Lands for the Province of Ontario, in his report for the year 1879, says:—

"The great loss sustained yearly by the Province and the revenue from waste of valuable material in the manufacture of square and waney pine, especially in connection with the former, which is hewn to a "proud edge," has for some time occupied my serious attention. It is estimated on good grounds, that one-fourth of every tree cut down to be made into square or waney timber is lost to the wealth of the country, and that the revenue suffers proportionally. When the tree is cut down, it is lined off for squaring, and the "round" outside of the lines is what is called *beaten* off on the four sides; the wood thus beaten or slashed off in preparation for hewing by the broadaxe is the prime part of the tree, from which the best class of clear lumber is obtained when the timber is taken in the round to a saw-mill. Besides the destruction of timber of the finest texture and greatest value, there is the upper portion of the tree, near to and partly into the top, which would yield lumber of an inferior quality, it is true,

* Sketches from America, by John White, Fellow of Queen's College (1870), p: 166.

but suitable either for domestic use or for export to the American market, where, during general business prosperity, large quantities of the lower grades of lumber are required for packing and other purposes connected with trade of all kinds, as much as 100,000,000 feet, it is stated, being sold annually by two or three firms in Brooklyn and New York, to be used as boxes for packages of petroleum alone; but the upper part of the tree is rejected by the square-timber manufacturer and left in the woods, with the fine wood beaten off, to rot and become material for feeding forest fires, by which more timber has been destroyed than has ever been cut down for commercial purposes.

" The following will show the estimated loss to the Province and the revenue from waste in getting out square pine from 1868 to 1877, both inclusive :—Total quantity taken from public and private lands during the ten years, 119,250,420 cubic feet; waste, one-fourth of each tree, equal to one-third of the total mentioned, viz., 39,750,140 cubic feet, or say, in round numbers, 477,000,000 feet, board measure, which may be valued, one-half at $10 per 1,000 feet, and one-half at $5 per 1,000 feet, representing relatively the prime timber beaten off and the inferior timber from the upper part of the tree, average value say $7.50 per 1,000 feet, equal to $3,577,500 loss to the Province for the ten years, or an annual loss in material wealth of $357,750.

" The quantity taken from public lands during the ten years is 87,620,135 cubic feet, the waste on which, on the basis given, being equal to 29,206,711 cubic feet, or 350,000,000 feet, board measure, subject to Crown dues at $750 per 1,000,000 feet, equal to $262,500 lost to the revenue during the ten years, or at the rate of $26,250 per annum.*

" The loss to the country and revenue from timber destroyed by fires, which might have been confined to a limited area, and possibly extinguished, before great damage had been done to the forest, had they not been fed by the *debris* of trees left to rot and dry, is incalculable.

" In 1877, I instructed the officer in charge of the woods and forests branch of the department to prepare a paper on the waste of timber referred to, for the purpose of submitting it to the department of Crown lands of Quebec, with the view of joint action by the two Provinces towards the discouragement of the further continuance of the square timber trade.

" On addressing himself to the task, he found that the lack of knowledge of the mode of dealing with the square timber, after its arrival in the old country in the square " log," was a great drawback to writing intelligently on the subject, as it was essential to know how the timber was disposed of at the great centres of import, such as Liverpool, London, Glasgow, etc.; who the parties were who ultimately acquired the handling of it; where it was cut into specification bills to meet the wants of those who put the product of the " logs," after they had been reduced to the required dimensions, to practical use, etc, so that the department might be in possession of facts, more or less important, when it undertook to show those who are engaged in the trade in Canada, that in abandoning it, and thereby stopping the supply of square timber, they would create a market for their material on the other side of the Atlantic in the shape of sawn lumber.

*If every part of the outside wood wasted in squaring timber could be used, the loss might be estimated at a much higher rate than is above estimated. If the area of a given circle be 1, the area of an enclosed square is 0.636 nearly. The loss is therefore about 0.364 in the outer wood alone, to say nothing of the tops left on the ground. But as there must inevitably be some loss in working, there could scarcely be realized more than 25 per cent. This, in the aggregate of large quantities, is a loss so immense that it should attract the attention of the manufacturers and lead to a thoughtful study into the means for its avoidance. In the careless way that lumber is manufactured, and with the wide-set saws too much in use, it could be easily shown that more than half the material of our forests is wasted, a considerable part of which might, with proper care, be saved.

" I have since procurred some information on the points referred to, from which I learn that the timber is imported directly by wealthy saw-mill proprietors either by the venture of individuals singly, in so many cargoes in each year, or the importation of a number of cargoes annually by several mill-men combined; or it is consigned by Canadian shippers to brokers or agents to be sold on commission; in the latter case, the timber is generally disposed of at auction, at which the saw-mill owners purchase it, and any surplus over what they require for their own establishments they sell in small quantities, sometimes a few pieces at a time, to builders and country dealers of limited means who have it sawn at small mills, and often by hand, at the villages in the interior for local wants. These saw-mill proprietors, having virtually a monopoly of the lumber and bill stuff produced from the timber imported or purchased by them at auction sales, are naturally opposed to the introduction of wood goods into the market they supply in any other shape than in the square log, as at present; but it is time that the Canadian lumberer engaged in the square-pine business should open his eyes to the alarming waste of a material, the value of which is increasing every year; that, in fact, he is stripping his limits and disposing of his timber frequently at a loss, or at best, during several years past, at a rate which seldom pays more than the cost of cutting down, squaring, drawing and taking to market, while at the same time he leaves in the woods as useless one-fourth of each tree he levels to the ground, one-half of the timber so left being the most valuable part of the tree; and see the necessity of his turning his attention to saw-milling operations as a more economical mode of manufacturing his timber, by which he would not only benefit himself by turning to profitable account what is now so wantonly wasted, but the Province generally by increasing the field of labor for its people, while the Provincial treasury would derive additional revenue from the material saved and utilized.

" It may not be out of place to mention here that saw-milling is, to a certain extent, a factor in the settlement of the country, from the fact that many of the employees, from their steady habits and value as workmen, are kept in permanent employment summer and winter in connection with the establishments, and are induced in consequence to take up lands in the vicinity, which are improved by the families of those having grown-up children, and by hired help in the case of unmarried men, till ultimately considerable sections in the neighborhood of the mills would become settled and cleared, with comfortable homes on the locations; while, on the contrary, the men employed in getting out square timber are generally without fixed homes or continuous employment. Their engagements terminate in the spring; in the interim, until they re-engage for the following winter, they too frequently remain idle, and spend their earnings in a reckless manner, and are penniless and often in debt when they return to the woods."

After noticing various available forest commodities for exportation from Canada, such as pit-props, mining timbers, telegraph poles, railway ties, etc., the forms and dimensions best suited to the English market, and suggestions as to their preparation, the commissioner refers to the topic he had previously been discussing with reference to the encouragement of sawn goods for exportation instead of the wasteful practise of getting out hewn timber. His suggestions have no local application, and are well suited to any region or country that has commodities to export.

" The characteristic of modern commerce is to seek out markets wherever they can be found, in which commodities to be disposed of can be sold to the best advantage whether natural products in a raw state where the means of profitable manufacture do not exist where they are produced, or in a manufactured state when such means are available, and in proportion to the energy and enterprise used in pressing forward and occupying every vantage ground in trade, is the measure of success which attends individuals and communities. It is not usual in these days to wait until a customer comes knocking at your door to find out what you have for sale; to succeed, it is necessary that such should be made known far and wide ; and to create a business of any magnitude the first object is to find out what is required not only at home but abroad, *and having the article,* to calculate whether or not the field can can be entered at a fair profit in furnishing what is wanted.

"In the Canadian timber trade there seems to have been no lack of energy ; but in my humble opinion it does not appear to have been accompanied by that kind of prudent enterprise which might be expected from the intelligent men who are engaged in it. The square-pine manufacturers have been contented from year to year to go on bargaining with a Quebec merchant to get out so many cubic feet of a certain average for a price agreed upon ; the merchant writes home to his agent or partner to effect sales, or goes himself or some one for him for that purpose, or frequently ships on his own account the timber which the lumberer has contracted for and delivered to him. Not unfrequently the lumberer possessed of means gets out his timber without advances in money or supplies having been made to him and takes it to Quebec to sell it at the best price he can obtain from the dealers there. Sometimes this has succeeded better than contracting ; but where the venture fails through a downward tendency in the market or a rise in freights, it becomes a serious matter to hold it over, as cove charges and other incidentals rapidly effect a shrinkage in the value of the article.

"But so it has gone on since the early days of getting out square pine ; the same well-trodden rut has been travelled ; the same traffic in the timber in the crude shape of the square " log " has been continued, the actual producer and *quasi* proprietor of the pine upon the timber limits reflecting on the waste of material, or the propriety and prudence of economizing it and turning it to more profitable account.

"Saw-mill owners, although they have had trying times during the past few years, are not generally so unfortunate as the operators in square pine, the trade in which is peculiarly fluctuating and uncertain. The former have always had more or less of the domestic trade : and, unless under extraordinary circumstances, such as the late prolonged depression, can depend on the United States for a market, with prices generally affording a reasonable profit, notwithstanding the American duty of two dollars per one thousand feet ; and with these markets, domestic and across the line, they have seemed to be satisfied without seeking a European opening for their lumber.

"I feel a delicacy in giving advice in this matter to parties who may very naturally say that they know their own business best ; but, nevertheless, I will venture to observe that those in Canada engaged or interested in the trade in timber which is next in value to agricultural products in the exports of the Dominion, viz., in 1878, $20,054,829, and $27,281,089, respectively, should acquire a knowledge of and endeavor to cultivate a transatlantic trade, and would suggest that a spirited effort should be made to extend the sawn-lumber business to countries which have hitherto imported the timber in a crude state and manufactured it to suit their purposes. Already have the European and other markets been successfully invaded by the produce of industries of various kinds, from the American continent, and there seems to be no reason why our great staple export should not meet with equal success.

"It may seem out of place in this report to indicate in anything like detail the steps which might be adopted to carry out what has been hinted at, but a preliminary step would seem to be for a few saw-mill proprietors to join together and send to the old country two or three practical men, having a thorough knowledge of lumbering, the different qualities of lumber produced in Canada, and the minutiæ of the working of saw-mills, who might be accompanied by one or two joiners or house carpenters to make technical observations as to the various uses and forms in which the lumber is applied. Let these parties visit the larger saw-mills in England, Ireland, and Scotland, and on the continent, if deemed expedient, with sufficient time allowed to inspect and report on the whole subject to their employers, having specially in view the required dimensions of boards and bill stuff in all forms, which would suit the several markets ; and also make inquiry as to freights, insurance, port charges, etc., and upon such report, and after due consideration, the parties interested would be in a position to come to a conclusion whether or not a fair paying business could be pushed in the direction indicated. The attempt seems to be worth making ; and if prepared assortments of Canadian lumber were exhibited in the principal markets of the old country, even although they may not take at first, which perhaps would be too much to expect, there is at least a prospect ot success through the exercise of sound judgment, patience, and perseverance."

EXPORTATION OF FOREST PRODUCTION OF CANADA.

From the earliest period of colonial trade the export of timber has been an important item of production for the British market, and much of the timber exported from the northern frontier of the United States has been shipped from Quebec, being generally rafted down the rapids of the St. Lawrence and placed upon vessels at Quebec. In later years the timber of the country bordering upon the upper lakes was brought in vessels to Clayton, N.Y. or to the foot of Wolfe Island, or to Garden Island, near Kingston, at which places for a long period, the principal business of making up rafts for the navigation of the rapids has been done. More recently the business of Clayton has much declined, while that of Wolfe Island and of Garden Island has increased.

This exportation of timber has been largely affected by the tariffs, which from political and financial reasons, the British Government thought it proper to impose.

From the report of a select committee of the House of Lords, appointed in, 1820 to inquire into the means of extending and securing the foreign trade of the country we learn that the encouragement afforded to the importation of wood from the British Colonies in North America by the imposition of heavy duties on wood from foreign States was of comparatively recent date, and that it had not formed a part of the commercial or colonial policy of the country before the then recent European wars. Till 1809 little or no duty had been imposed upon the various species of timber, but in that and the succeeding year, however, the nature of the political relations with the Baltic powers led to an apprehension that great difficulty might be found in deriving the usual supplies of timber from that quarter, not only for domestic use, but more particularly for the purposes of ship-building. The Canadian timber trade had not then been large in the aggregate, although relatively important to the country. There being some risk and uncertainty in a further expansion of the business, it was deemed expedient to give Canadian timber the benefit of an exemption from all duties on such as was fit for naval use, and a duty little more than nominal on other descriptions, while, at the same time, a considerable increase was made in the duty on wood from the north of Europe.

High permanent duties and a temporary war-duty were accordingly imposed upon all descriptions of wood imported from foreign countries.*

The Canadian merchants were never led to believe by the Government that these duties were to be permanent, but an expectation was held out that the duty of £2 1s. first imposed would be continued for some considerable time. No such expectation was fairly raised with respect to the war duty and the duty imposed in 1813, and the exemptions from duty on Canadian timber had always been temporary, and were limited to July 1820†

The protection thus begun was continued many years, and the two great monopolies of corn and timber—the first maintained for the assumed benefit of the possessors of land ; the second conceded to the clamour of a certain class of ship-owners—became through after years the object of attack by an energetic class of reformers, representing the more numerous but less organized class of consumers, until their efforts were finally crowned with success. After successive reductions from time to time the rates on timber from every country were reduced about 1859, to the uniform rate of 2s. per ton, and not long afterwards they were taken off altogether.

The following tables are derived from the reports upon Trade and Navigation reported annually by the Minister of Customs since the beginning of the Dominion Government, July 1, 1867, and previously by the Receiver-General of Canada. The column of years will be understood to be calendar years until the change made in 1864, when the fiscal year, beginning July 1, was substituted.

---

* 49 Geo. III, c. 93. These were doubled by the 50 Geo. III. c. 77, and afterwards partially increased by the 51 Geo. III. c. 93 and by the 52 Geo. III. c. 117 and a duty 25 per cent. upon the whole of the permanent duties were added by the 53 Geo. c. 33. The several duties above referred to were afterwards arranged and consolidated by the 53 Geo. III. c. 52.

The duty upon a load (50 cubic feet) of Baltic timber, which at the beginning of the wars in the early part of this century had been 6s. 8d. was raised by inconsiderable steps to 26s. 2d. in 1806, doubled in 1811, and in 1813 further advanced to 65s. Colonial timber, which had been admitted free of duty up to 1798, was then subjected to a duty of 3 per cent. *ad valorem*. From 1803 to 1809 the *ad valorem* rate was changed to a specific duty of about 2s. a load, and in the latter year this was removed. In 1821, in consequence of the report of the committee of the House of Lords, above cited, the system was changed by reducing the rate on European timber, while that from the colonies was made 10s. In 1849, 1s. 6d. per load was added to them respectively. In October, 1843, the duties were reduced to 25s. per load on foreign timber, and 2s. per load on that from British colonies.—G. R. Porter's *Progress of the Nation*, 1847, p. 380.

† *Parliamentary Papers*, 1820, vol. 3. (269), p. 4.

COMPARISON OF THE SEVERAL CLASSES OF FOREST PRODUCTS EXPORTED
FROM THE PROVINCE OF CANADA DURING THE TWELVE YEARS PRECEDING THE
FORMATION OF THE DOMINION GOVERNMENT.

(Quantities and Values.)

Years.	Ash.		Birch.		Elm.		Maple.	
	Tons.	Value.	Tons.	Value.	Tons.	Value.	Tons.	Value.
1856	2,589	$14,403	4,556	$40,200	36,453	$508,433	18	$169
1857	3,485	25,360	5,026	46,985	37,984	432,822	169	1,593
1858	2,378	16,999	4,005	30,339	19,451	163,389	37	285
1859	4,313	24,067	7,937	56,294	26,278	200,840	84	728
1860	2,478	14,976	12,508	100,759	25,629	207,297	249	1,996
1861	2,422	12,708	8,397	60,585	32,610	265,562	127	1,014
1862	2,496	12,770	4,159	32,424	27,689	202,573	139	882
1863	8,341	42,255	11,256	89,111	53,392	421,180	440	2,620
1864*	1,319	6,667	3,315	26,413	14,331	114,414	53	366
1865†	3,670	22,689	10,488	82,638	49,048	387,655	110	1,350
1866†	2,860	20,986	8,793	72,505	29,483	255,670	152	1,268
1867†	3,631	26,074	9,394	81,355	28,476	252,647	76	643

* Half year ending June 30.   † Year ending June 30.   Since 1864.

Years.	Oak.		White pine.		Red pine.		Tamarack.	
	Tons.	Value.	Tons.	Value.	Tons.	Value.	Tons.	Value.
1856	33,814	$377,190	361,046	$2,062,003	61,943	$471,691	2,117	$13,381
1857	48,539	576,630	500,781	2,821,320	61,323	526,458	4,571	28,471
1858	26,904	377,561	344,981	1,811,340	53,143	374,079	995	5,411
1859	34,300	359,731	395,694	2,249,006	43,643	363,567	2,185	11,382
1860	41,553	404,861	490,233	2,582,605	62,573	507,610	2,589	17,023
1861	55,979	526,997	523,112	2,594,388	71,381	508,609	1,802	11,116
1862	47,436	527,317	430,257	2,110,046	66,563	452,113	14,861	33,301
1863	73,327	754,328	650,483	3,304,903	103,329	745,642	19,591	124,955
1864*	28,387	205,546	194,822	918,323	32,653	230,831	5,326	34,994
1865†	118,313	1,089,417	606,300	2,963,534	108,877	761,037	10,681	65,332
1866†	64,026	710,861	450,950	2,324,063	85,638	593,134	11,266	71,938
1867†	62,895	696,461	413,036	2,118,754	78,792	499,858	5,411	36,015

* Half-year ending June 30.   † Year ending June 30.   Since 1864.

Exports to Great Britain.[*]

The timber trade of British North America is principally concentrated within the Provinces of Nova Scotia and New Brunswick, and throughout the St. Lawrence and Ottawa River basins, and north-western Ontario.

Up to about the year 1842—and partially afterwards—a very much lower duty was levied in Great Britain on wood which was the produce of Canada than on similar imports from European nations. This led to the fostering of a very large trade, especially in hewn timber, from Quebec and the lower ports on the St. Lawrence—a region which forty or fifty years ago was looked upon as Great Britain's principal source of supply. Large quantities of white pine and spruce as well as a small supply of red pine are still exported; the first being partly hewn and partly sawn, while the second is mostly in a sawn condition. A considerable quantity of oak, elm, ash and birch is likewise exported.

The following return of shipments will best show the movements of the Quebec export trade for the nine years ending with 1882.

In 1874, 854 timber carrying vessels of 636,672 tons cleared out.
" 1875, 642   "   "   478,441   "   "
" 1876, 786   "   "   624,110   "   "
" 1877, 796   "   "   670,620   "   "
" 1878, 476   "   "   399,833   "   "
" 1879, 433   "   "   364,628   "   "
" 1880, 634   "   "   555,451   "   "
" 1881, 459   "   "   380,186   "   "
" 1882, 426   "   "   359,925   "   "

Small parcels of sawn wood are shipped by steamer from Quebec, and these are not included in the above table; but such supplies although considerably larger than they were some years ago, are not sufficient to alter the fact that this export trade of Quebec is declining.

The following is a detailed statement in loads of the principal items of which that export was composed for the five years mentioned, namely :—

	1879	1880	1881	1882	1883
Hewn wood :					
Square and waney pine	106,009	231,051	182,038	158,243	208,540
Red pine	16,276	28,664	18,440	20,494	20,979
Oak	33,620	46,337	37,667	39,147	42,658
Elm	10,880	20,836	15,943	15,567	14,798
Birch	3,929	11,177	5,878	42,274	4,661
Sawn wood :					
Pine deals	122,602	266,900	177,659	144,315	183,016
Spruce deals	130,740	147,673	141,962	127,752	125,108

There are no means of ascertaining the exact amount of the Quebec shipments to the United Kingdom, but it is probably about five-sixths of the whole quantity despatched over sea. The total export of Quebec wood to Europe

[*] Robert Carrick, *Gefle, Sweden,* in Forestry and Forest Products (Edinburgh), 1884.

in 1883 had an approximate value of $7,802,550. A small but increasing business is likewise now being done with South America.

The city of Ottawa is the headquarters of the timber-manufacturing interest of Canada. The wood goods sawn here are principally pine boards for the United States market, and it is no exaggeration to say that the Ottawa valley production of sawn timber is at present the largest in the world. This reached a total of about 1,320,000 loads in 1882. Such a production far exceeds that of any district in Europe.

New Brunswick and Nova Scotia, however, must for the present be regarded as the seats *par excellence* of Great Britain's supply of American spruce. The chief ports that are engaged in shipping this class of goods, are St. John, Miramichi, Dalhousie, Richibucto and Bathurst, the bulk being despatched from the two first named.

In connection with this question, it is necessary to consider whether the Dominion of Canada can keep up or increase her present output, without endangering the future of her forests, and what proportion of that output will be required at home and in the United States.

In seeking an answer to these queries, only white pine, red pine and spruce need be considered, seeing that the quantities of oak, elm, birch and ash now being exported, are comparatively small.

So far as red pine timber is concerned, the supply has gradually diminished to a quarter of what it was in 1863. A substitute for it has been found in pitch pine, so that it is not much missed.

Hewn and sawn white pine is a most valuable description of timber, and when of the finest grades, is unrivalled for many purposes, such as house building, and other wood-work. It has been largely and continuously imported into the United Kingdom for more than fifty years, and a fully equivalent substitute will be difficult to find. The White Sea redwood approaches nearest to it in point of quality, but the latter in addition to its smaller dimensions and greater knottiness has other defects that diminish its value in comparison with the former.

The quantity of hewn white pine received at Quebec in 1876, was about 19,243,733 cubic feet, whereas in 1883, it was but 11,198,557 cubic feet, or taking the average for the years 1871 to 1875 inclusive, it exceeded 14,000,000 cubic feet per annum ; while for the five seasons, 1879 to 1883 inclusive, it was but 8,412,654 cubic feet. On the other hand the supply of white pine deals to Quebec has not decreased in so pronounced a manner, for although in 1876 it reached 278,363 loads, and only 147,979 loads in 1883, the average of the five years ending 1885, was 187,187 loads, against 238,731 loads on an average for the five years ending with 1880. Such figures in conjunction with the history of the quantity exported of late years, bear abundant evidence to the fact that a diminished quantity is available for export to Europe.

Reference may be made to the attempt which was made in connection with the tenth census of the United States to ascertain the quantity of mature white pine then existent in that country and ready for the axe. Professor Sargent, who had charge of this part of the census, reported in 1882 in these words :—" The entire supply (of white pine) growing in the United States and ready for the axe does not to-day greatly, if at all, exceed 80,000,000,000 feet ; and this estimate includes the small and inferior trees which, a few years ago, would not have been considered worth counting. The annual production of this timber is not far from 10,000,000,000 feet, and the demand is constantly and rapidly increasing."

According to this semi-official statement, there was, some years ago, only about eight years' supply, and now there is supposed to be only six or seven years' consumption ; there is, fortunately, reason to believe that the statement is inaccurate, and proof of the assertion that the United States consumes 5,050,000 of St. Petersburg standard hundreds, or 16,665,000 loads of white pine wood annually, will be awaited with interest. It must be taken into consideration that Canada supplies a good percentage of the present United States consumption, and that the precise minimum dimension has not been defined which the census assessors have put down as ready for the axe. The large number of trees, too, which, in the course of eight or ten years will reach the cutting size, although now a trifle under it, must also be considered. Taking credit for all these items, however, and assuming the available quantity to be as much underrated as the consumption is evidently over-estimated, the outlook is sufficiently alarming, and amply justifies compulsory replanting and reforesting enactments wherever the ground is best adapted for the production of forest.*

As regards American spruce, which competes in the British market with Swedish and Russian white wood, the question of its future supply becomes one of paramount weight. One of the most important facts to be borne in mind, is

* At a meeting of the Genessee Valley Forestry Association, held in the Chamber of Commerce Rooms at Rochester, on the 10th May, 1892, Mr. B. E. Fernow, Chief of the Division of Forestry, Department of Agriculture, is reported to have said : " Although there is still plenty of virgin timber in the country, the time when it will be comparatively exhausted is drawing near. We have in the United States about 500,000,000 acres in woodland. If all this were in good condition, with full grown timber on it—which is far from being true—there could not be found on it more than 1,500,000,000,000 cubic feet of wood, which is at the rate of 10,000 feet b.m. of saw-timber per acre in the average. Since we use annually from 20,000,000,000 to 25,000,000,000 cubic feet of wood, of which 30,000,000,000 to 40,000,000,000 feet b.m. is saw-timber, it appears that even with these extravagant assumptions regarding supplies, we would exhaust them in sixty or seventy years, assuming that new growth is consumed by increased requirements. That this increase takes place may be learned from my computation, according to which the values of forest products and wood manufactures during the census years 1860, 1870, 1880 and 1890 amount to $300,000,000, $600,000,000, $900,000,000 and $1,200,000,000, respectively, or an increase of thirty per cent. for every decade ; and since it takes at least sixty to seventy years to grow saw-timber—what we now cut is usually twice as old, or more—it would appear that whoever invests his money in forest culture to-day must be amply repaid by the crop, albeit his children will reap the profits really. Certain it is that it always requires time, and quite a long time, before the results of such management become visible, and that is largely why people are afraid to stake their money in the business.

" For profitable forest management, good, permanent roads are indispensable. Their money value may be judged from the experience of the little Dukedom of Brunswick, where, without any other changes, the building of a rational system of roads through its forest domain increased the income from the forest management by twenty per cent.

" In the government forests the annual net profits range from $4.11 per acre of forest in the highly cultivated and densely populated Kingdom of Saxony, to $1.19 in mountainous Bavaria, where the government forests comprise 2,300,000 acres on which the government spends $3,130,000 and gets in return $5,880,000, netting $2,730,000 every year, and giving besides employment to a large force of men. In Prussia the net annual profit for every acre of woodland was at the rate of $1.31 on $6,000,000 acres woodland, the expenditure last year being nearly $8,800,000 for the administration, and with prices in the woods of three dollars per cord of fire-wood in the average, and ten dollars and thirty-two cents per thousand feet b.m. of saw-timber, the returns were over $17,600,000, netting $8,835,119, and this is continuous, ever increasing revenue.

" Why should not the State of New York, now owning as State property twice the area which little Saxony owns in woodlands, and proposing to acquire additional lands so as to make the area between that of Bavaria and Prussia's government forests, why should not New York make its Adirondack forests, if not as profitable, yet fairly so, as those States have made their woodlands pay ?

" To bring this about, it is necessary to secure as soon as possible the acerage before it becomes more expensive, to place it under competent administration, to open up and make accessible the wilderness by a rational system of well-built roads, and inviting, not keeping out the railroads, under such restrictions to be sure as will properly guard the forest against danger from fires, for accessibility is the key-note of practical and profitable forest management."

the case, rapidity and spontaneous manner in which this tree reproduces itself.
Unlike white pine, which requires a century to grow a standard log, spruce is
almost irrepressible, and a spruce forest from which all the trees measuring
twelve inches and upwards in diameter have been taken, will, after a rest of two
decades, be ready for work again.

It does not appear that any systematic attempt has been made by the
authorities to reduce to figures the available quantity of mature spruce now growing
in New Brunswick, Nova Scotia, and the immense forest belt which lies between
the Ottawa river basin in the west and Mingan in the east. All these districts,
however, according to Mr. Joly, late Premier of the Province of Quebec, contain
immense quantities of spruce.*

The following tables exhibit the total number of pieces of squared white
and red pine and other woods, and of pine saw logs cut in the upper Ottawa
territories of Quebec and Ontario, on Crown lands, and also on private lands
from 1826 to 1881 (30th June) inclusive, that is before and since Confederation
as closely as can be learned from the records of the Crown Timber Office, Ottawa.†

FROM 1826 TO 1866 INCLUSIVE.

	Square timber.		Pine saw-logs (only).	
	Ontario.	Quebec.	Ontario.	Quebec.
White pine	3,048,382	1,739,094	4,084,258	2,230,056
Red pine	1,714,412	978,064	....	....
Other wood	255,950	146,019	....	....
Totals	5,018,744	2,863,177	4,084,258	2,230,056
Deduct from private lands	1,074,418	612,931	874,349	498,804
Cut on Crown lands	3,944,326	1,751,246	3,209,909	1,731,252

*The pitch-pine referred to in this chapter is the *Pinus Australis* of Michaux, otherwise the Southern
Pine, the trade in which has been entirely developed during the last twenty years. Its importation to
Great Britain is extending rapidly. It is shipped from the so-called Pitch Pine States, situated to the
east of the Mississippi River, namely, the Carolinas, Georgia, Florida, Alabama, and Mississippi.

The wood of the pitch-pine tree is the heaviest of the pine family, and, like all high resinous woods, is
supposed to possess great durability in climates of moderate severity. A very large industry has been
developed in the Southern States in connection with the gathering of its resinous products, such as tar,
turpentine, pitch, and the like. Turpentine is the raw sap of the tree, and the "boxing" or tapping necessary to collect this is supposed by some to be injurious to the "alburnum" or sapwood, although it has
probably little effect on the heartwood, or "duramen."

†A. J. Russell, Crown Timber Agent, Ottawa, in Reports on the Forests of Canada, London, (England) 1885.

From 1867 to 1881 Inclusive.

	Square timber.		Pine saw-logs (only).	
	Ontario.	Quebec.	Ontario.	Quebec.
White pine	1,925,247	1,119,382	17,920,850	17,277,103
Red pine	485,141	118,626	....	....
Other wood.........................	238,874	63,319	....	....
Totals	2,649,262	1,301,327	17,920,850	17,277,103
Deduct from private lands..................	615,879	244,895	3,380,275	2,679,412
Crown timber............................	2,033,383	1,056,432	14,540,575	14,597,691

Total Recorded Product, Upper Ottawa Agency, 1826 to 1881, 30th June

Ontario, 7,173,182 pieces pine ; 494,824 other woods ; 22,005,108 saw logs.
Quebec 3,955,166 " " ; 209,338 " ; 19,507,159 "

Total, 11,128,348 " " 704,162 " 41,512,267 "

In the foregoing table it is to be observed that square timber and saw-logs only are included, and all other wood goods are omitted, as unimportant for the object of the table. The dues accrued from them are, however included in the following equally brief exhibit of the revenue accrued from Crown timber dues in the upper Ottawa territories of Quebec and Ontario respectively from 1826 to 1881 inclusive, being from the remotest period of which there are any records in the Crown Timber Office, Ottawa.

Period.	Ontario.	Quebec.	Totals.	Years.
1826 to 1834	$ 169,078	$ 46,023	$ 215,101	9
1835 to 1851	$934,735\frac{60}{100}$	$460,643\frac{79}{100}$	$1,395,379\frac{69}{100}$	17
1852 to 1857	$453,058\frac{13}{100}$	$282,879\frac{21}{100}$	$735,937\frac{93}{100}$	6
1858 to 1866	$896,096\frac{27}{100}$	$609,861\frac{15}{100}$	$1,505,957\frac{42}{100}$	9
1867 to 1881	$3,279,538\frac{33}{100}$	$3,439,832\frac{84}{100}$	$6,719,371\frac{19}{100}$	15
1826 to 1881	$\$5,732,506\frac{47}{100}$	$\$4,839,240\frac{53}{100}$	$\$10,571,746\frac{89}{100}$	56

The collection of timber and slide dues being combined, greatly facilitates and tends to secure the accurate collection of both. The same timber and the same men are dealt with at the same time. The returns of the timber rangers of their inspections supported by sworn statements of cullers, afford the means of effectively checking the accuracy of the certificates the lumberers give the deputy slide masters, which through careless error or fraud might be quite erroneous, as to the numbers of saw-logs especially, and as to where they came from. On the other hand, the assistance the Dominion Officers of Inland Revenue and Customs, and their collectors of canal dues, are authorized to render at the request of the Crown timber agent, in preventing the departure of vessels, or the passage of barges loaded with lumber through canals is often of the greatest importance in securing the payment of timber dues. On the River Ottawa no boats or barges loaded with lumber are allowed to pass through the canals without permits from the Crown timber agent. Such permits are occasionally withheld for the enforcement of the payment of timber dues, and the detention of the boats (till released by direction of the Crown timber agent) is always promptly effected by the officers of the Dominion.

The Public Timber Lands of Canada.—Crown Lands.*

These lands belong to the Provincial governments within which they lie with the exception of those in the North-West Territories and the Province of Manitoba, which belong to the Dominion government. In Quebec and Ontario these lands are in charge of a Commissioner of Crown Lands ; in New Brunswick, of the Surveyor-General; in Nova Scotia, of the Attorney-General; and in British Colunbia, of the Chief Commissioner of Lands and Works.

It will be seen that hitherto almost no attention has been given in Canada to the reproduction of timber upon lands from which it has once been cut off ; but all the laws and regulations that have been established have reference only to the native forests of the country and to the securing of a revenue from the existing supply. The reservation of young timber, too small for profitable use, by the limitation of a size below which it should not be cut, has received attention only in the Province of Quebec, and only in respect to pine timber. Questions of conservation and restoration are, however, beginning to attract notice, as will be seen in the following pages, and it is earnestly to be hoped that a knowledge of the true interests of the country will lead to effectual measures for this end, before vested rights have been established to embarrass, and before the need of these measures has become urgent.

Former Timber Regulations in Canada.

The system of granting licenses to cut timber on the public lands in Canada was introduced in 1825, and since that time the right of renewal, upon compliance with regulations has been practically acknowledged. In 1845 an Order-in-Council was passed, making the licenses annual, but with the above understanding, and in the order of 1849 the lessee was permitted to transfer his limit by simple assignment. In 1851 a ground-rent system was introduced.

The branch of woods and forests in the department of Crown lands was organized under the former government of Canada in 1852. A system of local agencies was established, and reforms much needed had begun to be introduced at the time when the Dominion government was inaugurated. Among the abuses of earlier times was the monopolizing of immense tracts without using the privileges or paying an equivalent for them. A ground-rent system was at last adopted, which made reserved but unoccupied privileges unprofitable to hold, more especially as the rate increased in geometrical proportion as the penalty for not using.

As lessons of experience in questions of timber management always have value, it may be interesting to learn how this expedient, apparently so easy to enforce, and so effectual to control the mischief, was found to operate when put to the test of trial.

---

*Hough ; Report upon Forestry, (U.S.) 1878-79.

The Commissioner of Crown Lands of Canada, in his report for the year 1856, in describing the workings of this rule, says :—" It will be readily seen, however, that the operation of such a system would reach a climax within a limited period; that, although it could scarcely be said to be even a check in any degree upon monopoly, in the first instance, the increase in the annual rents on unoccupied tracts after the first few years, became so sudden and great that a crisis became inevitable.

" This crisis arrived in the year before last (1855), the rents of unoccupied berths having in many cases, reached a figure the preceding year which, if again doubled, with a certainty of being quadrupled in 1856, would have rendered the ground untenable.

" A general effort was therefore made by those interested to have the system suspended or rescinded. A new feature in the controversy arose on this occasion from the interference of a great body of the shipping merchants of Quebec, who submitted a counter-petition opposed to the views of those of the producing merchants who desired to be relieved from the accumulating ground-rents.

" The lumber trade being one of the principal resources of the country, the regulations by which it is governed must always be of great moment and worthy of the greatest consideration, and therefore I trust that the importance of the subject to the country at large may be deemed sufficient to warrant a pretty extended reference to the consideration bestowed upon it at the period of the crisis referred to, and which has resulted in establishing a degree of permanency in the institutions connected with it which was previously unknown.

" As the lumber trade is ordinarily conducted in this country, there are two distinct branches of it, viz., that in which the producer is engaged, and that which is carried on by the shipper. There are some firms who are engaged in both branches of the trade, but, although mutually dependent, they are always distinct from and sometimes antagonistic to each other. The principal feature in which they conflict is that it is the interest of the producer that the prices should rule high, as compared with the cost of production, while it is the interest of the shipper that they should rule low in the lumber markets of the country, as compared with the prices in England.

" This subject was very fully treated of in the evidence taken before the parliamentary committee in 1849, appointed to inquire into the causes of the ruinous state of the trade which had existed for some years previous to that date (see appendix P.P.P.P. of that year), which it may not be considered inopportune to refer to, as perhaps the greatest crisis the trade has ever had to contend with since it grew to anything like its present importance.

" By the evidence obtained by the committee on that occasion, it will be seen that, commencing with the year 1846, there was a supply in the Quebec market wholly disproportionate to the demand, originally caused by an unwisely forced production, and aggravated in the succeeding years by a diminished consumption arising from the general depression in commercial affairs which occurred in 1847. The important fact to be observed here, is, that in 1846, a year in which the statistics of the trade proved that all the elements of prosperity existed in the highest degree, the most wide-spread ruin occurred among the producers. The business of 1845 was most profitable to the country and to individuals engaged in the trade, while the business of 1846 was ruinous to individuals and a loss to the country. The demand and the shipments in 1845 exceeded those of previous years ; the demand and the shipments in 1846 were equally great, or even slightly in excess of those of the previous year. The reason of the prosperous state of the trade in the one year and its ruinous state in the other, is, therefore, to be found in the fact that in 1845 the supply was in just proportion to the demand, while in 1846 a supply was forced upon the market out of all proportion even to the great demand and shipments of that year; the result was that, in the one year, individuals realized a profit on their business and the country at large reaped a profit on the total export, while in the other year individuals had, from over-supply, to sell for much less, timber which (from over-stimulated production, enhanced price of labor, etc) had cost more, and were, therefore, in many instances, ruined, a loss being at the same time sustained by the country at large, which, in the total export of the season, parted with so much capital at something like half its value.

"The over-production of 1846 (which did not all reach market that year) continued to depress the trade for several years, the supply of square timber resulting from it in Quebec market having been as follows, viz.:—In 1845 there was a supply of 27,702,344 feet, to meet an export of 24,223,000 feet; but in 1846 a supply of 37,000,643, to meet an export of 24,242,689 feet; and in 1847 a supply (including the overstock of previous years) of 44,027,253 feet, to meet an export of 19,060,880 feet. Here then the distinctive interests of the different branches of the trade may be seen. The business of 1845, which was so profitable to the producers and the country, having been of but doubtful benefit to the shippers, who had to pay quite as high a price here as the prices in England would justify; while the business of 1846, which was so ruinous to the producers, who had to sell at less than the cost of production, was profitable to the shippers, who obtained the timber in Quebec at about half the price it had cost them the previous year, while there was not a corresponding diminution of price in the English markets, at least during that season, and those of them who had contracted realized the full benefit of their contract prices on the diminished rates they had to pay in Canada.

"It is needless to discuss the continued depression of the succeeding years, in which the general derangement of commercial affairs, which began in 1847, was the principal cause; but there can be no doubt that, so far as the lumber trade was concerned, the depression was aggravated by the enormous production of 1846, which continued to hang upon the market for years after. But it is important to observe that the cause of the over-production itself was shown by the parliamentary inquiry referred to, to have been in part indeed the natural stimulus arising from the successful operations of the previous years, but, in part also, the unwise course, at that time pursued by the government, of forcing production, as will hereafter appear upon explanation of the regulations.

"It is to the advantage of the shipping interest that production should again be forced; it is to the advantage of the producing interest that it should be limited. Shippers and producers are alike essential to the trade, and while it would be a mere waste of the labor and capital of the Province for the Government to *force* production, it may be safely assumed that the true course is to let the trade, as far as practicable, regulate itself, without interfering on the one side or the other.

"But it so happens that there must be some regulation to govern the cutting of timber on Crown lands, and it is an unavoidable incident of such regulations that they must exercise some influence upon the trade. The object the regulations should have in view, therefore, in this particular, is to exercise that influence to the least extent possible at the same time that they hold out equal facilities to all desirous of embarking in the trade, due protection to all in the rights acquired and full security for investments of capital necessary to be made, to render the resources of the timber territories available, but not to lock them up in unproductiveness.

"Such being the principles at stake and such the adverse interests involved, both parties memorialized the government, each endeavoring to secure the preponderance of their particular views.

" The memorial in the shipping interest did not, however, correctly represent the grounds upon which those who signed it really opposed the object sought for by the producing interest. I would indeed be sorry to accuse gentlemen of their standing and respectability of any intentional mis-statements, but yet, from being ignorant of that branch of the trade with which they were not connected and of the regulations by which it was governed they allowed themselves to be led into a train of argument which raised entirely false issues, some erroneous information or misconception having led to the result that every paragraph in their memorial conveyed either inferentially or directly some statement that could not be sustained by facts.

" They assumed in the first place that the ground rent was " a condition agreed to by the license holders when they obtained the privilege of cutting, etc.," which was not the fact as regards the great bulk of the trade, the timber berths having been obtained without any such condition, and the ground rents being an additional impost to which they have since been subjected. They next stated that " of late years the bulk of the timber limits of the Crown have been monopolized by a few houses," whereas, there had been no change by which this could have been effected, the only change introduced for several years, having been the very one they were seeking to maintain, establishing ground rents etc. as the most efficient check upon monopoly which had yet been found.

" I may here remark that the assumption that a great monopoly of the timber territory existed was at best a chimera, as proved by the fact that there are upon an average about nine hundred timber births under license in the hands of about five hundred persons. The assertion, therefore, that there is monopoly where there are five hundred competitors, each equally free to deal to a large or a small extent as he sees fit, or his means will allow, needs no further contradiction.

" There may indeed be some local monopolies, where persons of large means buy up the lesser establishments in their vicinity; but anything approaching a general monopoly in this trade, under existing regulations, is impossible; and, so far as any local monopoly exists, it is not by the government that it has been created or is sustained, but by the influence of capital, the application of which for the purposes of trade the government cannot control.

" The greatest local monopoly that has yet arisen in the trade was that which existed a few years on the Saint Maurice, and there it arose from the influence of capital at public competition, although the regulations on that occasion were specially calculated to throw the trade of the territory into the greatest number of hands possible. Capital, however, bore down all opposition for the moment, and it is due to the firmness with which the government resisted repeated, most urgent, and most influential appeals to relax the regulations that that monopoly was ultimately broken up.

" Indeed it may be truly said that the shipping branch of the trade, as carried on at Quebec, bears much more the character of a monopoly than the producing branch, the whole of the business arising from about five hundred competitors on public lands, and perhaps an equally great number of producers on private lands, being, so far as the business centres in Quebec, in the hands of about forty shippers, nine or ten of whom do more than three-fourths of the whole business. But this, in like manner, so far as it can be called a monopoly, is the result of capital, and is not influenced by government, which can as little interfere to limit the operations of the producer to one timber berth or a hundred timber berths as to limit the business of the shipper to one ship or a hundred ships.

" The memorialists also stated that the monopoly of which they complained was 'to the almost total exclusion of those whose means or influence was not so great as to obtain limits.'

" There was here a remarkable instance of men of high position descending to meddle with other people's affairs, and being thereby led to commit themselves to vulgar errors on matters of which they were themselves wholly ignorant.

" It will be seen that in the above they asserted two distinct grievances as the cause of the monopoly they complained of; first, that those without a certain amount of means could not obtain " limits " or timber berths; and, second, that (failing means) they might be obtained by influence. The first must indeed be admitted. Men of means will acquire timber berths, as well as houses and lands and ships, to the ' exclusion of those whose means are not so great as to obtain them '; it is an old grievance for which governments have not yet found a remedy.

" And even if, at the suggestion of these memorialists (who, by the way, were not of the class who usually advocate such a doctrine), the government had taken, or should yet take, some undefined way of throwing the timber berths into the hands of those who have not means to obtain them in a legitimate manner, those who possess means would (provided the tenure justified the investment) immediately buy them out, and then there would be the same cry for a repetition of the operation.

" With respect to the second grievance, it is sufficient to say that it is not to be found in the law or the regulations affecting the trade ; and as it could only exist in violation of both, the memorialists should have established the fact before they claimed credit for it as such, whereas they did not attempt to substantiate even one case of such violation.

" They suggested, in conclusion, that if the license holders were unable or unwilling to pay, etc., their timber berths should be thrown open to competition, and they, the memorialists, believed that, nothwithstanding the depressed state of the trade at that time, they would be readily taken by others without loss to the revenue.

" It is difficult to write seriously on such a proposition ; there can be no doubt that if the opportunity had occured and had been taken advantage of to submit to public competition, privileges which have already been in many cases dearly bought, and in the development of which on the whole, hundreds of thousands of pounds of private means have been expended (as shown by returns laid before Parliament in 1852), they would readily be taken without loss to the revenue, but it was an issue not more reasonable nor likely than that the ships of the memorialists would have been made available to the revenue if they had asked for a change in the navigation laws.

" Such was the false position assumed by the shipping interest at the period referred to, but the erroneous grounds upon which they opposed the prayer of the producing merchants of course made no argument either for or against the latter, which had to be dealt with upon its own merits.

" The memorial of the producing merchants was signed by some of the shipping merchants also, who are connected with or interested in the business of the producers and there appeared to be two or three firms, not known to the department to be connected with the producing interest, who signed, it is presumed, in a liberal view of what they conceived to be for the good of the country and the trade at large : some merchants and others at Ottawa had also joined in it, who are not personally engaged in the trade, but whose interests are bound up with those of the producers.

" The object of the memorialists as expressed, was to obtain a cessation for three years, or until the then existing depression had passed away, of the penalty imposed for non-occupation of timber berths. Although the object sought was professedly of a temporary nature, however, it would no doubt have been made a precedent for seeking government interference in every fluctuation of the trade thereafter. It would have been the first precedent that could be quoted since the adoption of the new system, and therefore I shall state the reasons that induced its rejection, as I conceive that upon the integrity of the system being maintained in the future depends much of the prosperity of the trade.

" It is to be observed that when the great depression occurred in the trade, which began in 1846, and from which it was about four years before it could be said to have recovered, the ground rent system was not in force. The license holders were at that time subject only to the payment of the amount of duty accrued on the quantities cut ; they were then as now obliged to occupy every year, but under pain of forfeiture of the right to renewal of license instead of the penalty of an increased payment.

" It was complained of this system that it favored monopoly, inasmuch as a berth could only be proved unoccupied at a very heavy expense, and then it was still subject to be repurchased by the former holder. The standard of occupation (that is the quantity required to be cut to constitute occupation) was in 1845-46 made too high, thereby having a tendency to force production. In obedience to the cry of monopoly, then prevalent, notice was also given by the department, about the same time—there being then no statute upon the subject—that all the larger timber berths would be sub-divided in three years ; this also, although never actually effected, had a tendency to force production, as license holders were naturally desirous of making the most of their berths by cutting off all the best timber in the interim.

" Parties differed in opinion as to the exact amount of influence these rules exercised upon the over-production, but it was generally admitted that they exercised some influence in that way. At all events the result of the ruinous state of the trade was that the government did afford relief in these particulars, the notice of sub-division was withdrawn, the standard of occupation was reduced, and finally the parties were allowed from year to year up to 1850 to hold their timber berths without any condition of occupation at all, and without any payment where they did not choose to occupy.

" The action of the government on the trade, during the periods of great prosperity and succeeding depression referred to, was thus in opposite extremes. It therefore became expedient that a better permanent system of regulations should be framed for the government of the trade, and the regulations of which the ground-rent system is a part were finally the result.

" By this system an annual ground-rent was imposed on timber berths, in excess of the duty, as a regular permanent charge, and as a check upon monopoly it was provided, by way of penalty, that the ground-rent should double upon each renewal of license on berths which had not been occupied during the preceding season, and continue doubling every year, so long as the berths continued unoccupied. Thus the rent paid for the largest size of berth the regulations permit—in excess of all other charges—is £6 5s., the same being payable annually.

" But upon non-occupation for one season the rent rises to £12 10s.—upon non-occupation for a second season to £25, for a third season to £50—and so on (as the system was first introduced) without limit, but reverting to the original rate of £6 5s. whenever occupation recommenced.

" For the first few years after the introduction of this system it could not force production to any very sensible extent; but the constant increase, in geometrical progression, at last comes to a point when the increase is so great and sudden that those who held any timber berths in reserve had either to occupy or relinquish them. Unfortunately as regards the great bulk of the license holders, the operation of the system had just reached the point (when they had either to produce more timber or relinquish that which they had already paid a series of rents for, and in some instances, otherwise laid out money upon, without return) at a moment when the trade was in a state of considerable depression, and required a decreased instead of an increased production. This state of depression, too, arose from causes wholly foreign to the internal management of the trade ; for it differed from the previous great crisis in the trade (that of 1846-47, etc.) in this, that it arose less from an excessive production than from a sudden cessation of demand—the result probably of the war then raging. It diff red also in degree, bearing only the character of a temporary embarrassment as compared with the wide spread-ruin which fell upon the trade on the former occasion. It was none the less necessary, however, to apply a remedy, if practicable, in time, and it was in this view that the producers sought to be temporarily authorized to suspend productions where the ordinary tendency of the regulations was to enforce it.

" It was not, therefore, as put by the opponents of the producing interest, a question of the holders of timber berths fulfilling or failing in their obligations ; and even if it had been so, the maintenance of the penalty in its full force would not, at least for some time, have compelled any considerable relinquishment of licenses ; on the contrary, the parties would have continued to hold them, and endeavored by extended operations to reduce to their original amounts the ground rents on such berths as the penalty had most accumulated upon, thus risking the consequences of increased production rather than abandon their licenses.

" The real question at issue, therefore, simply was, whether the penalty for non-occupation had been made too severe or not.

" But there was also the question of whether the exceptional circumstances then existing, arising out of the war or otherwise, were such as would justify the temporary suspension of the penalty.

" On the first head, as regards the penalty for non-occupation generally, it is to be observed that, if any regulation were to succeed in compelling the occupation of all the lands licensed, it would force a production far beyond the requirements of the trade ; no regulation could permanently have this effect, however, as the result of an excessive penalty would be to cause the relinquishment of a portion of the territory now under license, which (apart from the question of whether it would not afford, in every period of excitement, too great a facility for a rush into the trade) would leave a portion of the timber lands wholly unproductive, either in ground rents or duties, which now afford a very considerable revenue.

" The system of regulations for the granting of licenses to cut timber began by a course of trial and error and has gradually been perfected by experience.

" The ground-rent system was a trial ; it has proved a most happy and successful one, which has given general public satisfaction to the trade, but it would be too much to pretend that, in the first trial there had been no error, that it had been perfected at once without any experience of it practically.

" In the introduction of the system the then remote contingency was not provided for, that if no limit was set to the ultimate amount the ground-rent might reach, great hardships might in some cases be the result; such, for instance as might arise in case of several timber berths being taken up in a previously inaccessible locality, assuming in such a case that the license holders (joining together for that purpose) proceed to improve the stream (as is frequently done to the extent of many thousand pounds), lay out all the means they can command in the operation, and before the rents have reached an excessive amount are enabled to occupy the lower berths ; but some pressure then comes, they cannot push their improvements immediately to the upper berths, and the ground rents arrive at a point where they compel relinquishment, while they could not compete for the repurchase on equal terms with any new purchaser who would have the advantage of their outlay.

"It has been suggesetd that a remedy for this might be found by admitting improvements in lieu of occupation, which would be just in principle but practically extremely difficult of application.

"The cases urged upon the department from every part of the country would be numerous, the evidence to be adjudicated upon would be entirely *ex parte*, the exact nature of the improvements to be admitted would always be a matter of dispute, and, however honestly administered, the system would give rise to constant accusations of partiality and favor.

"Upon a full consideration, therefore, of all the circumstances it appeared that the difficulty might be met by a general rule calculated to perfect and give permanency to the system as a whole instead of impairing it.

"A rule was accordingly adopted which consists in limiting the extreme amount of ground rent on any berth to a sum equal to what the berth would produce in duty if duly occupied, the rent remaining at that rate per annum till occupation commences ; reverting then, of course, to the original rate as before. This, while it entails a heavy payment on those who reserve berths for future use, as much in fact as they would have to pay for the timber if they cut it, affords no public ground for complaint, for the public get the price of the timber annually while the timber itself remains, with the public interest in it, for future revenue, unimpaired ; at the same time it prevents the system from becoming oppressive, and therefore, inoperative, as all oppressive laws ultimately become.

"On the other head, with regard to the temporary suspension of the system the same issue as was then involved is now at stake and must continue to be so, It must be remarked, as a general rule, that any departure, for partial, local, or temporary causes, from the fixed laws affecting the trade, is bad in principle and calculated in every case to produce a bad effect.

"If, when a depression has arisen from over-production, or other causes, which the trade has brought upon itself, the government should once step in to affect the market or the supply, directly or indirectly, the same interference would be looked forward to again, and induce an over-speculative spirit in time of prosperity, sure to end in a similar result. If the government were at any time to relax the conditions it has seen fit to impose upon the holders of unoccupied timber berths without some other cause than the ordinary fluctuations of the trade, public confidence would be shaken either in the efficacy of the system itself or in the administration of it. Nothing but the strongest necessity, arising from causes foreign to the trade itself, could at any time justify an exception to this as a general rule, and the only question on this point worthy of consideration at that time was, whether the effects of the then state of war were such as to justify its being made an exceptional case.

"In considering this question it became necessary to take a retrospective view of the trade for some years, from which it appeared that there had not been any very excessive supply in the Quebec market as compared with the export. The supply was indeed somewhat excessive in 1852 and the stock of square timber on hand at the close of that year(18,151,750 feet) was also excessive, but the producers—profiting from the sad experience of 1846 and the embarrassments of succeeding years—having cautiously limited their operations, the supply was much less in 1853, and the stock on hand (12,632,929 feet) at the close of the season greatly reduced. But from the great demand these were years of great prosperity to the producing interest, and consequently an impetus was given to the supply produced in 1854, which was very great; but the export was also greatly increased and the stock in hand at the close of the season (13,465,602 feet) though large, yet with the more limited production for 1855 was not at all such as seriously to embarrass the trade had the usual demand existed. From whatever cause, however, the demand had greatly diminished, for at the time the subject was most strongly pressed upon the government, say 2nd July, 1855, the tonnage arrived in Quebec, from sea, was 121,778 tons against 240,021 tons to the same period of the previous year; and at the close of the year 346,449 tons against 580,323 tons the previous year; and in like manner, the quantity of square timber exported in 1855 was 15,389,774 feet against 25,346,800 feet in 1854. There is a defect in the present law which prevents the statistics being got so correctly in respect of deals. There is also a large quantity of timber usually absorbed in ship-building and exported in that shape, in which there had also been a falling off. The result of a full investigation of the subject, however, was to show that the trade was on the whole in a healthy condition, and that the depression at that period was only temporary, for although there had been no excessive production for some years previous, as compared with the export, the export itself had been great, having been gradually increasing till it produced a temporary glut, not in the Quebec market but in the English markets, which had precisely the same effect, and which was in some degree aggravated no doubt by a diminished consumption resulting from the war and the tightness of money matters consequent thereon.

"The prayer of the memorialists, therefore, to be authorized to suspend their operations for three years without incurring the penalty of increased rent, as provided by the regulations for non-occupation, was refused, for if even such an extreme case could arise, there did not then appear to be any cause operating to produce such permanent embarrassment as would have warranted the government in interfering with the integrity of a system which had, so far, been found to give stability to the trade and satisfaction to the public.

"The result has justified the course pursued; the export in 1856 having been nearly up to the average, or 3,919,378 feet, (equal to forty-six million inch board measure) in excess of the previous year. The season was in fact, upon the whole a very fair one, both for the producer and the shipper, and this without any such extreme measure on the one side or the other as the government had been asked the year before, to adopt for the safety of the trade.

"The only change adopted was one which had not an immediate effect; it consisted, as already stated, in making the ground rent on unoccupied berths cease to increase when it had reached the extreme amount which ground rent and the dues accruing on timber cut would both amount to upon a berth which was occupied. The public could scarcely ask more, as a protection against monopolizing timber berths, than that the parties who do so should be made to pay for the timber when they don't cut it the same as when they do cut and carry it to market.

"In former years more stringent laws were made against holding timber berths unoccupied, but the result was, as has already been seen, that when the crisis came, the government always gave way, thus proving that extreme measures are always the least effective, while they lead in matters of trade to uncertainty and fluctuation.

"I have entered thus at length into the circumstances attending the appeal of the opposing interests to the government in 1855, because there was then undoubt-

edly serious apprehension entertained by many that a time of great embarrassment and difficulty was at hand ; while a crisis had actually arrived in regard to
testing the efficacy of the by-laws by which the trade is governed, so far as it is
as a whole affected by the operations on public lands ; and because, therefore, the
action then taken has so far solved a difficult problem and is likely to exercise
a permanent influence on the trade."

As modified by experience, the management of the timber interests upon
the public lands in the later years of the former Canadian Government was in
charge of the Commissioner of Crown Lands, who was authorized to grant
licenses for cutting timber upon ungranted lands at such rates, and subject to
such regulations as might be established from time to time by the Governor-in-
Council and of which notice was given in the *Canada Gazette*. These licenses
were granted for a period not exceeding twelve months, and obliged the lessees
to make returns at the expiration of the lease, showing the number and kinds of
trees cut, and the quantity and description of saw-logs, or of the number and
description of sticks of square timber manufactured and carried away under
such license, which statement must be verified by affidavit before a justice of
the peace. The Crown dues were a claim upon the timber or any part thereof,
wherever found, and whether in the original logs or made into deals, boards or
other stuff, and which might be seized and detained wherever found until the
dues were paid.

Persons cutting, or causing to be cut, any timber on any of the Crown,
clergy, school, or other public lands, or removing, or inducing, or assisting in the
removal of timber thus cut without authority, acquired no right or claim for
cutting or preparing for the market, but the whole became forfeited, and if the
timber or saw-logs had been removed out of the reach of the officers of the
Crown lands department, or if it was found otherwise impossible to seize the
same, the person was liable in addition to the loss of his labor and disbursements,
to a forfeiture of $3 for every tree (rafting stuff excepted) that might be proved
to have been cut, to be recovered with costs of suit, in the name of the
Commissioner of Crown Lands or resident agent in any court having jurisdiction
in civil matters to the amount of the penalty. In all such cases it was incumbent
on the party charged to prove his authority to cut, and the averment of the
party seizing or prosecuting that he was duly employed under the timber act
was to be received as sufficient proof, thereof, unless the defendent proved to the
contrary.

Seizures might be made upon information supported by affidavit. If the
timber illegally cut had been mixed with other timber, the whole might be
detained until satisfactorily separated by the holder. Resistance to an officer or
authorized agent, by assault, force, or violence, or by threats of such, was made
a felony, and the carrying away of timber under seizure, whether openly or
secretly, and whether with or without force or violence, was deemed stealing and
rendered the person liable to punishment for felony. Whenever any timber was
seized for the non-payment of Crown dues, the burden of proof of payment, or
as to the land on which it was cut, was to rest on the claimant of such timber
and not on the officer making the seizure or the party bringing the prosecution.

Timber seized was to be deemed to be condemned at the end of thirty days
and publication of notice, unless the person claiming, sooner notified the nearest
officer or agent of the Crown land office that he intended to prove his claim.

Any judge of competent jurisdiction might order the release of timber under
seizure upon receiving from the alleged owner a bond with two good and
sufficient sureties, first approved by the agent, for double the value of the
timber in case of condemnation, such bond being taken in the name of the

Commissioner of Crown Lands, and to be delivered and kept by him until the claim was released or paid. Every person availing himself of any false statement or oath to evade the payment of Crown dues forfeited the timber on which dues were attempted to be evaded.

The malicious cutting or loosening of a boom, or the cutting loose or breaking up of a raft or crib, was made punishable by fine and imprisonment of not less than six months.

Such in brief was the system formerly in force. That it did not insure the forests upon the Crown lands from pillage and waste by lumbermen is sufficiently proved by the following statement made by the Commissioner of Crown Lands of the Province of Ontario, in 1877 in describing the system of supervision then in use and the abuses that had been formerly practiced.

"Previous to confederation, the guardianship of the forests as regards surveillance over the cutting of timber under license or in trespass on lands of the Crown was so ineffective or attended to with such laxity as to be in fact no guardianship at all, and pillage to a large extent was carried on almost with impunity; the seat of government was peripatetic,* and the agents of the Crown Lands Department for the collection of timber dues were located at certain points where returns were brought to them of such operations as parties chose to make, on which dues were paid, and the amount received with statement of timber, etc., on which it was paid, transmitted monthly to the department without any actual knowledge of or check on the extent of cutting; these returns and moneys were received at headquarters without comment or inquiry, and the one debited to the agent and the other placed to his credit."

AMOUNTS ACCRUED AND COLLECTED FOR TIMBER DUES, GROUND RENTS, AND BONUSES IN UPPER AND LOWER CANADA, DURING THE YEARS PRECEDING THE FORMATION OF THE DOMINION GOVERNMENT.

Years.	Accruals.			Collections.		
	Upper Canada.	Lower Canada.	Total.	Upper Canada.	Lower Canada.	Total.
	$ c.	$ c.	$ c.	$ c.	$ c.	$ c.
1857	135,310 64	120,797 96	256,108 60	94,921 15	114,023 53	208,944 68
1858	111,739 62	111,081 53	222,821 15	141,185 90	134,476 00	275,661 90
1859	140,409 96	142,071 97	282,481 93	136,189 33	145,745 59	281,934 92
1860	176,400 39	168,973 36	345,433 75	149,921 22	168,330 38	318,252 60
1861	156,253 57	154,101 38	310,354 95	127,995 88	127,849 10	255,844 98
1862	143,357 59	136,830 79	280,188 38	159,330 86	144,321 31	303,652 17
1863	170,160 12	157,484 72	327,644 84	197,093 73	189,562 80	386,656 53
1864	188,171 74	155,793 97	343,965 71	121,367 79	121,718 52	243,086 71
1865	146,079 67	151,034 24	297,113 91	183,380 75	160,035 23	343,415 98
1866	203,040 46	166,036 54	369,077 00	197,965 85	138,678 05	336,643 89
Total 10 years.	1,570,983 76	1,464,206 46	3,035,190 22	1,509,352 46	1,444,740 50	2,954,092 96

* For some years before the union of 1867 the seat of government of Canada alternated between Toronto and Quebec. It had previously been located at Montreal, and at a still earlier period at Kingston.

RECEIPTS FROM BONUSES AND GROUND RENTS ALONE, DURING THE UNION
OF THE PROVINCES OF CANADA, SO FAR AS THESE HAVE BEEN PUBLISHED.

Fiscal years.	Amount.
1856-'57	$244,112 90
1957-'58	203,263 59
1858-'59	276,741 16
1859-'60	316,983 35
1860-'61	290,933 04
1861-'62	283,383 31
1862-'63	309,252 15
1863-'64	325,294 51
1864-'65	324,535 61
1865-'66	300,486 18
1866-'67	369,800 53

RECENT AND EXISTING TIMBER REGULATIONS IN CANADA.

(a) *Dominion Lands.*—These lands, in Manitoba and the North-west Territories, are in charge of the Department of the State and a division thereof styled "The Dominion Land Office." The act under which they are administered was assented to April 14th, 1872. The surveys are conducted by the Surveyor-general and his deputies, and there are various agents concerned in the duties incident to this interest.

The system of surveys is by townships six miles square, sub-divided into sections of one mile square each, unless this arrangement is modified by the divergence of meridians, irregularities in previous surveys, or other causes. There is an allowance of one chain and fifty links between all townships and sections for roads. The townships are numbered northward from the international boundary, or the forty-ninth degree north latitude, and in Manitoba, east and west from a principal meridian, ran in 1869, that strikes this line of latitude about ten miles west of Pembina. The sections are numbered from one to thirty-six in each township, beginning at the south-east corner and running alternately from east to west and from west to east, so that the last number shall be in the north-east corner. In this the order of numbering is just the reverse of that employed on the surveys of public lands in the United States. Sections eleven and twenty-nine in each township are reserved for education.

The sections are divided into sixteen squares of forty acres each, numbered in the same way as the sections in townships, beginning at the south-east corner. The lines running north and south are designed to be true meridians, and those running east and west are chords intersecting circles of latitude passing through the angles of the townships.

The terms and conditions of the deed of surrender from the Hudson Bay Company stipulated a reservation of one-twentieth part of the portion described as the "fertile belt," which rendered it necessary to modify the general plan, and in the prairie region, where there are islands or belts of timber, a special mode of sub-division was provided, with the view of affording benefit to the greatest possible number of settlers, and for the prevention of petty monopolies. In these cases the woodlands are surveyed into lots of not less than ten nor more than twenty acres each, so as to afford one wood-lot to every quarter-section of prairie farm in each township. This, however, is not allowed to interfere with the sections set apart for schools, nor to those set apart and vested in the Hudson Bay Company. Each wood-lot is required to front on a section road-allowance.

In case an island or belt of timber come entirely within a quarter-section, or in several quarter-sections so that not more than twenty-five acres shall be included in each, it is not to be separately surveyed into wood-lots. These wood-lots are conveyed as homestead grants the same as other lands, but the grantee is not allowed to sell any of the timber on his lot to any saw-mill owners, or to any other than settlers for their own private use, under penalty of prosecution, as for trespass. Upon conviction they may be fined or imprisoned, or both, and they further forfeit their claims absolutely.

Any tract of land covered by forest timber may be set apart as timber lands, and reserved from sale and settlement; and except as it may be thought expedient by the secretary of State to divide a township into two or more timber limits, the several townships composing any such tracts shall each form a limit. The word " timber " is used to designate all lumber, and all products of timber, including firewood and bark.

Leases for cutting timber may be granted for twenty-one years, and upon the following conditions:—

1. The lessee to erect a saw-mill or mills in connection with such limit and lease, and subject to any special conditions which may be agreed upon and stated in the lease, such mill or mills to be of capacity to cut at the rate of a thousand feet, board measure, in twenty-four hours, for every two-a-half square miles of limits in the lease, or shall establish such other manufactory of wood goods as may be agred upon as the equivalent of such mill or mills, and the lessee to work the limit in the manner and to the extent provided in the lease within two years from the date thereof, and during each succeeding year of the term.

2. To take from every tree he cuts down all the timber fit for use, and manufacture the same into sawn lumber, or some other such saleable product as may be provided in the lease, or by any regulations made under this act.

3. To prevent all unnecessary destruction of growing timber on the part of his men, and to exercise strict and constant supervision to prevent the origin or spread of fires.

4. To make returns to the government monthly or at such other periods as may be required by the secretary of State, or by regulations under this act, sworn to by him or by his agent or employee cognizant of the facts, declaring the quantities sold or disposed of as aforesaid, of all sawn lumber, timber, railway-car stuff, ship-timbers and knees, shingles, lath, cordwood or bark, or any other product of timber from the limit, in whatever form the same may be sold or otherwise disposed of by him during such month or other period, and the price or value thereof.

5. To pay in addition to the bonus an annual ground rent of $2 per square mile, and further a royalty of five per cent. on his monthly account.

6. To keep correct books, of such kind and in such form as may be provided by his lease, or by the regulation under this Act, and to submit the same for the inspection of the collector of dues whenever required, for the purpose of verifying his returns aforesaid.

7. The lease shall describe the lands upon which the timber may be cut, and shall vest in the lessee during its continuance the right to take and keep exclusive possession of the lands so described, subject to the conditions hereinbefore provided or referred to, and such lease shall vest in the holder thereof all right of property whatsoever in all trees, timber, lumber, and other products of timber cut within the limits of the lease during the continuance thereof, whether such trees, timber, and lumber or products be cut by authority of the holder of such lease, or by any other person, with or without his consent; and such lease shall entitle the lessee to seize in replevin, revendication, or otherwise, as his.

property, such timber, where the same is found in the possession of any unauthorized person, and also to bring any action or suit at law or in equity against any party unlawfully in possession of any such timber, or of any land so leased, and to prosecute all trespasses thereon, and such other offenders as aforesaid, to conviction and punishment, and to recover damages, if any; and all proceedings pending at the expiration of any such lease may be continued and completed as if the lease had not expired.

8. Such lease shall be subject to forfeiture for infraction of any one of the conditions to which it is subject, or for any fraudulent return; and in such case, the secretary of State shall have the right, without any suit, or other proceeding at law or in equity, or compensation to the lessee, to cancel the same, and to make a new lease, or disposition of the limit described therein to any other party at any time during the term of the lease so canceled: provided, that the secretary of State, if he sees fit, may refrain from forfeiting such lease for non-payment of dues, and may enforce payment of such dues in a manner hereinafter provided.

9. The lessee who faithfully carries out the above conditions shall have the refusal of the same limits, if not required for settlement, for a further term not exceeding twenty-one years, on payment of the same amount of bonus per square mile as was paid originally, and on such lessee agreeing to such conditions and to pay such other rates as may be determined on for such second term

It was further provided that any ground rent, royalty, or other dues to the Crown not paid when falling due, should bear interest at six per cent. until paid, and be a lien on any timber cut within the limits. After three months' neglect, the Crown timber agent might seize so much of the timber cut as would be necessary to pay the claim and expenses, and sell the same at public auction, paying over to the lessee or owner of the timber any balance left after paying claims and costs.

In case the payment of the Crown dues was evaded by removal of the timber or products out of Canada, or otherwise, the amount due might be charged upon any other timber cut on Dominion lands by the same lessee or by his authority, or the claim might be recovered by action at law, in the name of the secretary of State, or his resident agent, in any court having jurisdiction in civil cases to the amount claimed.

The secretary of State was empowered to take bonds or promissory notes for any money due to the Crown, interest and costs, or for double the amount of all dues, fines and penalties, and costs, incurred or to be incurred, and he might then release any timber upon which the same would be leviable, whether under seizure or not; but the taking of such bonds was not to affect the lien and right of the Crown to enforce payment of such money on any other timber cut on the same limit, if the sums for which such bonds or notes were given should not be paid when due.

The penalties imposed for cutting timber without authority were forfeiture of the timber cut, and a fine not exceeding $3 for every tree cut or carried away, with costs. In such cases the burden of proof of authority to cut and take the timber was to be upon the party charged, and the averment of the party seizing or prosecuting that he is duly employed under this Act was to be sufficient proof thereof, unless the defendent proved to the contrary.

Upon information, supported by affidavit, that timber had been cut without authority on Dominion lands, and describing where the same can be found, or upon information to a Crown officer or agent as to such cutting without authority, the officer or agent was authorized to seize the timber and place it under custody until a decision could be had by competent authority.

If timber, cut without authority, has been made up with other timber into a crib, dram, or raft, or in any other manner mixed up with other timber, so that it cannot be identified, the whole of the timber so mixed is to be liable to seizure and forfeiture until satisfactorily separated by the holder. Timber held under seizure may be released upon sufficient security for the payment of its full value or of double the amount of all dues, fines, penalties, and costs, incurred or imposed thereon.

The penalties for resisting seizure, or removing timber after it was seized, were prescribed and proceedings therein specified. No sale or grant of Dominion lands was to give any title to any slide, dam, pier, or boom, previously erected upon it, unless expressly mentioned in letters patent, or other instrument establishing such sale or grant. The free use of such works was not to be interrupted, and the right of passing and repassing on either side, whenever necessary for use, and at portages, was reserved.

The Dominion Lands Act makes provision for military bounties, homestead entries, leases for grazing, and hay-cutting, mining, etc., and for direct sales of land.

(*b*) *Crown lands of Ontario.*—The timber Act now in force was passed in 1860, and is found as Chapter 28 of the Revised Statutes of Ontario, 1887. It is as follows :—

(1) An Act Respecting Timber in Public Lands.

Her Majesty, by and with the advice and consent of the Legislative Assembly of the Province of Ontario, enacts as follows :—

1.—(1) The commissioner of Crown lands, or any officer or agent under him authorized to that effect, may grant licenses to cut timber on the ungranted lands of the Crown, at such rates, and subject to such conditions, regulations and restrictions as may from time to time be established by the Lieutenant-Governor-in-Council, and of which notice may be given in the *Ontario Gazette*.

(2) No license shall be so granted for a longer period than twelve months from the date thereof ; and if, in consequence of incorrectness of survey, or other error, or cause whatsoever, a license is found to comprise lands included in a license of a prior date, the license last granted shall be void in so far as it interferes with the one previously issued, and the holder or proprietor of the license so rendered void shall have no claim upon the Government for indemnity or compensation by reason of such avoidance.

2. The licenses shall describe the lands upon which the timber may be cut, and shall confer, for the time being, on the nominee the right to take and keep exclusive possession of the lands so described, subject to such regulations and restrictions as may be established :—And the licenses shall vest in the holders thereof all rights of property whatsoever in all trees, timber, and lumber, cut within the limits of the license during the term thereof, whether the trees, timber, and lumber, are cut by authority of the holder of the license, or by any other person, with or without his consent ; and the licenses shall entitle the holders thereof to seize in revendication or otherwise, such trees, timber, or lumber, where the same are found in the possession of any unauthorized person, and also to institute any action against any wrongful possessor or trespassers, and to prosecute all trespassers and other offenders to punishment, and to recover damages, if any ; and all proceedings pending at the expiration of any license may be continued to final determination, as if the license had not expired.

3. Every government road allowance included in any Crown timber license, heretofore granted, or which may hereafter be granted, under section 1 of this Act, shall be deemed and taken to be and to have been ungrant d lands of the Crown, within the meaning of said section, and liable, as such, to be included in the license.

4. The licensee or nominee named in any license shall be deemed and taken to have, and to have had, all the rights in respect of every such road allowance, and the trees, timber, and lumber thereon, or cut thereon, as were, or, by the section 2 of this Act, may be conferred upon him, in respect of any other Crown lands embraced in such license, and the trees, timber, and lumber thereon, or cut thereon, except that he shall not be entitled to take or keep exclusive possession of such road allowance.

5. No by-law passed, or to be passed by any municipal council for preserving, selling, or otherwise appropriating or disposing of the timber or trees, or any part thereof, on a government road allowance or allowances included in any such license, shall be deemed or taken to have had or have any force or effect against any such license.

6. In case the council of any townships, organized as a separate municipality, or the council of any united townships, have passed, or hereafter pass, any by-law for preserving or selling the timber or trees on the government road-allowances within such township or within the senior township of united townships, and included in any such license, the corporation of such township or united townships shall be entitled to be paid out of the consolidated revenue fund of this Province a sum equal to two per centum of the dues received by Her Majesty for or in respect of the timber and saw-logs which, during the existence of the by-law, were cut within the township or united townships, under the authority of the license ; but no corporation shall be entitled to such percentage of the dues received for timber or saw-logs cut during the times or seasons when timber, or trees on any such road allowances were cut or removed, for which cutting or removal the corporation had, before the fifteenth day of February, one thousand eight hundred and seventy-one, obtained a verdict against any such licensee or nominee.

7. No municipal corporation shall be entitled to such payment as aforesaid, unless a certified copy of the by-law passed, or to be passed as aforesaid, accompanied by an affidavit of the clerk or reeve of the corporation, verifying the copy, and the date of the passing of the by-law, is filed in the Department of Crown Lands at Toronto within six months from the passing of the by-law ; and the affidavit may be made or taken before any person or officer who, under sections 42 or 43 of "The Public Lands Act," is authorized to take the affidavits in those sections mentioned.

8. All moneys to be paid as aforesaid, to any municipal corporation shall be expended in the improvement of the highways situate within the township or within the senior or junior township in respect of which such moneys were paid.

9. The percentage to which the junior township or townships of such united townships may be entitled, shall only be in respect of the dues received upon timber or trees which shall be cut after the 30th day of April, 1881. (See Rev. Stat. Ont., c. 25, ss. 13-15, as to the right of the Crown to grant timber licenses on Free Grant Lands.)

10. Every person obtaining a license shall, at the expiration thereof, make to the officer or agent granting the same, or to the Commissioner of Crown lands, a return of the number and kinds of trees cut, and of the quantity and description of saw-logs, or of the number and description of sticks of square timber manufactured and carried away under the license ; and the statement shall be sworn to by the holder of the license, or his agent, or by his foreman, before a Justice of the Peace ; and any person refusing or neglecting to furnish such statement, or evading or attempting to evade any regulation made by order-in-council, shall be held to have cut without authority, and the timber made shall be dealt with accordingly.

11.—(1) All timber cut under licenses shall be liable for the payment of the Crown dues thereon, so long as and wheresoever the timber or any part of it may be found in Ontario, whether in the original logs or manufactured into deals, boards or other stuff ; and all officers or agents intrusted with the collection of such dues may follow all timber and seize and detain the same wherever it is found until the dues are paid or secured.

(2) Nothing in this Act contained shall be construed to repeal the provisions of the section 4 of chapter 23 of the Consolidated Statutes of Canada, as regards timber removed into the Province of Quebec.

12. Bonds or promissory notes taken for the Crown dues either before or after the cutting of the timber, as collateral security, or to facilitate collection, shall not in any way affect the lien of the Crown on the timber, but the lien shall subsist until the dues are actually discharged.

13. If timber so seized and detained for non-payment of Crown dues remains more than two months in the custody of the agent or person appointed to guard the same, without the dues and expenses being paid, the Commissioner of Crown Lands with the previous special sanction of the Lieutenant-Governor in council, may order a sale of the said timber to be made after sufficient notice ; and the balance of the proceeds of the sale, after retaining the amount of dues and costs incurred, shall be handed over to the owner or claimant of the timber.

14.—(1) If any person without authority cuts or employs or induces any other person to cut, or assists in cutting any timber of any kind on the Crown, clergy, school or other public lands, or removes or carries away, or employs or induces or assists any other person to remove or carry away, merchantable timber of any kind so cut from the public lands aforesaid, he shall not acquire any right to the timber so cut, or any claim to any remuneration for cutting, preparing the same for market, or conveying the same to or towards market.

(2) When the timber or saw-logs made has or have been removed by any person out of the reach of the officers of the Crown lands department, or it is otherwise found impossible to seize the same, such person shall, in addition to the loss of his labor and disbursements, forfeit a sum of $3 for each tree (rafting stuff excepted) which he is proved to have cut or caused to be cut or carried away.

(3) Such sums shall be recoverable with costs, at the suit and in the name of the Commissioner of Crown lands or resident agent, in any court having jurisdiction in civil matters to the amount of the penalty.

(4) In such cases it shall be incumbent on the party charged to prove his authority to cut ; and the averment of the party seizing or prosecuting that he is duly employed under the authority of this Act, shall be sufficient proof thereof, unless the defendant proves the contrary.

15. Where satisfactory information, supported by affidavit made before a justice of the peace or before any other competent party, is received by the Commissioner of Crown Lands, or other officer or agent of the Crown Lands Department, that any timber or quantity of timber has been cut without authority on Crown, clergy, school or other public lands, and describing where the timber can be found, the Commissioner, officer or agent, or any one of them, may seize or caused to be seized in Her Majesty's name the timber so reported to be cut without authority, wherever it is found, and place the same under proper custody until a decision can be had in the matter from competent authority.

16. Where the timber so reported to have been cut without authority on the public lands has been up with other timber into a crib, dram, or raft, or in any other manner has been so mixed up at the mills or elsewhere as to render it impossible or very difficult to distinguish the timber so cut on public lands without license from other timber with which it is mixed up, the whole of the timber so mixed shall be held to have been cut without authority on public lands, and shall be liable to seizure and forfeiture accordingly, until satisfactorily separated by the holder.

17. Any officer or person seizing timber, in the discharge of his duty under this Act may, in the name of the Crown, call in any assistance necessary for securing and protecting the timber so seized.

18. Whenever any timber is seized for non-payment of Crown dues, or for any other cause of forfeiture, or any prosecution is brought for any penalty or forfeiture under this Act, and a question arises whether the said dues have been paid on such timber, or whether the timber was cut on other than the public lands aforesaid, the burden of proving payment, or on what land the timber was cut, shall lie on the owner or claimant of the timber, and not on the officer who seizes the same or the party bringing the prosecution.

19. All timber seized under this Act shall be deemed to be condemned, unless the person from whom it was seized, or the owner thereof, within one month from the day of the seizure, gives notice to the seizing officer or nearest officer or agent of the Crown lands office, that he claims or intends to claim the same; failing notice, the officer or agent seizing shall report the circumstances to the Commission of Crown Lands, who may order the sale of the said timber by the officer or agent, after a notice on the spot of at least thirty days.

20.—(1) Every judge having competent jurisdiction may, when he deems it proper, try and determine such seizures, and may order the delivery of the timber to the alleged owner, on receiving security by bond, with two good and sufficient sureties to be first approved by the agent, to pay double the value in case of condemnation.

(2) The bond shall be taken in the name of the Commissioner of Crown Lands to Her Majesty's use, and shall be delivered up to and kept by the Commissioner.

(3) If the seized timber is condemned, the value thereof shall be forthwith paid to the Commissioner of Crown Lands or the agent, and the bond cancelled, otherwise the penalty of such bond shall be enforced and recovered.

21. Every person availing himself of any false statement or oath to evade the payment of Crown dues, shall forfeit the timber on which dues are attempted to be evaded.

(2)- MANAGEMENT OF THE TIMBER LANDS OF ONTARIO.

Previous to June 13th, 1866, applications for license to cut timber on the Crown lands were made to the several Crown timber agents, who might grant such privileges upon payment at the rate of 2s. 6d. ($0.50) per square mile annually, payable in advance. These leases expired on the 30th of April in each year, and might be renewed before the 1st of July following. The changes since introduced are described by the Commissioner of Crown Lands in a statement prepared for the information of the then Premier of the Province of Quebec in 1877, a manuscript copy of which has been furnished us, as follows : On the 13th of June, 1866, prior regulations were superseded, and the clause respecting licenses to cut timber was modified, so that instead of agents granting them on application it was provided that such vacant berths as the Commissioner of Crown Lands saw fit should be offered at public auction, to be held half-yearly in each timber agency on the 10th of July and 10th of January, or such other dates as the Commissioner might think proper to fix by public notice, at an upset price of $4 per square mile, or such rate as he might fix by such notice, the berths to be awarded to the highest bidder, etc., in addition to the yearly ground rent of fifty cents per mile and tariff dues on timber when cut, the Commissioner or agent in the intervals between sales to grant licenses on application on payment of the bonus and ground rent mentioned.

The Regulations of 1851 and those of 1866 imposed a fine for non-occupation of timber berths as follows : If a berth in surveyed territory had not been occupied, i.e., worked upon during the season for which license was granted or renewed, or in unsurveyed territory the year after granting or renewal of license, the ground rent of fifty cents was doubled, and so on in case of non-occupation until the ground rent reached 23s. 4d. ($4.67), or maximum charge per square mile, at which rate it stood till the berths had been worked upon, on which the rent fell again to fifty cents per mile ; the making of an average of 500 feet of square timber, or twenty saw-logs to the mile, being admitted as due occupation. The object of compulsory occupation or the payment of an increased ground rent was to prevent large areas of country from falling into the hands of capitalists, to the exclusion therefrom of men of smaller means ; but the penalty of additional charge for rent was easily evaded, seeing that the holders of limits had only to cut, or pretend to have cut, 357 pieces of square timber or 1,000 logs, to have a fifty mile limit maintained at fifty cents per mile rent, or reduced thereto had the rent been advanced.

After Confederation, compulsory occupation in Ontario was dispensed with, and the ground rent increased from fifty cents to $2 per square mile, and by the third clause of existing regulations it is made imperative that all new timber berths should be sold by public auction to the bidder of the highest amount of bonus per square mile ; that berths should be offered for sale at such time and place as the Commissioner thought fit, instead of at any particular date or place ; and that in the interim between sales no new licenses be granted, as under the regulations of 1866.

The duty of the Commissioner of Crown Lands with respect to disposing of timber berths, would seem clear and simple, inasmuch as he is by the auction system relieved from the necessity of acting on individual applications for licenses ; but the fact is, that the management of public forests in Ontario is surrounded by many difficulties, not the least of which is the settlement of the country, which is extensively and rapidly taking place, in territory held under timber license, where lumbering operations are being carried on simultaneously with the location of the lands.

The management of timber on lands under license in unsurveyed territory, or in surveyed lands where settlement has not yet penetrated, is comparatively easy ; all that is required being a close inspection of operations by wood rangers. But in old settled townships, where licenses granted many years past still obtain, and where settlers who had, prior to 1st July, 1867, purchased lots out of limits, being actual residents on their lots with certain improvements, are allowed to cut and sell the timber on their lands under the "settler's license regulations," the dues on the timber so sold being applied towards payment of the purchase money due the Crown, less ten per cent. for collection : and in newly surveyed townships in free grant territories covered by license, where locations have been or are being made under the Free Grants Act, as well as lands sold under the Land Act of 1860 within or adjoining timber limits, subject to the Pine Tree Regulations under Order-in-Council of 27th May, 1869, there is great care required in guarding against imposition and fraud upon the revenue, by passing timber cut on lands of the Crown in trespass as cut under authority of settler's license or general timber license, or in process of clearing the lands for cultivation under the 10th section of the Free Grants Act, and the Order-in-Council of 27th May, 1869, with respect to lands sold under the Land Act of 1860. To watch the interest of the revenue and at the same time avoid apparent harshness in dealing with settlers on the public lands demands the greatest circumspection by the department, and zeal and vigilance on the part of its employees on the ground ; yet, notwithstanding the exercise of every care and precaution, the conflicting interests arising between lumber operators and settlers are frequent and perplexing.

The Free Grant Townships in the Muskoka, Parry Sound, and Nipissing Districts are being rapidly settled upon, the lands being in many cases selected and large improvements made before they were open for location or sale under the Act ; in view of this fact, and that it would be impolitic to assume the attitude of retarding the settlement of the country, the question of dealing with the pine timber on the lands before they were formally located, so that the timber might be utilized in the public interest, instead of allowing it to be destroyed by fires, incidental to the clearing of the land, was somewhat embarrassing, seeing that the sawn lumber and square timber trade was in such a state of depression as had never before been experienced, and that in consequence the result of selling the townships situated as described, as timber berths, it was anticipated would be anything but satisfactory in a revenue point of view ; however, as settlement could not be kept back, it became imperative that the right to cut the timber on the lands should be disposed of, so that as much as possible might accrue to the public chest. Accordingly, eight or nine townships, in the condition referred to were inspected as to the pine timber thereon, and reports examined with regard to the quantities in different parts of the townships, and berths of various areas from four to twenty-six square miles each were prepared so as to have the several groups of pine distributed over the respective berths and thereby as far as possible insure sales ; through the careful management in the laying out of the berths, the sale, which took place was very successful, the amount realized giving an average of $200 per square mile.

In April, 1869, new regulations were introduced of which the following is a copy. They took the place of those established by Order-in-Council dated June 12, 1866, and published in the *Canada Gazette* of June 23, 1866, and enforced from that date :—

(3) CROWN TIMBER REGULATIONS.

(Established under Chapter 23 of the Consolidated Statutes of Canada by order of His Excellency the Lieutenant-Governor-in-Council, dated the 16th April, 1869).

1st. The Commissioner of Crown Lands may, at his discretion, cause the limit lines of any timber berths under license, which have not been already surveyed, to be properly surveyed and run, the costs of such survey to be paid by the holder of the license ; and where two or more licenses are interested in the survey, the Commissioner shall determine what portion of the costs of the survey shall be paid by each, and such costs of survey shall be a charge upon the timber berth, to be paid with the ground-rent before renewal of the license.

2nd. The Commissioner of Crown Lands, before granting any licenses for new timber berths in the unsurveyed territory, shall, as far as practicable, cause the section of country where it is intended to allot such berths to be run out into townships, and each township, when so surveyed, shall constitute a timber berth, but the Commissioner of Crown Lands may cause such townships to be subdivided into as many timber berths as he may think proper.

3rd. The berths or limits, when so surveyed and set off, and all new berths or limits in surveyed territory, shall be explored and valued, and then offered for sale by public auction at the upset price fixed by such valuation, at such time and place, and on such conditions, and by such officer, as the Commissioner of Crown Lands shall direct by public notice for that purpose, and shall be sold to the highest bidder for cash at the time of sale.

4th. All forfeited timber berths may be offered for sale on the second Tuesday in August in each year by public auction, at such upset price, and at such place as the Commissioner of Crown Lands may fix and appoint by public notice, or at such other rate as he may fix by such notice, and shall be awarded to the highest bidder making payment at the time of sale, but should the said timber berth not be then sold, the same may be granted to any applicant willing to pay the said upset price and ground-rent or on such other terms as the Commissioner of Crown Lands may direct.

5th. License holders who shall have complied with all existing regulations shall be entitled to have their licenses renewed on application to the Commissioner of Crown Lands, or to such local agent as he may appoint for that purpose.

6th. The Commissioner of Crown Lands shall keep a register of all licenses granted or renewed, and of all transfers of such licenses; and a copy of such register with a plan of the licensed limits, shall be kept by the Crown timber agent of the locality, and open to public inspection.

7th. All transfers of timber berths shall be made in writing, but shall be subject to the approval of the Commissioner of Crown Lands, to whom they shall be transmitted for approval or rejection, and they shall be valid only from the time of such approval, to be expressed in writing.

8th. Timber berths are to be described in new licenses as " not to interfere with prior licenses existing or to be renewed in virtue of regulations." When the description of any berth or boundary, as given by any license, clashes with the description of any other licensed berth or territory, the license of more recent origin (tracing back only to the time when such license or any previous license, of which it is a renewal, was first granted) shall give way, and the Commissioner may amend or cancel such license wholly or in part, and substitute another in place thereof, so as to correct the description of the berth or limit intended to be licensed ; and in all cases where any license has issued in error or mistake, or is found to be inconsistent with any other license, or inconsistent or incompatible with the regulations under which it was granted, the Commissioner of Crown Lands may cause it to be cancelled or amended, or he may refer all matters in dispute with reference to the boundaries and position of timber limits to arbitration, each of the contending parties to choose one arbitrator and the Commissioner of Crown Lands shall appoint an umpire, naming a day on or before which the award of such arbitrators or of such umpire shall be made and delivered to the parties and such award shall be binding on them.

9th. Timber cut on limits for which license has been suspended or held in abeyance shall be considered as having been cut without authority, and treated accordingly.

10th. Occupants, locatees, or purchasers of pubic lands, who have not completed all the conditions of sale or location, shall not, unless under settler's license, or for clearing, fencing, or building purposes on the said land, be permitte l to cut timber or logs thereon, or to dispose of it to others. Persons found doing so shall be subject to the penalties established by law for cutting timber on the public lands without authority.

11th. All timber licenses are to expire on the 30th of April next after the date thereof, and all renewals are to be applied for and issued before the 1st of July following the expiration of the last preceding license, in default whereof the right to renewal shall cease, and the berth or berths shall be treated as forfeited.

12th. No renewal of any license shall be granted unless or until the ground-rent and all costs of survey, and all dues to the Crown on timber, saw-logs, or other lumber, cut under and by virtue of any license, other than the last preceding, shall have been first paid.

13th. All timber berths or limits shall be subject to an annual ground-rent of $3 per square mile, payable in advance, before the issuing of any original license or renewal.

14th. All timber, saw-logs, wood, or other lumber, cut under any license now in force, or under any license which may be hereafter granted, shall be subject to the payment of the following Crown dues, that is to say :—

Black walnut and oak, per cubic foot	$0 03
Elm, ash, tamarac and maple, per cubic foot	0 02
Birch, basswood, cedar, buttonwood and cottonwood, and all boom timber, per cubic foot	0 01¼
Red and white pine timber (per O.C. 27th April, 1887), per cubic foot	0 02
All other woods	0 01
Basswood, buttonwood and cottonwood, saw-logs, per standard of 200 feet board measure	0 15
Red and white pine saw-logs and boom timber, per standard of 200 feet B. M., (per O.C. 27th April, 1887)	0 20
Walnut, oak and maple saw-logs, per standard of 200 feet board measure	0 25
Hemlock, spruce, and other woods, per standard of 200 feet board measure	0 10
All unmeasured culled saw-logs, to be taken at the average of the lot, and to be charged for at the same rate.	
Staves, pipe, per mille	7 00
Staves, West India, per mille	2 25
Cordwood (hard) per cord	0 20
Cordwood (soft) per cord	0 12½
Hemlock tan-bark, per cord	0 30

Railway timber, knees, etc., to be charged 15 per cent. *ad valorem.*

15th. The duties on timber shall be charged upon the quantity shown by the specification of measurement at the office of the supervisor of cullers, at Quebec, or that of the deputy-supervisor of cullers, at Sorel or Montreal, or by other reliable measurement, but where such actual measurement cannot be obtained, each stick of white pine timber shall be estimate l as containing seventy cubic feet, red pine as containing thirty-eight cubic feet, oak, fifty feet, and elm, forty-five feet, and all other wood as containing thirty-four cubic feet.

16th. All licensees, or occupants of timber berths, shall furnish, through themselves, their agents, cullers, and foremen, to such agent or agents as the Commissioner of Crown lands may appoint for that purpose, and at such time and place as such agent or agents may require, satisfactory proof upon oath as to the exact locality where all the timber, saw-logs, and other lumber in his or their possession were cut, giving the number of pieces and description of timber, saw-logs, and other lumber, cut by themselves and others to their knowledge upon each of the timber berths held or occupied by him or them, respectively, designating what quantity, if any, had been cut on settlers' lands, giving the names of such settlers, the name of the township, and the number of each lot and concession, exhibiting at the same time, for the inspection of such agent or agents, the books of count and measurement of such timber, saw-logs, and other lumber, under his or their control, respectively ; and shall moreover furnish such agent or agents all required information and facilities to enable him or them to arrive at a satisfactory determination as to the quantity and description of timber, saw-logs, and other lumber, made by him or them, or held in his or their possession, respectively, on which government dues are chargeable ; and in the event of such agent or agents deeming it expedient to cause such timber, saw-logs, and other lumber to be counted or measured, the said licensee, or occupier of such timber berth, and his or their agent, cullers and foremen, shall aid and assist in such count and measurement, but should such licensee or occupier, or his or their agents, fail to comply with these conditions, such licensee shall forfeit all right to a renewal of his license, and the berth and limit shall become vacant. And to enable persons who sell their timber under settler's license to obtain their refund of dues, and timber cut on patented lands, to pass duty free, it will be necessary for the parties interested to prove on oath, taken before such agent or agents, and to his or their satisfaction, the number of pieces and description of timber and saw-logs cut on each lot respectively. And in the event of such proof being deemed unsatisfactory, the said agent or agents may determine the same by causing a strict count of the stumps to be made, and then certifying according to such count.

17th. The Commissioner of Crown lands, or any authorized agent, shall at all times have free access to and be permitted to examine the books and memoranda kept by any licensee, showing the quantity of lumber in board measure, sawn by him from logs cut on his timber berth or berths, and failing to produce such books and memoranda when required so to do, will subject such licensee to a forfeiture of his right to a renewal of his license. .

18th. When any license holder is in default for, or has evaded the payment of dues to the Crown on any part of his timber or saw-logs, such dues may be levied on any other timber or saw-logs belonging to such defaulter, cut under license, together with the dues thereon.

19th. Before moving any raft or parcel of timber, lumber, or saw-logs, from the agency in which it has been cut, the owner or person in charge thereof shall report the same to the Crown timber agent, making, if required, declaration upon oath, as to where the said timber was cut, the number of pieces and description of each kind of wood contained in such raft or parcel of timber, and the number of cribs, stating at the same time the number and description of pieces cut on private lands, also on lands under settler's license, giving the names of the owners or licensees of such land, with the names of the townships and number of each lot and concession ; and should such Crown timber agent not be satisfied with the correctness of such report, he shall cause a strict count to be made of the timber in such raft ; and on being satisfied of the correctness of such report of count, the Crown timber agent may grant a clearance, in due form, for such raft, stating the number of pieces and description of timber contained therein, distinguishing the timber cut on private lands and under settler's license from that cut on the Crown domain.

20th. The owner or holder of any such raft or parcel of timber shall, within twenty-four hours after the same shall have arrived at its destination at Quebec, Sorel, Montreal, or other port of sale or shipment, report the arrival of such raft to the collector of Crown timber dues, or if at Sorel or Montreal, to the deputy-supervisor of cullers ; and should the said raft be found by the specification of measurement to contain a greater number of pieces of timber than is noted in the clearance, the surplus number of pieces, if not satisfactorily explained, shall be held as having been cut on Crown lands without authority, and subject to the payment of dues accordingly.

21st. Parties omitting to obtain their clearance at such agency, or omitting to report the arrival of such raft at its destination, as above mentioned, may be refused further license, and may be subject to forfeiture of the timber for evasion of regulations, as provided in Cap. 23, of the Consolidated Statutes of Canada.

22nd. Persons evading or refusing the payment of timber dues, or the final settlement of bonds or promissory notes for the payment of such dues, or in default with the Crown timber office or agent ; also persons taking forcible possession of disputed ground, before obtaining decision in their favor, and persons refusing to comply with the decision of arbitrators or of the umpire, as provided by the 8th section of these regulations, or with the regulations established by Order-in Council, or who forcibly interrupt surveyors in the discharge of their duty, shall be refused further licenses, and their berths shall be forfeited at the expiration of the then existing license.

23rd. Dues of all kinds on timber cut under license, remaining unpaid on the 30th November following the season in which it was cut, shall be subject to interest from that date, but without prejudice to the power of the Crown to enforce payment of such outstanding dues at any time the Commissioner of Crown Lands may think proper.

(4) On the Various Forms of Timber Licenses in Use.

There are four forms of timber license in use in the Province of Ontario; two for what is called the " Western Timber District," and the " Belleville District," one containing the right to cut timber on road allowances and the other not, and neither of them granting the right to cut rafting stuff on lands of the Crown. Two forms of licenses are used for the " Ottawa Agency," one having a stipulation concerning road allowances, and the other not, but both conferring the right to cut rafting stuff from the Crown lands.

The reason why the right to cut rafting stuff is confined to the Ottawa agency, is because, on the Ottawa timber and logs come from a long distance up the river, and from different tributary streams, and have to be rafted, broken up, and re-rafted in some cases several times before the timber and logs reach their destination : whereas on the rivers in other parts of the Province, no rafting takes place, the timber and logs being driven down the streams loosely till they reach the large waters of the lakes or the River St. Lawrence, on the shores of which rafting stuff can be cut or purchased.

The following copy of the simpler form of license used in the Western Timber District, will, with its notes, give an idea of these different licenses :

(5) Form of a Timber License in the Western Timber District.

By authority of Chapter 26 of the Revised Statutes of Ontario, and the Crown Timber regulations, dated the 16th day of April, 1869, and for and in consideration of the payments made and to be made to Her Majesty :

I do hereby give unto   and unto
                              agents or workmen, full power and license to cut every description of timber on lands or lots unlocated and unsold at the date of this License, or sold or located during the time this License is in force, and pine trees on lands or lots sold under Orders in Council of 27th May, 1869, or sold or located under the Free Grants and Homesteads Act of 1868 or amendment of the said Act, by Chapter four of the Statutes of Ontario of 1880, and pine and cedar trees, when reserved, on lots sold under Order in Council of 3rd April, 1880, prior to the date of this License, and pine trees on lots patented under said Chapter four, or patented as mining lands, under the General Mining Act, or patented or leased under Statute 54 Victoria, Chapter eight, upon the location described on the back hereof by   and to hold and occupy the said location to the exclusion of all others, except as hereinafter mentioned from   to thirtieth of April, 18   , and no longer ; with the right of conveying away the said timber   through any ungranted, uncleared, or waste lands of the Crown :—

And by virtue of this license, the said licensee   ha   right by the said statute to all timber cut by others during the term of this license in trespass on the ground hereby assigned, with full power to seize and recover the same.

But this license is subject to the following conditions, viz.:—

To the withdrawal therefrom of lots located or sold under the Free Grants and Homesteads Act of 1868, prior to the passing of Chapter four of the Statutes of Ontario of 1880, and for which patent may be granted on the ground that five years had elapsed from the date of such location or sale, and that the conditions of settlement had been complied with prior to thirtieth April preceding the date or issue of the license.

That any person or persons may at all times make and use roads upon, and travel over the ground hereby licensed.

That nothing herein shall prevent any person or persons from taking from the ground covered by this license, standing timber of any kind (without compensation therefor) to be used for the making of roads or bridges or public works, by or on behalf of the Province of Ontario, the authority of the Department of Crown Lands having first been obtained.

That persons settling under lawful authority or title within the location hereby licensed, shall not in any way be interrupted in clearing and cultivation by the said licensee   , or any one acting for   or by permission.

That the Commissioner of Crown Lands, under Order in Council of 27th April, 1885, may at any time during the currency of this license, cancel the right to cut timber other than pine upon any lots, included in the description in this license, which may have been sold or located subsequent to the date hereof, or upon any lots in said description which may have been squatted upon with the *bona fide* intention of location or purchase.

And further : under condition that the said license or   representatives shall comply with all regulations that are or may be established by Order in Council, and shall submit all the timber, saw logs or other lumber cut under this License to be counted or measured, and settle for the duties chargeable thereon, when required by me or any officer thereunto authorized,—otherwise the said timber will be forfeited to the Crown, and the said licensee be subject to such other penalties as the Act provides.

Given under my hand, at Toronto, the   day of   in the year of our Lord, one thousand eight hundred and ninety-   in duplicate.

Commissioner.

The stipulation in regard to road allowances found in two of the forms, is as follows :—

And every government road allowance or parts thereof, embraced within the boundaries of the tracts or parcels of land above mentioned or described, and all such portions of any government road allowance as border upon any tract, lot, or parcel of land above mentioned or described, and lie between the side-lines or between the front and rear-lines, or between a side-line and a front or rear-line, or between different parts of any line of said tracts, lots, or parcels of land produced across such road allowance ; provided, however, that when any portion of a road allowance is found to be included in any two licenses covering lands on opposite sides of such road allowance, then each license is to extend only to the centre line of such road allowance; and provided also, that all disputes arising out of any conflict of licenses covering government road allowances shall be decided by the Commissioner of Crown Lands, who may define what portion of any road allowance is included in each license, and his decision shall be binding.

This license not to interfere with prior licenses.

(G) SYSTEM OF WOOD RANGING.—EFFORTS TO PREVENT WASTE.

The Commissioner of Crown Lands, in the communication already cited describes the operation of these regulations and the system of wood-ranging was then introduced. This is admitted as at first crude and experimental, but it has since gone on with modifications as suggested by experience, until it is deemed at present as perfect as can practically be carried out.

A staff of from twenty to thirty experienced and reliable rangers are employed each season, some of them being engaged from December till the 30th of April, and a few of the supervising rangers up to the end of October. The result has been satisfactory in the highest degree, the revenue having increased in the several agencies immediately after the inception of the system to the extent of from fifty per cent, and in one agency even 400 per cent.

Instead of agents dealing with accounts for timber dues as formerly, all returns, together with ranger's reports, are transmitted to the department, where the timber limit operations and cutting on special lots of land are checked, and all accounts made up, and transmitted to agents for the collection of the dues and transmission of the same to headquarters as collected.

Wood-rangers have standing instructions to report generally on any wanton or special waste, when such has been observed in connection with lumbering operations, and in cases of licensees allowing standing pine through which fire has passed to become lost instead of utilizing it before it is destroyed by what is termed the " boring worm." A few cases of waste transpired some years ago by licensees arranging with jobbers to cut saw-logs on their timber limits, the logs, by agreement, to be up to a certain standard of quality—all logs falling short of the standard fixed being rejected and left in the woods—and an attempt made to leave the rejected timber out of the returns; but through the vigilance of the wood-rangers of the department, such transactions were nipped in the bud, and abandoned when parties found that they had to account for and make payment to the Crown on every tree cut down. The only real waste of timber in lumbering is in connection with the manufacture of square pine and board, (or octagonal) pine timber, especially the former, in squaring which and in the rejection of the upper portion of the tree where the limbs begin; fully one-third of the tree is wasted, viz., one-sixth of the best of the timber in siding off to reach the square, and one-sixth of the upper part of the tree which is left in the woods, but which if drawn, would be valuable at a saw-mill, where it could be cut into various qualities of lumber, either fit for domestic use or export. The waste referred to has been noticed by this department for years past, but under the regulations past and present and the tenure under which licenses to cut timber are held, and have been held for many years, it is found difficult to uproot a system which has obtained so long, and in which there are so many vested interests and so much capital involved.

PRODUCTION OF TIMBER IN CANADA, 1890.

The following table taken from the Statistical Year Book of Canada (1891), p. 14, gives the production of timber in the whole of Canada during the year 1890 :

Timber.	Ontario.	Quebec.	New Brunswick.	British Columbia.	Nova Scotia.	Manitoba and N.W.T.
Saw logs B. M.	522,524,283	495,449,000	108,569,122	79,177,055	**78,603,742	30,605,906
Square timber, c. ft.	3,392,629	2,151,791	16,818	....	....	....
Boom " pieces	150,361	5,240	7,375	....	....	....
Hardwood, c. ft.........	12,527	67,428	†	....	....	....
Railway ties, No..........	672,410	139,550	79,488	....	....	....
Cordwood, cords.........	29,971	8,747	1,356			
Telegraph poles	468	635	3,163			
Cedar, lin. ft.	162,346	....	4,716,201			
Cedar posts, tanbark and bolts, cords	4,147	10,769	258		....	....
Pile timber, B. M...	11,664	....	....		....	....
Shingles, M..............	....	3,331	615	....		....
Battens, knees, etc., No...	....	1,230	14,787	....	....	1,449,916
Posts and rails, No.	* 1,225	....	6,820	....	....	§ 156,402
Staves, poles, etc., M.	....	....	‡ 63	....	....	....
Dues received, $..........	878,772	806,052	112,475	29,678	....	102,951

* Traverses.   * *Trans-Atlantic shipments only.   † Included in square timber.   ‡ Rafting pins.
Pulp and bobbin wood included.   § Laths.

THE ONTARIO FIRE ACT.

The following Act (Cap. 213, R.S.O.) was passed by the Legislature of Ontario in 1878, with the view of preventing the occurrence of the fires which have wrought so much devastation among the forests of the Province :—

An Act to preserve the Forests from destruction by Fire.

WHEREAS large quantities of valuable timber are an- nually destroyed by fires which are in many instances the result of negligence and carelessness, it is therefore necessary to provide stringent regulations for the prevention of such fires. *Preamble.*

Therefore Her Majesty, by and with the advice and consent of the Legislative Assembly of the Province of Ontario, enacts as follows :—

1. The Lieutenant-Governor may, by proclamation to be made by him from time to time, issued by and with the advice and consent of the Executive Council, declare any portion or part of the Province of Ontario to be a fire district. *Lt.-Governor may proclaim a fire district.*

2. Every proclamation under this Act shall be published in the *Ontario Gazette*, and such portion or part of the Province as is mentioned and declared to be a fire district in and by the said proclamation, shall, from and after the said publication, become a fire district within the meaning and for the purposes of this Act. *Publication of fire district.*

3. Every such portion or part of the Province mentioned in such proclamation shall cease to be a fire district upon the revocation by the Lieutenant-Governor in Council of the proclamation by which it was created. *Revocation.*

4. It shall not be lawful for any person to set out, or cause to be set out or started, any fire in or near the woods within any fire district, between the first day of April and the first day of November in any year, except for the purpose of clearing land, cooking, obtaining warmth, or for some industrial purpose; and in cases of starting fires for any of the above purposes, the obligations and precautions imposed by the following sections shall be observed. *Fires not to be started except for certain purposes and in certain periods.*

5. Every person who shall, between the first day of April and the first day of November, make or start a fire within such fire district for the purpose of clearing land shall exercise and observe every reasonable care and precaution in the making and starting of such fire, and in the managing of and caring for the same after it has been made and started, in order to prevent such fire from spreading or burning up the timber and forests surrounding the place where it has been so made and started. *Precautions to be taken in case of clearing land.*

6. Every person who shall, between the first day of April and the first day of November make or start within such fire district a fire in the forest, or at a distance of less than half-a-mile therefrom, or upon any island for cooking, obtaining warmth, or for any industrial purpose, shall— *Precautions in case of cooking, etc.*

1. Select a locality in the neighbourhood in which there is the smallest quantity of vegetable matter, dead wood, branches, brushwood, dry leaves, or resinous trees ;

2. Clear the place in which he is about to light the fire by removing all vegetable matter, dead trees, branches, brushwood and dry leaves from the soil within a radius of ten feet from the fire ;

3. Exercise and observe every reasonable care and precaution to prevent such fire from spreading, and carefully extinguish the same before quitting the place.

Precautions in case of matches, burning substances, etc. **7.** Any person who shall throw or drop any burning match, ashes of a pipe, lighted cigar, or any other burning substance, or who shall discharge any fire-arm within such fire district shall be subject to the pains and penalties imposed by this Act, if he neglect completely to extinguish before leaving the spot the fire of such match, ashes of a pipe, cigar, wadding of the fire-arm, or other burning substance.

Act to be read to employees by heads of surveys, lumberers, etc. **8.** Every person in charge of any drive of timber, survey or exploring party, or of any other party requiring camp fires for cooking or other purposes within such fire district, shall provide himself with a copy of this Act, and shall call his men together and cause said Act to be read in their hearing, and explained to them at least once in each week during the continuance of such work or service.

Precautions as to locomotives. **9.** All locomotive engines used on any railway which passes through any such fire district or any part of it, shall, by the company using the same, be provided with and have in use all the most approved and efficient means used to prevent the escape of fire from the furnace or ash-pan of such engines, and that the smoke stack of each locomotive engine so used shall be provided with a bonnet or screen of iron or steel wire netting, the size of the wire used in making the netting to be not less than number nineteen of the Birmingham wire gauge, or three sixty-fourths parts of an inch in diameter, and shall contain in each inch square at least eleven wires each way at right angles to each other, that is in all twenty-two wires to the inch square.

Duty of engine drivers. **10.** It shall be the duty of every engine driver in charge of a locomotive engine passing over any such railway within the limits of any such fire district, to see that all such appliances as are above-mentioned are properly used and applied, so as to prevent the unnecessary escape of fire from any such engine as far as it is reasonably possible to do so.

Penalty for non compliance with this Act. **11** Whosoever unlawfully neglects or refuses to comply with the requirements of this Act in any manner whatsoever, shall be liable upon a conviction before any justice of the

peace to a penalty not exceeding fifty dollars over and above
the costs of prosecution, and in default of payment of such fine
and costs, the offender shall be imprisoned in the common gaol
for a period not exceeding three calendar months ; and any
railway company permitting any locomotive engine to be run
in violation of the provisions of the ninth section of this Act
shall be liable to a penalty of one hundred dollars for each
offence, to be recovered with costs in any court of competent
jurisdiction.

12. Every suit for any contravention of this Act shall be
commenced within three calendar months immediately following such contravention. *Time for bringing action.*

13. All fines and penalties imposed and collected under this
Act shall be paid one-half to the complainant or prosecutor
and the other half to Her Majesty for the public use of the
Province. *Disposal of fines.*

14. It shall be the special duty of every Crown Land agent,
Woods and Forests agent, Free Grant agent, and bush ranger,
to enforce the provisions and requirements of this Act, and in
all cases coming within the knowledge of any such agent or
bush ranger to prosecute every person guilty of a breach of
any of the provisions and requirements of the same. *Government agents to enforce this Act.*

15. Nothing in this Act contained shall be held to limit or
interfere with the right of any party to bring and maintain
a civil action for damages occasioned by fire, and such right
shall remain and exist as though this Act had not been
passed. *Act not to interfere with right of action for damages occasioned by fire.*

DESCRIPTION OF "FIRE DISTRICTS" UNDER CAP. 213 OF THE STATUTES OF ONTARIO.

District No. 1.—Commencing at a point on the north shore of Lake Huron
where Provincial Land Surveyor Albert P. Salter's meridian line between ranges
numbers twenty-one and twenty-two west intersects the water's edge, said
point being the south-west angle of the Township of Plummer ; thence easterly,
following the turnings and windings of the shore along the water's edge of Lake
Huron and the Georgian Bay to the mouth of French River ; thence south-easterly,
along the easterly shore of the Georgian Bay, and taking in Parry Island, to the
north-west angle of the Township of Matchedash ; thence south-easterly along
the westerly boundaries of the Townships of Matchedash and North Orillia to
the south-west angle of North Orillia ; thence north-easterly along the southerly
boundary of North Orillia to the waters of Lake Couchiching ; thence easterly
across said lake to the south-west angle of the Township of Rama ; thence easterly along the south boundaries of the Townships of Rama, Dalton, Digby and
Lutterworth to the north-west angle of the Township of Galway ; thence southerly along the westerly boundaries of the Townships of Galway and Harvey to
the south-west angle of Harvey ; thence easterly along the south boundaries of
the Townships of Harvey, Burleigh, Methuen, Lake and Tudor, to the north west
angle of the Township of Elzevir ; thence southerly along the west boundary of

14 (F.)

Elzevir to the south-west angle of said township ; thence easterly along the south boundaries of the Townships of Elzevir, Kaladar, Kennebec, Olden, Oso and South Sherbrooke, to the south-east angle of the Township of South Sherbrooke ; thence north-westerly along the easterly boundaries of the Townships of South and North Sherbrooke to the southerly boundary of the Township of Lavant; thence north-easterly along the southerly boundaries of the Townships of Lavant and Darling, to the south-easterly angle of the Township of Darling; thence north-westerly along the easterly boundaries of the Townships of Darling and Bagot, to the north-easterly angle of the Township of Bagot; thence south-westerly along the northerly boundaries of the Townships of Bagot and Blithefield, to the easterly boundary of the Township of Brougham ; thence north-westerly along the easterly boundaries of the Townships of Brougham, Grattan, Wilberforce and Alice, to the waters of the Upper Allumette Lake ; thence north-westerly, following the water's edge of said lake and the Ottawa River to the head of Lake Temiscamingue ; thence due north along the boundary between the Provinces of Ontario and Quebec, to the northern boundary of the Province of Ontario ; thence westerly along the said northern boundary to its intersection with the production northerly of Provincial Land Surveyor Albert P. Salter's meridian line between the said ranges numbers twenty-one and twenty-two west, and thence southerly along said meridian line produced to the place of beginning.

District No. 2.—All that part of the said Province lying west of Provincial Land Surveyor Albert P. Salter's meridian line between ranges twenty-one and twenty-two west, near Bruce Mines, in the District of Algoma, and west of the said meridian line produced to the northern boundary of the Province, the said meridian line being the western boundary of the Fire District established by the Proclamation of March 27th, 1878.

THE ONTARIO FIRE-RANGING SYSTEM.

In 1888 the Department of Crown Lands (Ontario), inaugurated a system of fire ranging, explained in the circular-letter to limit-holders given below, the cost of which is borne in equal parts by the Province and the lumbermen. It is very generally adopted by limit-owners and is believed to have been instrumental in greatly reducing the annual loss through forest fires :—

Sir,—The Commissioner of Crown Lands, feeling the importance of creating some better organization for preventing the destruction of the forest by fire, has approved of a scheme, the principal points of which are herein stated to you, so that you may, should the position of your limits make it desirable, avail yourself of its advantages.

It is proposed that during the dangerous period, say from the first day of May to the first day of October in each year, there shall be placed on such limits as are exposed to danger a man or men who will be empowered and instructed to use every endeavor to prevent and suppress fires in every way possible, and the ranger who is placed in charge of a limit will be authorized to engage whatever help may be necessary to cope with a dangerous fire where prompt action is necessary ; these men will be supplied copies of the " Fire Act," and instructed to post them up in public and conspicuous places, to visit each person resident on the limit and give them, if thought advisable, a copy of the Act, explaining to them its provisions, penalty for its infraction, etc., and to endeavor to enlist their assistance and sympathy to make the Act effective.

The department will leave the limit holder to suggest the number of men who should be placed on his limit, and as it is of all things necessary that practical bushmen of good judgment and well acquainted with the limit should be selected, he, the limit holder, will nominate the man to be placed in charge of the limit and his subordinates, if any, the department reserving the right to limit the number of men to be employed on any limit and also to reject or remove any man whom it finds unfitted to discharge the duties of the position.

It is hoped that limit holders will recognize the necessity of recommending men of good judgement and cool temper who, while fully discharging their duties, will not harass or annoy settlers or others, as, if an animus is created in the breasts of the settlers the scheme will undoubtedly fail to effect the result expected. Limit holders will be expected to exercise supervision over these men and see that they thoroughly and effectually perform their duties.

With respect to remuneration the department thinks that the man in charge of a limit should be paid   dollars a day, which should cover board and ordinary expenses, and where subordinates are required, that suitable men can be obtained at   dollars per day, which should also cover board and ordinary expenses; the men will be appointed-bush and fire rangers and instructed from here so as to clothe them with authority under section 14 of the Fire Act, and a copy of the instructions will be furnished each limit holder.

As the limit holder is reaping a large proportion of the benefit, it is intended that he should bear one-half of the cost of men and expenses which may be incurred under this scheme.

The department will pay wages and expenses and charge to each limit holder his proportion, which will be a charge upon the limit and an account will be rendered at the close of the season, when prompt payment must be made.

Should you desire to avail yourself of this scheme you will at once address a letter to the department to that effect, stating the limits you wish protected, the number of your license for current season, the number of men you would recommend to be employed, and submit a list of those you would recommend for appointment on your limits.

AUBREY WHITE,
Assistant Commissioner.

DEPARTMENT OF CROWN LANDS,
(WOODS AND FOREST BRANCH,)
April, 1888.

AN ACT RESPECTING THE ROCKY MOUNTAINS PARK OF CANADA.

Chap. 32 of 50-51 Vict. setting apart a national park at Banff in the Rocky Mountains reads thus :—

Whereas it is expedient in the public interest that a national park and sanatorium should be set apart and established in the North-west Territories ; Therefore, Her Majesty, by and with the advice and consent of the Senate and House of Commons of Canada, enacts as follows :—

1. The tract of land comprised within the limits hereinafter set forth, that is to say, commencing at the easterly end of Castle Mountain station grounds, on the Canadian Pacific Railway, as shown on a plan of right-of-way filed in the Department of Railways and Canals by the Canadian Pacific Railway Company, thence on a course about south thirty-five degrees east, ten miles, more or less, to a point in latitude seven minutes, six seconds and ninety-six hundredths of a second south of the point of commencement, and in longitude seven minutes, fifty-four seconds and ninety-eight hundredths of a second east of the point of commencement ; thence on a course about north fifty-five degrees east, twenty-six miles, more or less, to a point in latitude five minutes, forty-six seconds and twenty hundredths of a second north of the point of commencement, and in longitude thirty-seven minutes, twenty-three seconds and thirty-one hundredths of a second east of the point of commencement ; thence on a course about north thirty-five degrees west, ten miles, more or less, to a point in latitude twelve minutes, fifty-three seconds and ninety-one hundredths of a second north of the point of commencement, and in longitude twenty-nine minutes, thirty-two seconds and thirty-eight hundredths of a second east of the point of commencement ; thence on a course about south fifty-five degrees west, twenty-six miles, more or less, to the place of commencement, containing by admeasurement two hundred and sixty square miles, be the same more or less, so far as the title to the said tract of land in whole or in part, is now vested in the Crown, is hereby withdrawn from sale, settlement and occupancy under the provisions of "The Dominion Lands Act" or any regulations made under the said Act or any other Act with respect to mining or timber licenses or any other matter whatsoever.

2. The said tract of land is hereby reserved and set apart as a public park and pleasure ground for the benefit, advantage and enjoyment of the people of Canada, subject to the provisions of this Act and of the regulations hereinafter mentioned, and shall be known as the Rocky Mountains Park of Canada.

3. No person shall, except as hereinafter provided, locate, settle upon, use or occupy any portion of the said public park.

4. The park shall be under the control and management of the Minister of the Interior, and the Governor-in-council may make regulations for the following purposes :—

(a) The care, preservation and management of the park and of the watercourses, lake, trees and shrubbery, minerals, natural curiosities and other matters therein contained.

(b) The control of the hot springs situate in the said park, and their management and utilization for purposes of bathing and sanitation and in every other respect.

(c) The lease for any term of years of such parcels of land in the park as he deems advisable in the public interest, for the construction of buildings for ordinary habitation and purposes of trade and industry, and for the accommodation of persons resorting to the park.

(d) The working of mines and the development of mining interests within the limits of the park, and the issuing of licenses or permits of occupation for the said purposes; but no lease, license or permit shall be made, granted or issued under this or the next preceding paragraph of this section which will in any way impair the usefulness of the park for the purposes of public enjoyment and recreation.

(e) Trade and traffic of every description.

(f) The preservation and protection of game and fish, of wild birds generally, and of cattle allowed to pasture in the park.

(g) The issuing of licenses or permits for the pasturage of cattle, and the management of hay lands.

(h) The removal and exclusion of trespassers.

(i) And generally for all purposes necessary to carry this Act into effect according to the true intent and meaning thereof.

(2) The Governor-in-council may, by the said regulations, impose penalties for any violation thereof, not exceeding in each case the sum of fifty dollars, or, in default of payment with costs, imprisonment for not more than three months.

5. Every regulation made as aforesaid, shall, after publication for four consecutive weeks in the *Canada Gazette,* and in any other manner that may be provided thereby by the Governor-in-council, have the like force and effect as if it was herein enacted, and such regulations shall be laid before Parliament within fifteen days after its first meeting thereafter.

6. Nothing in this Act contained shall affect the obligations of the Government (if any) arising out of the conditions of the acquisition of the North-west Territories.

7. This Act may be cited as "Rocky Mountains Park Act, 1887."

REGULATIONS RESPECTING ROCKY MOUNTAINS PARK.

By Order-in-Council of Monday, 30th June, 1890, under authority of " The Rocky Mountains Park Act," 50-51 Vict., chap 32, s. 4, the Order-in-Council of the 27th day of November, 1889, establishing regulations for the control and management of the Rocky Mountains Park of Canada was cancelled, and the following Regulations, approved in point of form by the Minister of Justice, were substituted after the 1st day of July. 1890, for the regulations established by the said Order-in-Council.

(1) No person shall, without permission from the Minister of the Interior, reside permanently within other portions of the park than those sold or leased.

(2) The superintendent of the park (hereinafter called the superintendent) may issue permits to visitors for camping upon such ground as he may designate; any one camping without such permit shall be considered a trespasser, and the fee for such permit shall be one dollar per month per tent; provided, however, that no such permit shall be granted for camping in any portion of the park situated south of Bow River.

(3) The defacement of any object at any of the hot springs, or of any of the natural rock formations, or timber, by written inscription, or otherwise, is strictly forbidden ; as is also the throwing of any stones, sticks or other substances whatsoever into any of the springs or streams in the Park.

(4) No advertisements, other than those issued or permitted by the Minister of the Interior, shall be posted or displayed within the park, except on leased property in the town site of Banff, or property in the village of Anthracite.

(5) No live stock shall be permitted to run at large, nor shall pigs, sheep or goats be brought into or kept within the park, provided, however, that licensed butchers may bring in and keep for a period not exceeding thirty days, and at such places and in the manner to be prescribed by the superintendent, animals to be slaughtered for food purposes.

(6) The superintendent shall from time to time select and designate pasturing grounds within the park, upon which leaseholders may pasture not in excess of two milch cows and two horses for each lot leased ; but leaseholders availing themselves of this regulation shall make provision satisfactory to the superintendent for herding the animals and driving them to and from the pasture grounds.

(7) All stock found pasturing, except where authorized, may be impounded and held until a proper guarantee be given that the trespass will not be repeated, and until a fine be paid sufficient to cover the expenses of impounding such stock, feeding them while so impounded, and advertising. Failure to give the necessary guarantees and to pay the fine within thirty days shall render the stock liable to be sold by the superintendent, and the proceeds of such sale, after paying thereout the fine, cost of maintenance, advertising and sale, shall be paid by the superintendent to the owner of the stock. The superintendent may authorize any person to act as pound-keeper, the rates of remuneration to be settled by the Minister of the Interior.

(8) The superintendent shall, upon application, furnish each owner of a dog or bitch, upon payment of a fee of one dollar in the case of a dog and two dollars in the case of a bitch, with a license authorizing him to keep such dog or bitch ; such licenses shall expire on the thirtieth day of June in each year, and shall then be renewed ; and any unlicensed dog or bitch may be impounded or destroyed, at the discretion of the superintendent.

(9) No person shall cut or remove any timber, growing or dead, or remove or displace any mineral deposits or natural curiosities, unless by written permission of the superintendent.

(10) No rubbish or any matter of an offensive nature shall be deposited, except in such places and at such times and under such conditions as the superintendent shall designate.

(11) No person shall ride or drive on or over any bridge within the park faster than a walk ; furious riding or driving on public roads is also prohibited.

(a) Horses driven with sleighs shall be provided with bells.

(b) No person shall ride or drive across or on any side-walk, boulevard, vacant lot, or common within the park, without the written permission of the superintendent. Horse-racing is also prohibited, except in such places as may be set apart for the purpose by the superintendent

(c) Horses in use or attached to any vehicle shall not be allowed to stand without being tied or in charge of some grown person.

(12) The waters of the hot springs shall be controlled by the superintendent, and shall be supplied to licensed bath-houses at such rental per annum as may be fixed from time to time by Order-in-Council, and the superintendent may at any time shut off the supply of the said water, after two weeks' notice in writing, from any such bath-house, the lessee of which may be in arrears for rent or who

may have in any way infringed any of the provisions of this or the next succeeding clause ; and no person shall in any way interfere or tamper with any spring, pipes, valves, traps, tanks, or any other apparatus connected with the supply and distribution of the said water.

(13) The superintendent or his authorized agent shall have free access for inspection at all reasonable times to any bath-house or building using the water of the springs, or to any pipe leading to or within such bath-house or building.

(14) The Minister of the Interior shall have power to cause such portion of the park as from time to time he may designate to be surveyed and laid out in building lots, for the construction thereon of buildings for ordinary habitation and purposes of trade and industry, and for the accommodation of persons resorting to the park, and may issue leases for such lots for any term not exceeding forty-two years, with the right of renewal, at rentals to be from time to time fixed by him : also, to set apart such portions of the park as he may think proper for the sites of market-places, jails, court-houses, places of public worship, burying-grounds, benevolent institutions, squares, and for other similar public purposes.

(15) The location, design, and general character of any buildings to be erected as dwelling-houses or for purposes connected therewith, or fences, shall be subject to the approval of the superintendent and to the sanction of the Minister of the Interior.

(a) No timber on any lot leased for residential purposes, except so much as is actually necessary to be removed to make room for the building and reasonable access thereto, shall be cut or removed, except by permission of the superintendent.

(16) The Minister of the Interior may issue licenses of occupation for the working of mines and the development of mineral interests within the limits of the park, subject, however, to the approval in each instance by the Governor-in-Council, of the terms, conditions, and duration of such licenses of occupation.

(17) All leases or licenses of occupation shall be in such form as may be approved by the Minister of the Interior and the Minister of Justice.

(18) No bar-room or saloon shall be permitted within the park.

(19) The following restrictions on the sale of intoxicating liquors in the park shall be imposed and enforced, in addition to the restrictions imposed by the North-west Territories Act. The sale of intoxicating liquors even under the special permission granted under section ninety-two of the said Act, is strictly prohibited, except in hotels, and there it shall only be allowed to hotel guests for table use. Nor shall any person after obtaining such special permission, sell, exchange, trade or barter, or have in his possession within the park, even for hotel use under this regulation, any intoxicating liquor, until his special permission issued in accordance with section ninety-two of the said North-west Territories Act has been countersigned by the Minister of the Interior or his deputy, for which countersigning a fee of fifty dollars shall be charged in each case ; and no permit for a hotel shall be so countersigned unless such hotel shall have at least twenty bed-rooms of a size and to be furnished in a manner satisfactory to the superintendent.

(20) If at any time during the continuance of the permit the superintendent reports that the accommodation hereinbefore specified is not maintained, or if it is proved to the satisfaction of the Minister of the Interior that the hotel is not being conducted in an orderly and proper manner, the permit may be revoked and cancelled by the Minister of the Interior and the permittee shall have no claim to have repaid to him any portion of the fee paid for countersigning such permit.

(21) No person shall do business as a peddler in the park or act as guide therein without a license from the Minister of the Interior, who shall have power

to revoke such license in his discretion; and no guide shall be entitled to charge for his services more than fifty cents per hour for six hours or under, and not more than three dollars for any day not exceeding ten hours.

(22) All slaughter-houses, butcher-shops, fish-stalls, and any other business which from its nature is or may become offensive or obnoxious, shall be carried on only at such places as the superintendent may designate in the license for the establishment of such business, and shall be subject at any time, on sixty days' notice in writing delivered to the owner or lessee in person, or left at his place of residence or place of business, to removal to such other place as the superintendent may designate. Every license issued under this clause shall be subject to revocation at any time upon thirty days' notice to the licensee, and the business shall entirely cease on the revocation of a license.

(23) The Minister of the Interior may issue a license to any person or persons undertaking to place a steam yacht or other vessel, or vessels, suitable for the conveyance of passengers, and in all respects complying with the Steamboat Inspection Act or Acts regulating steam and other vessels, on any waters within the park, to date from the first day of April in each year. The maximum fare which may be charged for the conveyance of passengers in such boats shall not exceed when running on regular trips up to eight miles, fifty cents; above eight and up to twelve miles, seventy-five cents; over twelve miles, one dollar.

(24) Licenses to carry on livery stables may be issued by the Minister of the Interior, the fee for which shall be ten dollars per annum for each vehicle drawn by two or more horses, and six dollars for each vehicle drawn by one horse; and no person shall keep horses or conveyances for hire without first having obtained such license. The rates which may be charged for the hire of carriages or other vehicles, and saddle horses, shall not exceed the following:—

(a) For the conveyance of one passenger from or to the railway station to or from any licensed hotel or boarding-house within a radius of one-and-a-half miles of the station, fifty cents; to all points beyond one-and-a-half and within three miles of the railway station, one dollar.

(b) For the conveyance of one passenger when there are at least four passengers in the vehicle, from any one point within one mile of the Bow River bridge, at the end of Banff avenue, to and from Devil's Lake, two dollars.

(c) For conveyance in any vehicle, drawn by two horses, and carrying not more than four persons, for one passenger, one dollar for the first hour, and twenty-five cents an hour for each additional passenger for the first hour; and for every subsequent hour, fifty cents for one passenger and twenty-five cents for each additional passenger.

(d) For conveyance in any vehicle drawn by two or more horses and carrying more than four persons, seventy-five cents an hour for each person for the first hour, and twenty-five cents an hour for every subsequent hour.

(e) For conveyance in any vehicle drawn by one horse, one dollar an hour for one person for the first hour, fifty cents an hour for an additional person for the first hour, and fifty cents for each person for every subsequent hour.

(f) For saddle-horses, three dollars for a whole day, two dollars for a half day or by the hour seventy-five cents for the first hour and fifty cents for each subsequent hour. In calculating half-a-day, one o'clock, p.m., shall be the hour of division; the maximum time allowed for a half day shall be five hours; and twenty-five cents may be charged for each subsequent hour.

(g) The rates for cartage of freight or general merchandise shall be subject to agreement between the parties interested.

(25) The tires on waggons used for freighting purposes on the roads constructed by the government within the park shall be at least two inches and

a-half in width ; all vehicles shall be provided with brakes ; and it shall be the duty of the superintendent to condemn and prohibit the use of any vehicle which in his opinion is unsafe.

(26) All drivers of public vehicles shall be licensed ; the fee therefor shall be one dollar; and such license may be revoked and cancelled at any time by the superintendent if it is proved to his satisfaction that the holder thereof has been guilty of incivility, insobriety, or misconduct, while discharging his duties.

(27) No person shall keep a pool, billiard, or bagatelle-table, or bowling-alley for use by the public, without a license ; such license shall be for one year from the first day of May in each year, and the fees for such license shall be the following :—

(a) For one billard or pool-table, twenty dollars and for each additional table ten dollars.

(b) For one bagatelle, Mississippi, pigeon-hole or other table or board with balls, twenty dollars and for every additional table ten dollars.

(c) For a bowling-alley, ten dollars.

(28) Every description of gaming and all playing of faro, cards, dice, or other games of chance for stakes of money or other things of value, and all betting and wagering on any such games of chance, are strictly forbidden and prohibited within the park, and no person shall play at or allow to be played on his premises, or assist, or be engaged in any way in any description of gaming, as aforesaid.

(29) The shooting at, wounding, capturing, killing, or in any manner injuring any wild animal or bird within the park, is hereby prohibited, excepting, however, mountain lions, bears, wolves, lynxes, wolverines, coyotes, wild cats and hawks. Fishing with nets in any of the waters of the park is also prohibited.

(30) The outfits of all persons found hunting, or fishing with nets, or having in their possession game or fish killed within the park in contravention of clause 29 of these regulations shall be subject to seizure and confiscation.

(31) Permission to cut hay within the park shall be obtained from the superintendent and shall be subject at all times to his supervision and control.

(32) No person shall take or use any stone, sand, gravel, or other material in the park without a permit from the superintendent, and the the following fees shall be paid to the superintendent for such materials :—

Sand . . . . 10 cents, per load.
Stone . . . . 25 " "
Gravel . . . . 25 " "

(33) Persons desiring to burn lime or manufacture brick within the park shall obtain a permit from the superintendent, defining the location of the kiln or brick yard, and pay a royalty of one cent and a half per bushel for all lime burnt, and, for all brick manufactured, a rate per thousand to be fixed by the Minister of the Interior.

(34) The use of fire-arms within the park, except under permit from the superintendent, is strictly prohibited.

(35) If any offence is committed under any of the provisions of these regulations, such offence shall be prosecuted, under the "Summary Convictions Act," before the superintendent of the park, who for the purposes hereof shall be, ex-officio, a justice of the peace, with jurisdiction anywhere within the park or before any officer of the North-west mounted police force empowered by law to sit and act as a justice of the peace.

(36) Except as hereinafter specially provided, every one who violates any provision of any of these regulations, shall be liable to a penalty not exceeding

twenty dollars and costs, and in default of payment to imprisonment for a term not exceeding one month.

(37) Every one who violates any of the provisions of clause number nineteen of these regulations, which relates to the sale of intoxicating liquors within the park shall be liable to a penalty not exceeding in each case the sum of fifty dollars and costs, and in default of payment thereof to imprisonment for a term not exceeding three months ; and a moiety of every penalty imposed and collected under the provisions of this clause of these regulations shall belong to Her Majesty and the other moiety to the person laying the information.

(38) Every one who violates any of the provisions of clause twenty-eight of these regulations, which relates to gaming, shall be liable to a penalty not exceeding in each case the sum of fifty dollars and costs, and in default of payment thereof to a term of imprisonment not exceeding three months ; and a moiety of every penalty imposed and collected under the provisions of this clause of these regulations shall belong to Her Majesty and the other moiety to the person laying the information.

(39) In order the more effectually to repress the offences specified in clauses numbers nineteen and twenty-eight of these regulations, every officer of the park, or officer of the North-west mounted police force, or constable of the North-west mounted police accompanied by or acting under the orders of a commissioned officer of the said force is hereby authorized, by force, if necessary, and without the necessity of any intervention or process of law, to enter any suspected place, to arrest therein on view any person or persons found committing any of the offences aforesaid, and to bring him or them before any of the officers who, by these regulations, are empowered to sit and act as justices of the peace within the Park to be dealt with according to law ; and also to seize any tables and other instruments and money, securities for money, liquor and vessels and appliances used in connection therewith, used in contravention of the said clauses ; and upon the conviction of such person or persons or any of them of such offence, in addition to any penalty imposed in respect thereof, the said table or tables and other instruments shall be forfeited and sold, or in the discretion of the convicting justice, destroyed, and the money so seized as aforesaid shall be forfeited and applied, together with the proceeds of sales towards the revenues of the park in a manner hereinafter provided.

(40) The revenues derived from every source under any of the provisions of these regulations shall be deposited forthwith to the credit of the Receiver-General on account of the park, except as otherwise herein specially provided.

(41) Printed copies of these regulations, to be furnished by the Department of the Interior for that purpose, shall be posted and kept in a conspicuous place in every government office, and in every hotel, boarding-house, bath-house and livery-stable within the park,

(42) For the control and management of the park in any matter whatsoever not specially provided for by the Rocky Mountains Park Act, 1887, or by any other Act of the Parliament of Canada applicable to the park, or by the foregoing regulations, any existing ordinances of the North-west council in that behalf shall be in force.

(43) Wherever in these regulations the expression " the superintendent of the park " or " the superintendent " is used it shall mean the officer holding that office at the present time under appointment by the governor-in-council, or any person who may hereafter be so appointed to the said office.

MINERAL LANDS IN THE PARK.

The following Order-in-Council was adopted 12th October, 1892, respecting mineral lands within the Park :—

Whereas section 5 of the Act 55-56 Victoria, chaper 15, amending the Dominion Lands Act, provides that lands containing coal or other minerals, including lands in the Rocky Mountains Park, shall not be subject to the provisions of this Act respecting sale or homestead entry, but the Governor-General-in-Council may, from time to time, make regulations for the working and development of mines on such lands, and for the sale, leasing, licensing or other disposal thereof; provided, however, that no disposition of mines or mining interests in the said park shall be for a longer period than twenty years, renewable, in the discretion of the governor-in-council, from time to time, for further periods of twenty years each, and not exceeding in all sixty years.

His Excellency, in virtue of the provisions of the above cited Act, and by and with the advice of the Queen's Privy Council for Canada, is pleased to to make the following regulations to govern the issue of licenses of occupation for the working of mines and minerals within the Rocky Mountains Park of Canada :—

(1) Licenses to mine coal from lands within the park shall be disposed of by public competition only, and the Minister of the Interior shall, from time to time, as he may find expedient in the public interest, survey, lay out, and offer for disposal by auction or by tender, locations for the mining of coal under such licenses.

(2) The duration of such licenses shall be twenty years, unless sooner terminated by consent of the Crown and the licensee, or cancelled for non-fulfilment of conditions, and such licenses shall be renewable in the discretion of the Governor-in-Council for further periods of twenty years each and not exceeding in all sixty years, on such terms and conditions as may at the time of renewal be agreed upon by the government and licensee.

(3) The ground rent shall be $1.20 per acre per annum, payable half yearly in advance

(4) A royalty of ten cents per ton shall be paid by the licensee on all coal, taken out of the mine. Returns under oath shall be furnished quarterly to the Minister of the Interior by the licensee, showing the quantity of coal taken out and the royalty shall be paid at the time of making such returns. If the royalty which is due for one half-year equals the rental paid for that half-year, then the amount paid for rent shall be credited to such royalty.

(5) The area to be licensed to one person shall not exceed three hundred and twenty acres, and the licensee shall not make any transfer or assignment of his license without the consent in writing of the Minister of the Interior.

(6) The boundaries beneath the surface of the location shall be the vertical planes or lines in which their surface boundaries lie.

(7) The license shall be subject to the general regulations for the control and management of the Rocky Mountains Park of Canada, dated the 30th June, 1890, and to such farther and other regulations as may be made from time to time in that behalf by the Governor-in-Council.

THE ADIRONDACKS.

FOREST LEGISLATION.

A desolation like that which has overwhelmed many once beautiful and fertile regions of Europe awaits important parts of America, and other comparatively new countries over which civilization is now extending its sway, unless prompt measures are taken to check the action of destructive causes already in operation. It is almost in vain to expect that mere restrictive legislation can do anything effectual to arrest the progress of the evil, except so far as the State is still the proprietor of extensive forests. Woodlands which have passed into private hands will everywhere be managed upon the same economical principles as other possessions, and every proprietor will, as a general rule, fell his woods unless he believes that it will be for his pecuniary interest to preserve them. In France, law has been found impotent to prevent the destruction or wasteful economy of private forests.

Fortunately for the immense economical and sanitary interests involved in this branch of rural and industrial husbandry, public opinion is thoroughly roused to the importance of the subject. Plantations of a certain extent have been made, and a wiser system is pursued in the treatment of the remaining native woods.

The people of the far west have thrown themselves into the work with much of the passionate energy which marks their action in reference to other modes of physicial improvement. California has appointed a State forester with a liberal salary, and made such legal provisions and appropriations as to render the discharge of his duties effectual. The hands that built the Pacific Railway at the rate of miles in a day are busy in planting belts of trees to shelter the track from snow-drifts, and to supply at a future day timber for ties and fuel for the locomotives. The settlers on the open plains, too, are not less actively engaged in the propagation of the woods.

It was not till 1869 that the legislature of the State of New York turned its attention to the subject of tree-planting, when it passed a law to encourage planting trees by the sides of public highways, and in 1872 created by enactment a "Commission for State Parks," whose duty was to enquire into the expediency of providing for vesting in the State the title to the timber regions lying within the Counties of Lewis, Essex, Clinton, Franklin, St. Lawrence, Herkimer, and Hamilton, and converting the same into a public park. The commission known as the "park commission" made a report recommending that no more lands lying in the counties named should be sold, but that as lands were acquired by the State through tax sales, they should be held for future forest management. The methods recommended by the commission were not acted upon until 1883, when a law was passed prohibiting further sales of lands in the counties named in the Act, and also in the counties of Saratoga and Warren. During the interval the sale of State lands had been continued, and in 1883 the State had a much less acreage in its possession than it would have had if it had adopted the recommendations contained in the commission's report as soon as they were made.

At the session of the legislature in 1884 there was appropriated $5,000 to be used by the comptroller of the State in "the employment of such experts as he may deem necessary to investigate and report a system of forest preservation." In July of that year the comptroller (Hon. Alfred C. Chapin) appointed as such experts, Prof. Charles S. Sargent, of Cambridge, Mass.; D. Willis James, Esq., of New York City; Hon. William A. Poucher, of Oswego, and Edward M. Shepard, Esq., of Brooklyn. The committee, so constituted, reported the result of their investigations, coupled with their recommendations as to future policy, to the comptroller in January, 1885, and in forwarding their report to the legislature Comptroller Chapin said: "The problem in its fulness affects the welfare of many sister commonwealths, and of the nation at large. It is eminently fitting that in its solution the Empire State should lead the way."

As a sequel to, and the result of the recommendations of the Comptroller's Committee, the Bill establishing the Forest Commission was passed in the following May, as follows :—

THE FOREST COMMISSION ACT.

Chapter 283, Laws of 1885 ; as subsequently amended..

An Act to establish a Forest Commission, and to define its powers and duties, and for the preservation of forests, passed May 15th, 1885.

The people of the State of New York, represented in Senate and Assembly, do enact as follows :—

COMMISSIONERS, HOW APPOINTED.

Section 1. There shall be a forest commission, which shall consist of three persons, who shall be styled forest commissioners, and who may be removed by the governor for cause. The forest commissioners shall be appointed by the governor by and with the advice and consent of the Senate.

TERMS BY LOT.

§ 2. At the first meeting of the forest commission they shall divide themselves by lot, so that the term of one shall expire in two years, one in four years and one in six years from the first day of February next ensuing. Except as to the three terms of office thus determined, the term of office of a forest commissioner shall be six years from the first day of February on which the preceding term expires.

HOW APPOINTED AFTER 1888.—VACANCIES.

§ 3. During the month of January, in the year 1888, and in every second year thereafter, the governor, by and with the advice and consent of the Senate, shall appoint one forest commissioner. Vacancies that may exist in the office of a forest commissioner after the commencement of a term of office, shall be filled by the governor's appointment, subject to the confirmation of the Senate at its next session, for the unexpired portion of the term in which the vacancy occurs.

COMMISSIONERS, COMPENSATION OF.

§ 4. The forest commissioners shall serve without compensation, except that there shall be paid them their reasonable expenses incurred in the performance of their official duties.

EMPLOYEES, AND PAY OF.

§ 5. The forest commission shall have power to employ a forest warden, forest inspectors, a clerk, and all such agents as they may deem necessary, and to fix their compensations, but the expenses and salary of such warden, agents, clerk, inspectors, and assistants, shall not exceed in the aggregate, with the other expenses of the commission the sum therefor appropriated by the legislature.

ROOMS, ETC.

§ 6. The trustees of public buildings, under Chap. 349, laws of 1883, shall provide rooms for office for the forest commission, with proper furniture and fixtures and with warming and lights.

FOREST PRESERVE DEFINED.

§ 7. All the lands now owned, or which may hereafter be acquired by the State of New York within the Counties of Clinton (except in the Towns of Altona and Dannemora) Delaware, Essex, Franklin, Fulton, Hamilton, Herkimer, Lewis, Oneida, St. Lawrence, Saratoga, Warren, Washington, Greene, Ulster and Sullivan, shall constitute and be known as the Forest Preserves, except all such lands within the limits of any incorporated village or city, and except all such lands, not wild lands, as have been, or may hereafter be, acquired by the State of New York, upon or by foreclosure of or sale pursuant to any mortgage upon lands made to the Commissioners for loaning certain moneys of the United States, usually called the United States Deposit Fund, and all such excepted lands acquired by the State of New York may be sold and conveyed as provided by law.

STATE LAND CANNOT BE LEASED. CONDITIONS UNDER WHICH IT CAN BE SOLD OR EXCHANGED. APPRAISERS APPOINTED. DUTY OF COMPTROLLER.

§ 8. The lands now or hereafter constituting the forest preserve shall be forever kept as wild forest lands, and shall not be sold, nor shall they be leased or taken by any person or corporation, public or private, except that whenever any of the lands now constituting the forest preserve or which may hereafter become a part thereof, owned by the State within any county specified in section seven of the act hereby amended, shall consist of separate small parcels or tracts wholly detached from the main portions of the forest preserve and bounded on every side by lands not owned by the State, then it shall be lawful, and the comptroller shall have power to sell and convey such separate tracts or parcels, or the timber thereon, to such person or persons, corporation or association as shall have offered the highest price therefor; but no such tracts or parcels of land or the timber thereon, shall be sold by the comptroller except upon the recommendation of the forest commission or a majority thereof, together with the advice of the attorney-general in behalf of the State. Such separate tracts or parcels of land may be exchanged by the comptroller for lands that lie adjoining the main tracts of the forest preserve upon the recommendation of the forest commission or a majority thereof, together with the advice of the attorney-general on behalf of the State; but the values of said lands so exchanged must be first appraised by three disinterested appraisers sworn to faithfully and fairly appraise the value of said lands, and the difference if any, between the values of such parcels so proposed to be exchanged shall be paid by the party so exchanging with the State into the State treasury, but the State shall not pay the amount of any such difference. Two of said appraisers shall be nominated and appointed by the county judge of the county in which said lands proposed to be exchanged are situate or in case such lands are situate in two counties, then the county judge of each county shall nominate and appoint each one appraiser. The two appraisers so appointed shall select a third appraiser, and they shall report to the comptroller the result of said appraisal, before such lands shall be exchanged as aforesaid. The said appraisers so appointed shall receive the same compensation for their services as is provided for appraisers of decedent's estates, to be paid by the party so proposing to exchange lands with the State. It shall be the duty of the comptroller annually to report to the legislature all sales or exchanges of lands made under the provisions of this act, together with all bids and the amounts received therefor, and in said report shall be included the reports of appraisers of lands exchanged in accordance with the foregoing provisions. The proceeds of all lands so sold, or the receipts from all exchanges so made, shall be invested by the comptroller, with the approval of the forest commission, in the purchase of forest land adjoining great blocks of the forest preserve now owned by the State.

FORESTS TO BE PROTECTED AND PROMOTED.—POWERS OF LAND OFFICE AND COMPTROLLER TRANSFERRED TO FOREST COMMISSION.—RULES GOVERNING FOREST PRESERVE.

§ 9. The forest commission shall have the care, custody, control and superintendence of the forest preserve. It shall be the duty of the commission to maintain and protect the forests now on the forest preserve, and to promote as far as practicable the further growth of forests thereon. It shall also have charge of the public interests of the State, with regard to forest and tree planting, and especially with reference to forest fires in every part of the State. It shall have as to all lands now or hereafter included in the forest preserve, but subject to the provisions of this act, all the powers now vested in the commissioners of the land office and in the comptroller as to such lands as are now owned by the State. The forest commission may, from time to time, prescribe rules or regulations, and may from time to time alter or amend the same, affecting the whole or any part of the forest preserve, and for its use, care and administration ; but neither such rules or regulations, nor anything herein contained, shall prevent or operate to prevent the free use of any road, stream or water as the same may have been heretofore used, or as may be reasonably required in the prosecution of any lawful business.

OFFICERS MAY ARREST, WHEN.

§ 10. The forest warden, forest inspectors and other persons acting upon the forest preserve, under the written employment of the forest warden or of the forest commission, may, without warrant, arrest any person found upon the forest preserve violating any of the provisions of this act; but in case of such arrest, the person making the arrest shall forthwith take the person arrested before the nearest magistrate having jurisdiction to issue warrants in such case, and there make, or procure to be made, a complaint in writing, upon which complaint the magistrate shall act as the case may require.

ACTION TO RECOVER DAMAGES AND FOR TRESPASS.—TRESPASS DEFINED.—ATTORNEYS, HOW EMPLOYED.—UNDER DIRECTION OF ATTORNEY-GENERAL.

§ 11. The forest commission may bring, in the name or on behalf of the people of the State of New York, any action to prevent injury to the forest preserve or trespass thereon, to recover damages for such injury or trespass, to recover lands properly forming part of the forest preserve, but occupied or held by persons not entitled thereto, and in all other respects for the protection and maintenance of the forest preserve, which any owner of land would be entitled to bring. The forest commission may also maintain, in the name or on behalf of the people of the State, an action for the trespass specified in section seventy-four, article fifth, title five, chapter nine, part one of the revised statutes, when such trespass is committed upon any lands within the forest preserve. In such action there shall be recoverable the same penalty, and a like execution shall issue, and the defendant be imprisoned thereunder without being entitled to the liberties of the jail, all as provided in sections seventy-four and seventy-six of the said article ; and in such action the plaintiff shall be entitled to an order of arrest before judgment as in the cases mentioned in section five hundred and forty-nine of the code of civil procedure. The trespass herein mentioned shall be deemed to include in addition to the act specified in the said section seventy-four, any act of cutting or cause to be cut, or assisting to be cut any tree or timber standing within the forest reserve, or any bark thereon, with intent to remove such tree or timber, or any portion thereof, or bark therefrom, from the said forest preserve. With the consent of the attorney-general and the comptroller, the forest commission may employ attorneys and counsel to prosecute any such action, or to defend any action brought against the commission, or any of its members or subordinates, arising out of their or his official conduct with relation to the forest preserve. Any attorney or counsel so employed shall act under the direction of and in the name of the attorney-general. Where such attorney or counsel is not so employed, the attorney-general shall prosecute and defend such actions.

INJUNCTIONS.

§12. In an action brought by or at the instance of the forest commission, an injunction, either preliminary or final, shall upon application be granted restraining any act of trespass, waste or destruction upon the forest preserve.

PARTITION.—ACTION FOR, HOW BROUGHT.—EFFECT OF.—AGREEMENT FOR.

§13. Whenever the State owns or shall own an undivided interest with any person in any land within the counties mentioned in section seven of this act, or is or shall be in possession of any such land as joint tenants or tenants in common with any person who has an estate of freehold therein, the attorney general shall, upon the request of the forest commission, bring an action in the name of the people of the State of New York for the actual partition of the said lands according to the respective rights of the parties interested therein; and upon the consent in writing of the forest commission, any such person may maintain an action for the actual partition of such lands, according to the respective rights of the parties interested therein, in the same manner as if the State were not entitled to exemption from legal proceedings, service of process in such actions upon the attorney-general to be deemed service upon the State. Such actions, the proceedings and the judgment therein, and the proceedings under the judgment therein shall be according to the practice at the time prevailing in actions of partition and shall have the same force and effect as in other actions, except that no costs shall be allowed to the plaintiff in such actions, and except that no sale of such lands shall be judged therein. The forest commission, may without suit, but upon the consent of the comptroller, agree with any person or persons owning land within the said towns jointly or as tenants in common with the State for the partition of such lands and upon such agreement and consent, the comptroller shall make on behalf of the people of the State any conveyance necessary or proper in such partition, such conveyance to be forthwith recorded as now provided by law as to conveyances made by the commissioners of the land office.

INCOME.

§14. All incomes that may hereafter be derived from State forest lands shall be paid over by the forest commission to the treasury of the State.

EXPENSES.

§15. A strict account shall be kept of all receipts and expenses, which accounts shall be audited by the comptroller, and a general summary thereof shall be reported annually to the legislature.

ANNUAL REPORT.

§16. The forest commission shall in January of every year, make a written report to the legislature of their proceedings together with such recommendations of further legislative or official action as they may deem proper.

SUPERVISORS PROTECTORS OF STATE LANDS, EXCEPT IN FOREST PRESERVE.—
DUTY OF DISTRICT ATTORNEY.—SUPERVISORS TO REPORT.—DISTRICT AND
FOREST GUARDS.

§17. The supervisor of every town in this State in which wild or forest lands
belonging to the State are located except within the counties mentioned in
section seven of this act, shall be by virtue of his office the protector of these
lands subject to the instruction he may receive from the forest commission. It
shall be his duty to report to the district attorney for prosecution any act of
spoliation or injury that may be done and it shall be the duty of such district
attorney to institute proceedings for the prevention of further trespass, and for
the recovery of all damages that may have been committed with costs of
prosecution. The supervisors shall also report their proceedings therein to the
forest commission. In towns where the forest commission shall deem it necessary,
they may serve a notice upon the supervisor, requiring him to appoint one or
more forest guards, and if more than one in a town, the district of each shall be
properly defined. The guard so appointed shall have such power, and perform
such duties, and receive such pay, as the forest commission may determine.

INSTRUCTIONS IN FORESTRY IN SCHOOLS.

§ 18. The forest commission shall take such measures as the department of
public instruction, the regents of the university and the forest commission
may approve for awakening an interest in behalf of forestry in the public
schools, academies and colleges of the State, and of imparting some degree of
elementary instruction upon this subject therein.

METHODS OF PUBLIC INSTRUCTION.

§ 19. The forest commission shall, as soon as practicable, prepare tracts or
circulars of information, giving plain and concise advice for the care of woodlands
upon private land, and for the starting of new plantations upon lands that have
been denuded, exhausted by cultivation, eroded by torrents, or injured by fire,
or that are sandy, marshy, broken, sterile or waste, and unfit for other use.
These publications shall be furnished without cost to any citizen of the State
upon application, and proper measures may be taken for bringing them to the
notice of persons who would be benefited by this advice.

SUPERVISORS TO ACT AS FIREWARDENS, WHEN.—DISTRICT MAPS.—FIREWARDENS
PAID BY TOWNS.—FIREWARDENS IN FOREST PRESERVE.—AUTHORITY OF
FIREWARDENS.

§ 20. Every supervisor of a town in this State, excepting within the counties
mentioned in section seven of this act, shall be *ex-officio* firewarden therein.
But in towns particularly exposed to damages from forest fires, the supervisor
may divide the same into two or more districts, bounded, as far as may be, by
roads, streams of water, or dividing ridges of land or lot lines and he may in
writing, appoint one resident citizen in each district as district firewarden
therein. A description of these districts, and the names of the district firewardens thus appointed, shall be recorded in the office of the town clerk. The
supervisor may also cause a map of the fire district of his town to be
posted in some public place with the names of the district firewardens appointed.

15 (F.)

The cost of such map, not exceeding five dollars, shall be made a town charge ; and the services of the firewardens shall also be deemed a town charge, and shall not exceed the sum of two dollars per day for the time actually occupied in the performance of their duties as such firewardens. The compensation for services of the persons who may assist in extinguishing forest fires, shall be a town charge, and shall not exceed the sum of one dollar per day for each person employed; but all bills for such services must be approved by the firewarden of the town in which the fire occurred before payment shall be made. It shall be the duty of the board of auditors in each town to examine, audit, and allow promptly all reascnable bills presented to them for services and disbursements under this section. Within the counties named in section seven of this act, such persons shall be firewardens as may from time to time be be appointed by the forest commission. The persons so appointed shall act during the pleasure and under the direction of the forest commission ; and there shall be applicable to them all the provisions of this act, with reference to supervisors and district town wardens. Upon the discovery of a forest fire, it shall be the duty of the firewarden of the district, town or county, to take such measures as shall be necessary for its extinction. For this purpose he shall have authority to call upon any person in the territory in which he acts for assistance, and any person shall be liable to a fine of not less than five nor more than twenty dollars for refusing to act when so called upon.

OFFICERS, POWERS AND DUTIES OF, IN CASE OF FIRE.

See Revised Statutes, Chap. 20, title 14, Part 1, Vol. 3, p. 2086.

§ 21. The forest commission, the forest warden, inspectors, the foresters, and any other person employed by or under the authority of the forest commission, and who may be authorized by the commission to assume such duty, shall within the counties mentioned in section seven of this Act, whenever the woods in any such town shall be on fire, perform the duties imposed upon, and in such case shall have the powers granted to the justices of the peace, the supervisors, and the commissioners of highway of such towns, by title fourteen, chapter twenty, of part one, of the revised statutes, with reference to the ordering of persons to assist in extinguishing fires, or stopping their progress ; and any person so ordered by the forest commission, the forest warden, the forest inspectors, the foresters, or any of them, or any other person acting or authorized as aforesaid, who shall refuse or neglect to comply with any such order, shall be liable to punishment prescribed by the said title.

NO ACTION FOR TRESPASS, WHEN.

§ 22. No action for trespass shall be brought by any owner of land for entry made upon his premises by persons going to assist in extinguishing a forest fire, although it may not be upon his land.

PRIVILEGE, CASE OF FIRE.

§ 23. The fire-warden, or the supervisor, where acting in general charge, may cause fences to be destroyed or furrows to be ploughed to check the running of fire, and in cases of great danger back fires may be set along a road or stream, or other line of defence, to clear off the combustible material before an advancing fire

Reports of Supervisors and Firewardens.—Summary to be Reported to Legislature.

§ 24. The supervisor of every town of which he is firewarden, as aforesaid, and in which a forest fire of more than one acre in extent has occurred within a year, shall report to the forest commission the extent of the area burned over, to the best of his information, together with the probable amount of property destroyed, specifying the value of timber as near as may be, and amount of cordwood, logs, bark or other forest product, and of fencing, bridges and buildings that have been burned. He shall also make inquiries and report as to the causes of the fires, if they can be ascertained, and as to the measures employed and found most effectual in checking their progress. A consolidated summary of these returns by counties, and of the information as to the same matter otherwise gathered by the forest commission, shall be included in the annual report of the forest commission.

What Railroad Companies Must Do.—Remove Grass, etc.

§ 25. Every railroad company whose road passes through waste or forest lands, or lands liable to be over run by fires within this State, shall twice within each year cut and burn off or remove from its right-of-way, all grass, bush, or inflammable material, but under proper care, and at times when the fires thus set are not liable to spread beyond control.

Locomotives.—Spark Arresters.—Engineers and Firemen.

§ 26. All locomotives which shall be run through forest lands shall be provided within one year from the date of this Act, with approved and sufficient arrangements for preventing the escape of fire from their furnace or ash-pan and netting of steel or iron wire upon their smoke stack to check the escape of sparks of fire. It shall be the duty of every engineer and fireman employed upon a locomotive to see that the appliance for the prevention of the escape of fire are in use and applied, as far as it can be reasonably and possibly done.

Not to Deposit Ashes, and to Report Fires.

§ 27. No railroad company shall permit its employees to deposit fire, coals or ashes upon their track in the immediate vicinity of wood lands liable to be overrun by fire, and in all cases where any engineers, conductors, or trainmen, discover that fences along the right-of-way, on woodlands adjacent to the railroad, are burning, or in danger from fire, it shall be their duty to report the same at their next stopping place and the person in charge of such station shall take prompt measures for extinguishing such fires.

Railroad Companies.—In Case of Fire.

§ 28. In season of drought, and especially during the first dry time in the spring after the snows have gone and before vegetation has revived, the railroad companies shall employ a sufficient additional number of trackmen for the prompt extinguishment of fire. And where a forest fire is raging near the line of their road, they shall concentrate such help and adopt such measures as shall most effectually arrest its progress.

RAILROAD COMPANIES.—PENALTY FOR NEGLECT.

§ 29. Any railroad company violating the provisions or requirements of this Act shall be liable to a fine of one hundred dollars for each offence.

RULES OF COMMISSION TO BE POSTED.

§ 30. The forest commission shall, with as little delay as practicable, cause rules for the prevention and suppression of forest fires to be printed for posting in school-houses, inns, saw-mills and other wood-working establishments, lumber camps and other places in such portions of the State as they may deem necessary. Any person maliciously or wantonly defacing or destroying such notices, shall be liable to a fine of five dollars. It shall be the duty of forest-agents, supervisors, and school-trustees, to cause these rules, when received by them, to be properly posted, and replaced when lost or destroyed.

INCENDIARIES.—PENALTIES.

§ 31. Any person who shall wilfully or negligently set fire to, or assist another to set fire to any waste or forest land belonging to the State or to another person, whereby the the said forests are injured or endangered, or who suffers any fire upon his own lands to escape or extend beyond the limits thereof, to the injury of the wood-lands of another, or of the State, shall be liable to a fine of not less than fifty dollars, nor more than five hundred dollars, or to imprisonment of not less than thirty days, nor more than six months. He shall also be liable in an action for all damages that may be caused by such fires ; such actions to be brought in any court of this State having jurisdiction thereon.

APPROPRIATION.

§ 32. Fifteen thousand dollars is hereby appropriated out of any moneys in the treasury, not otherwise appropriated, for the purposes of this act. And no liabilities shall be incurred by said forest commissioners in excess of this appropriation.

§ 33. This Act shall take effect immediately.

CHAPTER 37.—LAWS OF 1890.

AN ACT TO AUTHORIZE THE PURCHASE OF LANDS LOCATED WITHIN SUCH COUNTIES AS INCLUDE THE FOREST PRESERVE.

The people of the State of New York, represented in Senate and Assembly, do enact as follows :—

Section 1. The forest commission, with the approval and concurrence of the commissioners of the land office, may purchase lands so located within such counties as include the forest preserve, as shall be available for the purposes of a State park, at a price not to exceed one dollar and fifty cents, such approval and concurrence to be endorsed on a copy of the resolution of the said forest commission authorizing such purchase, and certified to by the clerk of said commissioners of the land office.

§ 2. The forest commission may have such lands appraised by one or more appraisers, not to exceed three in number, to be appointed by that commission. The expenses of such appraisal shall be a per diem allowance to the appraisers, not to exceed three dollars per day for the time actually employed and the necessary expenses incurred in each case, such expenses to be audited by the comptroller and paid out of the funds appropriated by the Legislature for the purposes of this act; but no purchase of lands shall be made in excess of previous appropriations for that purpose.

§ 3. The sum of twenty-five thousand dollars or so much thereof as may be necessary, is hereby appropriated out of any moneys in the treasury, not otherwise appropriated, for the purpose of this Act; and no liabilities shall be incurred by said forest commission in excess of this appropriation.

§ 4. This Act shall take effect immediately.

In returning the foregoing bill to the legislature with his signature, the governor filed therewith the following memorandum :—

State of New York.

Executive Chambers, Albany, March 11, 1890.

Memorandum filed with Senate Bill No. 91, entitled : "an Act to authorize the purchase of lands located within such counties as include the forest preserve."

Approved. There is no objection to this Act. The criticism which possibly may be urged against it is that it is good enough so far as it goes, but that it is wholly inadequate to meet the requirements of the situation. It is not a broad and comprehensive measure, providing for the establishment of an Adirondack park such as is imperatively demanded by the best interest of the State, but is simply a slight step in the right direction. The authority conferred is very inadequate, the amount appropriated is quite limited, the restrictions upon the prices to be paid are likely to produce unsatisfactory results, the provisions in regard to apportionment of lands are incomplete and somewhat unnecessary, and in many other respects the measure falls short of what it was hoped the legislature might enact. The bill must be regarded as a mere temporary expedient, and, as such, can do no harm; and although it will not afford a proper and complete solution of the Adirondack Park question, it encourages the hope that in the near future a more substantial and adequate measure may be passed to fully accomplish the object recommended in my recent message to the legislature relating to this subject.

I cheerfully approve the bill, in the expectation that its enactment may lead to such a result.

DAVID B. HILL.

CHAPTER 556.—LAWS OF 1890.

An Act Further to Amend Chapter Four Hundred and Twenty-Seven of the Laws of Eighteen Hundred and Fifty-Five, Entitled "An Act in Relation to the Collection of Taxes on Land of Non-Residents, and to Provide for the Sale of Such Land for Unpaid Taxes."

The people of the State of New York, represented in Senate and Assembly, do enact as follows :—

Section 1. Section seventy-four of chapter four hundred and twenty-seven of the laws of eighteen hundred and fifty-five, entitled "An Act in relation to the collection of taxes on the lands of non-residents and to provide for the sale of such lands for unpaid taxes," as amended by chapter four hundred and fifty-three of the laws of eighteen hundred and fifty, is hereby further amended so as to read as follows:—

§ 74. The occupant of any such lot, or any other person, may, at any time before the service of said notice by the purchaser or the person naming under him, and within three years from the expiration of the two years allowed by law for the redemption thereof, redeem any land so occupied, by filing in the office of the comptroller satisfactory evidence of the occupancy required, and by paying to him the consideration money for which the land to be redeemed was sold, and thirty-seven and one-half per centum thereon, together with the sum paid for the deed, if any, and such amount as may have been paid to the State for subsequent taxes thereon, or for the redemption from subsequent tax sales thereof, and, in addition thereto, providing such lot has been legally exempt from taxation for one or more years subsequent to the sale in question, of a sum that would represent the gross amount of taxes and interest that would have been due thereon, provided it had been taxed during each of the years it may have been exempt, on its assessed valuation and at the rate per cent. of taxation thereon, for the year when last returned to the comptroller's office. In all cases of tax sales heretofore made by the comptroller, where the land sold was in the actual occupancy of any person at the expiration of the two years allowed for the redemption thereof, and the purchaser or the persons naming under him shall have failed to serve notice of such sale upon the occupant or occupants thereof, and to file evidence of such service in the comptroller's office as provided by section sixty-eight of this Act, and the occupant or any other person shall fail to file in the comptroller's office within one year after this Act shall take effect a written notice of such occupancy, together with an application for the redemption of such lands, and to furnish the comptroller with satisfactory evidence of the occupancy required, and make such redemption within two years after this Act shall take effect, then and in all such cases the said tax sales of such land, and the conveyance thereof by the comptroller shall become absolute, and the occupant and occupants, and all other persons interested in the said land, shall be forever barred from all right and title thereto.

§ 2 All Acts and parts of Acts inconsistent with this Act are hereby repealed.

§ 3 This Act shall take effect immediately.

FIREWARDENS.

The organization of the force of firewardens in the State of New York has not been effected without considerable care and labor, it having involved the careful consideration of 281 separate appointments.

The maintenance of this system entails also a large amount of office work in the way of correspondence and in forwarding to each firewarden the necessary packages of posters, rules, and other supplies. In addition to the 281 firewardens in the counties of the forest preserve, there are the 430 supervisors in the other counties who are firewardens ex officio, with each of whom a correspondence is maintained.

No salary is attached to the position of firewarden; still the commission has succeeded in filling the places with good citizens, intelligent men who have shown themselves equal to the responsibilities devolving upon them, and have evinced a zeal and efficiency which argues well for the future care and preservation of our forest lands.

Each firewarden, when appointed, is furnished with a warrant bearing the signature and seal of the Department.

This warrant reads as follows :—

STATE OF NEW YORK.

THE FOREST COMMISSION, ALBANY, N. Y. 189

Esq.,

Town of   County of

The Forest Commision hereby appoints you a firewarden in and for your town in accordance with the provisions of the "Act to establish a Forest Commission" etc., passed May 15, 1885.

It will be your duty, as firewarden, whenever a forest fire occurs within the limits of your town, whether it be on State or other lands, to promptly notify a sufficient force to assist you ; to go to the place where the fire is burning, and to take charge of and to direct the work necessary for extinguishing it. All persons in the territory whom you may order to render you such assistance, are required by law to obey your order, and any person who may refuse to act in obedience to your order is, by statute liable, to a fine of not less than five nor more than twenty dollars.

If a forest fire occurs in your vicinity, although it may be in the adjoining town, it will be as much your duty to go immediately to the place of such fire as if it were in your own town ; and, in the absence of the fire warden of the town within the limits of which such fire may be, to assume the same authority and to discharge the same duties that you are empowered to assume and discharge in your own town, until the arrival of the fire warden of that town, upon which you will turn over all charges of the fire to him.

The same diligence and exertion must be used for the extinction of forest fires on private lands as on lands of the State. The public welfare requires that all forests should be protected from fire, no matter to whom they may belong.

After a forest fire has occurred in your town you must make a report of the same to the forest commission, stating the date and place of the fire, the number of acres burned over, the amount and nature of the damage, and the cause of the fire if known.

Your attention is called to the provisions of the twentieth section of the Forest Commission Act, for dividing your town into fire districts. Action theron, is left to your own discretion ; but if taken, you should report it to the commission.

It is essential that the rules and regulations of this commission, governing the methods of preventing and extinguishing forest fires, should be made fully known to the public. To that end you will be required to post, and keep posted, the cards containing the printed rules throughout your town, conformably to the provisions of section 30 of the before mentioned Forest Commission Act, and wherever you may judge it to be necessary in order to accomplish a complete public notification. Such posted cards, as may, at any time, be missing should be replaced at once. The cards will be furnished to you, and you can always be supplied with them on application to the commission.

A fire warden is required, by law, to be a resident of the town for which he is appointed. If you do not reside in the town, herein named, or if you should hereafter change your residence to another town, please notify this commission at once.

The office of a firewarden is distinct from that of a forester. Firewardens are not required to discharge any duties except those necessary for the prevention and extinction of forest fires, as before explained, and such other duties for a like purpose as may be, from time to time, assigned to them by this commission.

It is provided, by statute, that the pay of a firewarden for his official services shall not exceed the sum of two dollars a day for the time that he may be actually employed ; and also, that the bills of firewardens shall be paid by their respective towns. You are to render all your bills for services to your town and if you have any difficulty in having such bills audited and paid you should notify the forest commission.

This appointment is tendered to you in reliance upon a recommendation in which this commission places confidence. Should you accept the appointment, you are expected to discharge the duties of your office zealously, faithfully, in full compliance with the letter and spirit of the forest commission Act, and of the rules of the commission (both of which you are asked to read carefully), and in a manner at once honourable to the forest commission and yourself.

Be kind enough to inform the commission immediately, whether you accept or decline your appointment ; and in case that your acceptance is not forwarded within thirty days from date, you will be understood as declining.

By order of the Forest Commission.

(L.S.)   Secretary.

RULES AND REGULATIONS.

These are printed on heavy cards 12x15 inches. Latterly they have been printed on white muslin. as this material has proved more durable, the most of the placards being posted in the woods, or on fences, school-houses and mills, where they are exposed to the weather. Over 15,000 of these rules have been posted by the foresters and firewardens throughout the preserve counties, and the commission believes that much of the immunity from fire is due to their general distribution. They have been an important aid in warning the careless, and in educating the people in this particular.

Much of the force and value of these regulations is lost because there is no penalty attached to their violation ; and the commission is not authorized to add any clause in this respect.

RULES FOR THE PREVENTION AND SUPPRESSION OF FOREST FIRES AND FOR THE PROTECTION OF FOREST TREES.

(Established by the Forest Commission.)

1. All persons intending to light fires for the purposes of clearing or improvement, must give notice of their intention to the nearest firewarden before such fire is lighted, They must also give notice to all owners or occupants of adjoining lands, at least forty-eight hours before lighting such fires, which will be permitted only when the wind is favourable.

2. No fires, of the character before specified will be permitted until the trees are in full leaf. After such fires are lighted competent persons must remain to guard them until the fire is completely extinguished.

3. Fires will be permitted for the purpose of cooking, warmth, and insect smudges ; but before such fires are kindled, sufficient space around the spot where the fire is to be lighted must be cleared from all combustible material ; and before the place is abandoned, fires so lighted must be thoroughly quenched.

4. All fires other than those hereinbefore mentioned, are absolutely prohibited.

5. Hunters and smokers are cautioned against allowing fires to originate irom the use of firearms, cigars, and pipes; and all persons are warned that they will be held responsible for any damage or injury to the forest which may result from their carelessness or neglect.

6. Felling trees, and girdling or peeling bark from standing trees, are prohibited. Fallen timber only, may be used for firewood and camp construction.

7. Foresters and firewardens are instructed, and all citizens are requested, to report to the forest commission immediately all cases of damage or injury to forest trees arising from a violation of these rules which may come to their knowledge.

Beneath these 'rules, and on the same placard are printed sections 10, 30 and 31 of the Forest Commission Act, and the following:—

Section 74 (as amended by chapter 256, Laws of 1889):—

Every person who shall trespass on any lands belonging to the people in this State, or on any Indian lands, or who shall trespass upon any other lands within the bounds of the forest preserve, or which may hereafter be included in the forest preserve, by cutting or carrying away timber growing thereon, shall forfeit and pay the sum of twenty-five dollars for every tree that shall be cut or carried away by him or under his direction.

FROM THE PENAL CODE.

414. A person who, having been lawfully ordered to repair to the place of fire in the woods and assist in extinguishing it, omits, without lawful excuse, to comply with the order, is guilty of a misdemeanor, and shall forfeit the sum of fifty dollars and be liable to a fine and imprisonment.

640. *Malicious injury and destruction of property.*—A person who wilfully cuts down, destroys or injures any wood or timber standing or growing or which has been cut down and is on lands of another or of the people of the State; or cuts down, girdles, or otherwise, injures a fruit, shade, or ornamental tree, standing on the lands of another or of the people of the State, is punishable by imprisonment not exceeding three months, or a fine not exceeding two hundred and fifty dollars, or both.

REASONS FOR ESTABLISHING ADIRONDACK PARK.

The following account of the reasons which induced the people and Legislature of the State of New York to undertake the work of establishing a great Park among the Adirondack mountains. and of the means which have so far been adopted, is condensed from the reports of the New York Forest Commission for 1890 and 1891 :—

In pursuance of a resolution of the Senate adopted January 20th, 1890, the committee on finance recommended the adoption of the following concurrent resolutions :—

Resolved, (if the Assembly concur) "That the forest commission be and hereby is, directed to take into consideration the message of the Governor, addressed to the legislature, calling attention to the subject of establishing a State park in and about the headwaters of the rivers having their sources in the Adirondack wilderness, and after thoroughly investigating the possibilities of such an undertaking, to report to the legislature its conclusion thereon and its recommendations as to the most effective methods to be employed to accomplish that end—either by bill, or otherwise—together with any pertinent facts within the knowledge of the commission relating to the general subject of forest preservation or extension, and further to report the number of acres or square miles of land essential to fulfilling the requirements of a suitable reservation or park, and the probable cost thereof, and to report also in regard to the other subjects referred to in said message of the Governor."

The concurrent resolution reported by the committee was passed by the Senate, March 5th, 1890, and by the Assembly, April 4th, 1890.

The commission of State Parks made a report in 1873, from which the following is an extract, " It has been shown that the forests protect and preserve the springs and streams among them; and when we find individuals managing their property in a reckless and selfish manner, without regard to the vested rights of others, it becomes the duty of the State to interfere and to provide a remedy. Here, by ruthless destruction of the forest, thoughtless men are depriving the country of a water supply which has belonged to it from time immemorial and the public interests demand legislative protection. The canal interests of the State are very great, and are already suffering from this wrong. The water supply of the Champlain canal is entirely obtained from the streams of this wilderness, and the Erie canal, from Rome to Albany, is almost entirely supplied from the same watershed. In the Hudson, near Troy and Albany, navigation at midsummer has become very difficult. The mill-owners at Glen's Falls and at other points find that their water supply is failing; and the farming lands throughout the State suffer from storms and droughts of increasing severity. It is of no consequence, that, through ignorance of the natural law governing rain and rivers, men have hitherto permitted without protest, the injustice which they felt but the cause of which they did not understand. The State must apply the remedy, and to protect their interests preserve the forest. The great Adirondack forest has a powerful influence on the general climatology of the State; upon the rainfall, winds and temperature, moderating storms, and equalizing throughout the year the amount of moisture carried by the atmosphere, controlling and in a measure subduing the powerful northerly winds, modifying their coldness and equalizing the temperature of the whole State."

The commissioners say, in concluding their report:—

" There is no need for any expenditures save possibly in the improvement of a few of the principal roads leading to the settlements. The forest is in itself a natural park, and it would be improper to think of enclosing and fencing it for it should be a common unto the people of the State. The question before your Commission is one of great importance to the State, and requires their further consideration. For the present we deem it advisable and recommend that the wild lands now owned and held by the State be retained until that question be decided."

Under the resolution passed by the Legislature of 1890, the forest commission found itself confronted by four main topics for its examination and decision.

First—Is the establishment of a State park in the Adirondack wilderness feasible?

Second—If it be, what shall be the area of the park?

Third—What lands shall be embraced within the park?

Fourth—How shall the lands that ought to be included within the park, and not owned by the State, be acquired?

A survey of the actual condition of affairs showed that the region popularly known as the Adirondacks is diversely estimated at widely different areas. Taking the most reliable data, the gross area of the Adirondack wilderness proper is shown now to be about 5,600 square miles, or 3,600,000 acres This includes the area of water (lakes, ponds and rivers), overflowed lands, clearings, farms and some villages, or settlements. This area is by no means a compact tract, but lies in widely separated parcels, varying in extent from one-quarter of an acre to 70,000 acres, interspersed among tracts held by individuals and corporations (mostly lumbermen and paper manufacturers), an unknown number of clubs or other associations and persons who have established private preserves and parks in the woods for purposes of pleasure and recreation and of hotel sites. How to consolidate the lands necessary to form a park became a serious question. Only two methods were suggested. One that the State should condemn and take the land necessary to form the park by the exercise of its right of eminent domain; the other that the State should acquire the land by purchase.

A leading representative of the lumber interest, who has made the subject of forestry and timber supply and consumption a matter of study both in this country and in Europe, said to the commissioners: "It seems to me there is a practical side to this subject that should have some consideration. I have tried this summer to make a sufficient study of the way they have been managing forests in general. I have studied the German system, and have become very much interested in it. If it is possible to raise the money there is no question but what it would be the better way if the State could buy the lands outright and own them, but it is a great question whether the sum of money required for this could be raised at once, and my hope is that some plan will be devised by which the land can be bought at a low price, say about three dollars an acre, and allow the spruce to be taken out down to twelve or fourteen inches. If that was done the State would acquire the lands at a comparatively low price. The spruce below the twelve inch limit in a few years would grow up so that second cutting could be done, and within fifteen or twenty-five years I think the whole lands purchased by the State could be paid for with interest and cost the State nothing. The German forests that I have visited, paid last year some six or seven million dollars to the German government. I see no reason why our lands cannot be treated in the same way. The trouble with us is we are always afraid of planning out for anything seventy-five years ahead. If you went to the legislature and said you wanted to spend several millions of dollars, and political questions came up, you would kill the whole thing. What I am afraid of is that the railways are bound to come in. The land is not owned by the State. Suppose you own the land, you are able to prevent any roads going through. It might be that if they went through they would increase the value of your land. If the land is not owned by the State the railroad cannot be stopped. There is no question but that railroading does hurt the Adirondacks. The only advantage of their going through is that it would bring the whole of the Adirondacks closer together. This would be for the advantage of the rich and the poor alike. It would give the poor clerks and poor people an opportunity to go up there and live at low rates. As it is now, they have got to have guides and go to a large expense. My hope is that something will be done in the way of buying the land at low prices. There are 25,000 people in the Adirondacks who are depending for their living upon the different operations in lumber and who are in favor of this park, and will work in the interest of this scheme if they think they will not lose their livelihood."

The commissioners go on to say that a misunderstanding has prevailed to some extent with regard to the attitude of forestry to the lumber interests of private owners. It is, however, generally understood now that the true interests of the lumbermen are not incompatible with forest preservation, and it has been declared to be one of the objects of the forestry movement in this country "to harmonize the interests of the lumberman and the forester, and to devise for the lumbering interest such protection as is not given at the cost of the forests." Forestry is not opposed to having trees cut down in the proper way. They must be cut to supply the world with timber. Civilization could hardly exist without it. It is from trees, and from trees only, that our needs for wood are supplied through the timber dealer and the lumberman.

The phrase, "lumbered land," is a somewhat misleading one. It does not imply that such land is cleared, devastated, or even stripped of timber. The term is used locally to describe lands from which the "soft wood" (spruce, hemlock, pine and tamarac, one or all) has been taken, leaving the hard wood (birch, cherry, maple, beech, etc.) standing.

Generally, there is so much of this hardwood left on a "lumbered" tract that an inexperienced eye glancing over it would scarcely detect the work of the axe. The woodman expects to see such land covered with spruce again, large enough to be marketable, in about fifteen years.

Even the denuded forest lands to which reference is made are usually sufficiently well covered with a light growth of poplar and shrubs of various kinds to play a serviceable part in the purposes of forestry, and they will largely, if preserved from fire, be reclaimed by the forest.

The questions as to the area of the park and the lands that ought to be embraced within its limits have received much consideration from the commission, and as in the case of other questions a decided difference of opinion has been found to exist upon them.

The suggestion of the governor in his messsage of January, 1890, was that the area of the park should be "from fifty to seventy miles square." The smaller area mentioned would contain 2,500 square miles, or 1,600,000 acres, and the larger area 4,900 square miles, or 3,136,000 acres. The minimum area for the park that was suggested to the commission was 1,600 square miles, or about 1,000,000 acres, while most have urged larger areas, ranging from 2,500,000 to 4,000,000 acres.

The objects to be gained by establishing the park are stated by its various advocates in varying language, although perhaps agreeing in substance. One of the purposes is alleged in general terms to be "the preservation of the forests." The benefits derivative from forest preservation are stated as the maintenance of our timber supplies, the conservation of the sources of our rivers by the protection of watersheds, the protection and preservation of fish and game, and the founding of a permanent public resort for those seeking pleasure and rest, and which shall also be a sanitarium for invalids.

The commissioners reported that the loss to the State of New York which would be entailed by the destruction of the Adirondack forest, taking into account the manufacturing and canal interests involved, could only be counted by millions of dollars, and this, without taking into consideration the loss to the health of the citizens by the removal of the most valuable of all sanitariums, and the destruction of the valuable game preserves of the Adirondacks, and they unanimously recommended that the legislature should enact the necessary and suitable laws for the establishment and management of a park in the Adirondack wilderness, for the reasons and upon the general basis set forth in their report.

THE ADIRONDACK PARK ACT.

The Act which was passed by the State Legislature, 20th May, 1892, respecting the Adirondack Park, is as follows:—

The people of the State of New York represented in Senate and Assembly do enact as follows:—

Section (1) There shall be a State park established within the Counties of Hamilton, Herkimer, St. Lawrence, Franklin, Essex and Warren, which shall be known as the Adirondack park, and which shall, subject to the provisions of this

Act, be forever reserved, maintained and cared for as ground open for the free use of all the people for their health or pleasure, and as forest lands necessary to the preservation of the headwaters of the chief rivers of the State, and a future timber supply.

(2) For this purpose the forest commission shall have power, as herein provided, to contract for the purchase of lands situated within the County of Hamilton, the Towns of Newcomb, Minerva, Schroon, North Hudson, Keene, North Elba, St. Armand and Wilmington, in the county of Essex; the Towns of Harrietstown, Santa Clara, Altamont, Waverly and Brighton, in the County of Franklin; the Town of Wilmurt, in the County of Herkimer; the Towns of Hopkinton, Colton, Clifton and Fine in the County of St. Lawrence; and the Towns of Johnsburgh, Stony Creek and Thurman, in the County of Warren.

(3) In any case where lands are situated within the towns specified in section two, the purchase of which lands will in the opinion of the forest commission, be advantageous to the State, but which cannot as shall appear to the satisfaction of the forest commission, be bought on advantageous terms unless subject to leases or restrictions, or to the right to remove certain timber as hereinafter mentioned, the forest commission may make a contract for the purchase of such lands, providing that the contract and the deed or deeds to be made in pursuance thereof, shall be subject to such leases, restrictions or right. But no lands shall be so purchased subject to any right to remove hard-wood timber, or any trees of soft-wood with a diameter of less than ten inches at the height of three feet from the ground, or subject to any rights, leases or restrictions, or the right to remove any timber after the period of ten years from the date of the conveyance.

(4) The forest commission shall have power, from time to time, due notice having been given, to contract to sell and convey any portion of the lands within so much of the forest preserve as is now, or hereafter may be situated within the Counties of Clinton, Fulton, Lewis, Oneida, Saratoga, Washington, St. Lawrence, Franklin (except the Town of Harrietstown), Herkimer (except the Town of Wilmurt), Essex (except the Towns of Newcomb and North Elba), the Town of Hope in the County of Hamilton, and the County of Warren (excepting, however, therefrom, all islands in Lake George, and all land upon the shore thereof), the ownership of which by the State is not, in the opinion of the forest commission, needed to promote the purpose sought by this Act, or by chapter two hundred and eighty-three of the laws of eighteen hundred and eighty-five. The proceeds of all such sales, as in this section provided, shall be paid to the treasurer of the State, and shall be held by him in a separate fund and as a special deposit, which shall at all times be available to the forest commission for the purpose of purchasing lands situated within the towns mentioned in section two of this Act at such price per acre as may be determined by the forest commission and approved by the commissioners of the land office as hereinafter provided.

(5) All conveyances of lands belonging to the State which are to be delivered in pursuance of any contract authorized by section four, shall be executed by the comptroller and may contain such restrictions, reservations or covenants as the forest commission shall deem to be promotive of the purposes sought by this Act, or by chapter two hundred and eighty-three, laws of eighteen hundred and eighty-five. No contract made in pursuance or under the authority of this Act shall take effect until the same shall have been approved by the commissioners of the land office, such approval to be appended to the copy of the resolution of the forest commission authorizing such contract and certified by the clerk of the commissioners of the land office.

(6) Every conveyance executed in pursuance of this Act shall be certified by the attorney-general to be in conformity with the contract, and shall otherwise be approved by him as to form before the acceptance or delivery thereof. Every conveyance to be received by the forest commission, and executed in pursuance or under the authority of this Act, shall be made to the people of the State of New York as grantee, and shall be recorded in the proper county or counties, and shall after such record, be delivered by the forest commission to the commissioners of the land office to be treated as part of their archives.

(7) Payment for the purchase of land authorized by this Act, shall be made upon the certificate of the forest commission and the audit of the comptroller from moneys appropriated by this Act for the purchase of land or from moneys received from the sale of lands as provided in section four. Such expenses as may be necessarily incurred by the forest commission in the preliminary examination of lands purchased or sold under the authority of this Act, or in the examination of title of lands purchased under this Act, and all other expenses incidental to the conveyances and purchases so made shall be paid by the forest commission from the appropriations made from time to time for the purpose of such purchases, or from the fund established from the proceeds of the sale of lands as provided in section four.

(8) All lands now owned, or which may hereafter be acquired by the State within the towns mentioned in section two of this Act (except such lands in border towns as may be sold in accordance with the provisions of section four), shall constitute the Adirondack park. The forest commission shall have the care, custody, control, and superintendence of the same, and shall have within the same and with reference thereto and every part thereof, and with reference to any acts committed thereon and persons committing the same, all the control, powers, duties, rights of action, and remedies now belonging or which shall hereafter belong to the forest commission or the commissioners of the land office, within, or with reference to, the forest preserve or any part thereof, or with reference to any acts committed therein, or persons committing the same. The forest commission shall have power to prescribe and to enforce ordinances or regulations for the government and care for the Adirondack park, not inconsistent with the laws of the State of New York, or for the licensing or regulation of guides or other persons who shall be usually engaged in business thereon; to lay out paths and roads in the manner prescribed by law; to appoint the superintendent, inspectors, foresters and all other officers or employees who are to be engaged in the care or administration of the park and to fix their compensation, the same to be payable, however, only out of the appropriations made from time to time for the expenses of the forest commission.

(9) The forest commission shall have power to lease from time to time, as it may determine, tracts of land within the limits of the Adirondack park, not exceeding five acres in any one parcel to any person for the erection of camps or cottages for the use and accommodation of campers, such leases to be general in form except as to the term and amount of rental, and the term not to exceed five years, and the leases to contain strict conditions as to the cutting and protection of timber, the prevention of fires, and a reservation of a right of passage over the same for travellers at all proper and reasonable times, and to contain a covenant on the part of the lessee or lessees to observe the ordinances or regulations of the forest commission, theretofore prescribed or thereafter to be prescribed, as the same may be from time to time. No exclusive fishing or hunting privilege shall be granted to any such lessees.

(10) Except as in this act otherwise provided, the Adirondack park shall for all purposes, be deemed a part of the forest preserve. All laws for the pro-

tection of the forest preserve shall be applicable to the Adirondack park, except as in this act otherwise provided; and the forest commission may conduct the same prosecutions and institute and maintain the same proceedings, which it is, or shall be entitled to conduct, institute or maintain with reference to any portion of the forest preserve : and all acts forbidden upon the forest preserve are, and shall be deemed forbidden within the Adirondack park except as herein otherwise provided ; and all violations of law upon the Adirondack park shall be subject to the same punishments and penalties as if such violation were committed upon any part of the forest preserve.

(11) The foresters and other employees of the forest commission shall, when so directed by the forest commission act as game and fish protectors ; and as such they shall have all powers within the Adirondack park which game and fish protectors have or shall have, under chapter five hundred and seventy-seven of the laws of eighteen hundred and eighty-eight, and any law hereafter to be enacted, and they shall from time to time make such report to the commissioners of fisheries as that board may require. Nothing in this act contained shall be construed to permit any violation within the Adirondack park of the game and fish laws of the State heretofore or hereafter to be enacted, or to restrict or alter as to such park any of the prohibitions or penalties prescribed or hereafter to be prescribed by such fish and game laws. It shall be the duty of the forest commission with the concurrence and approval of the commissioners of fisheries, to provide for the enforcement within the Adirondack park of such fish and game laws by such means as the forest commission shall deem wise, in addition to such other means as are or shall be provided by law.

(12) The forest commission shall include in its annual report an account of its proceeding with reference to the Adirondack park, and shall make such recommendations with reference thereto as it shall deem wise. The forest commission shall state also in its annual report the number of acres purchased and sold during the year under the provisions of this act, the locality of the same, the prices paid or received, and all other information of importance connected with such transfers ; and shall state the amount of money required in the next fiscal year for the purchase of lands and for the expenses of the park.

(13) Chapter four hundred and seventy-five of the laws of eighteen hundred and eighty-seven, and all acts and parts of acts inconsistent with this act are, so far as they are so inconsistent, hereby repealed.

(14) This act shall take effect immediately.

PUBLIC UTILITY OF THE PARK.

The Commissioners in their report for 1891, go on to say :—

" At this late day, after all that has been written and said on the subject, we do not propose to discuss the necessity of forest preservation and the acquirement of the Adirondack wilderness by the people. Throughout the entire State there is a demand for this which is continuously urged by the newspapers, the local forestry associations, and by a general, widespread expression of public sentiment. Among all the demands and arguments in favour of this project, a dissenting voice has not yet been heard. If there have been such, they arose from questions of detail only.

" The original idea called for forest preservation, with reference only to protecting the head waters of our rivers, and providing a future economic and perpetual timber supply. But lately the acquisition of this territory has been urgently demanded by the public for the purposes of a health and pleasure resort, and the original movement has become largely subordinate to the latter one.

"It is immaterial whether it be called a park or a forest preserve, and its use as a pleasure or sanitary resort need not interfere with its management for forestry purposes. The friends of the forestry movement, therefore, view with pleasure the agitation in favour of the Adirondack park, and welcome the promoters of that enterprise as needed allies in the work of acquiring the necessary land.

"Section three of the Act provides for the purchase of land on which there is a growth of merchantable soft wood. Such lands are, for the most part, owned by lumbermen who will not—in fact, could not, part with such lands without abandoning their business. The State cannot acquire any tracts of this character except by the exercise of the right of eminent domain, an arbitrary measure which should not be resorted to until all other methods have failed. Moreover, there is no reason to believe that a bill authorizing the condemnation of the property of the lumbermen in northern New York could be passed. But the lumbermen have, so far as we can learn, expressed a willingness to turn over their lands to the State at a low price, provided they could have the privilege of removing the small proportion of trees comprising the merchantable soft wood. Now, the lumbermen will certainly remove such timber from their lands, sooner or later, and the forest commission see no possible way of preventing it. The only question is : Shall we secure these lands now, subject to the removal of the soft wood, or wait until it is removed and then attempt to buy the land ?

"After several years' observation and experience in this very matter, we believe the delay will only result in the State paying higher prices for these same lands, the soft wood having been removed in the meantime just the same. Six years ago prominent lumbermen called at the office of the forest commission and offered to organize a syndicate that would furnish the State half a million acres of Adirondack forest land for the nominal sum of one dollar (not one dollar per acre), provided that they could have the privilege of removing the soft wood and be relieved of the taxes. We are now willing to pay $750,000 for the same land which was offered to us then for one dollar, and pay it subject to the same conditions. Further delay in this matter will only result in the State paying higher prices and under more stringent conditions. It is the old story of the Sybilline books. There seems to be some misapprehension as to the result of the clause permitting the removal of the soft wood. The trees which would be removed under the sanction of section three of the proposed act would not exceed, on an average, eight trees to the acre. Their removal would not affect the general appearance of the forest, would not diminish the area of foliage, or lessen its value as a protection to the watershed of our rivers. The hard wood trees and young evergreens would still remain. There would be many lots on which scarcely a tree would have been removed under this clause ; although there might be, here and there, a few lots on which, by reason of the spruce growing in thick masses or so called "clumps" there would be a perceptible thinning. But even in the latter case the young trees would in a few years attain a growth which would cover all traces of previous operations, and under a properly conducted forestry management, furnish future revenues to the State.

"This commission does not consider it necessary to argue in favour of a policy which has already received the sanction of the best thought of the country. In 1873 a commission, headed by the late Governor Horatio Seymour, made a report strongly urging the reservation of the Adirondack wilderness.

"On January 22, 1890, Governor David B. Hill forwarded to the legislature a special report urging the establishment of an Adirondack park and the purchase of lands necessary to such purpose, his message outlining substantially the provisions adopted in the foregoing Act. In that message he urged that the limits, within which lands are to be obtained by the State for this purpose, should be

settled and defined ; and that all the lands out i e these limits would be subject
to sale. He recommended that the area f the state park should be from fifty
to seventy miles square, the mean area of 2,847,000 acres corresponding with the
amended boundaries now proposed by the committee. He also recommended that
the State give permission to persons desirous building summer camps upon
State lands ; and that small parcels should be let under suitable restrictions at
a moderate rental for such purposes, arguing that such occupants would have an
interest in preserving the forests in all their beauty, and would be the best of
firewardens or foresters; and that the wilderness would thus afford a summer
home to persons of moderate means as well as the wealthy.

" In the same year a committee of the State Senate reported that there could
be no dissent from any of these propositions ; and that 'it is admitted that the
creation of such a park would be of incalculable benefit to the State and the
whole country. The claims of the tourist and the summer resident for a great
public reservation, where the worker, the traveller and the lover of nature may
find rest, recreation and recuperation, great as they are, sink into insignificance
beside the acknowledged fact that the preservation of the Adirondack forest, which
can only be accomplished by the liberal action of the State, promptly taken, is a
sanitary necessity. Consideration for the invalid, who, according to eminent
authority, finds here a refuge of unsurpassed value, and regard for the public
health and protection of vast commercial and manufacturing industries depending
upon the Adirondack forests for even and seasonable distribution of the rainfalls
of that region, alike counsel action in this matter immediate liberal and comprehensive.'

" The report of this committee was signed by the Hons. George B. Sloan,
Charles T. Saxton and J. S. Fassett. Further than this, the various forestry
associations of the State of New York, and the boards of trade in various prominent cities, together with the entire press of the State, have unanimously and
persistently urged that the legislature no longer delay action in this matter, and
that their action, when taken, should be comprehensive and proportionate to the
magnitude of the interests involved.

" Section 5 provides for the purchase of lands in the Catskill Mountains. The
law of 1885, establishing the forest preserve, contemplated a reservation in the
Catskill as well as in the Adirondacks, and not without good reason.

" As one of the primary objects of forest preservation is the protection of the
watersheds of our chief rivers, the wooded slopes of the Hudson watershed
demand special consideration. The four counties of Greene, Ulster, Delaware and
Sullivan contain mountains whose forests protect the head waters of important
streams that flow to the Mohawk, the Hudson, and the Delaware. The Schoharie
creek, which takes its rise in the Catskill Mountains, is a large stream that flows
northward and joins the Mohawk River at Tribes Hill, its waters flowing thence
to the Hudson. This stream is also utilized as an important feeder to the Erie
Canal. The Esopus Creek also rises in the Catskill Mountains and flowing to the
north and east pours its waters into the Hudson at Saugerties. This stream is
valuable for its water power, which is used to advantage by the manufactories
situated near its mouth. The east and west branches of the Neversink, and the
east branch of the Delaware also rise here. The State has already about 50,000
acres of forest land in these four counties, the bulk of which is situated on and
near Slide Mountain, in Ulster County, the highest peak in the Catskills.

" But there are more important reasons for the establishment of a forest park
in the Catskills. This portion of the preserve is in close proximity to the great
cities of New York and Brooklyn, and the many cities along the Hudson. It is
easily accessible to three-fourths of the population of the State, and receives

16 (F.)

annually a far greater number of summer visitors than the famous Adirondack region. It is a favorite spot with the vast populations of New York and Brooklyn on account of its accessibility, cheap railroad fare, and desirable accommodations for people of moderate means.

" Still, aside from all such reasons, the acquisition of forest lands in the Catskill Mountains is necessary for forestry purposes, and for the preservation of an important watershed. '

FOREST FIRES.

Under this heading the Commissioners in their report for 1891 say :—

" Forest fires in this State during the past year have done but little damage as compared with the destructive conflagrations which raged within the wooded districts several years ago, prior to the organization of the present firewarden system. That the succeeding reports show a greater number of fires than last year, is due to the fact that extra effort has been made this season to secure reports from every town in the State. A large number of fires reported this year, it will be noticed, were checked soon after they started and before any serious damage was inflicted ; and the large number of such cases is encouraging evidence of the value and efficiency of the organization. In gathering these reports an effort was made to make the statistics on this matter perfect and complete. To this end a correspondence was maintained with the 900 firewardens, including the town supervisors who act as such within the towns outside the forest preserve, all of the woodlands thoughout the entire State, whether private or public, being under the charge of the forest commission in this respect. This correspondence was pushed until every town in the State had reported, the firewardens being directed to report the absence of fire as well as cases where they had occurred. It will be noted with interest that the principal causes of these fires were not the ones which had hitherto been claimed as such. In no case have they been reported as originating in lumbering operations, or in the " slash " left by log choppers. The most frequent source is found in petty farming operations, and the burning of fallows ; and, next the railroads and locomotives were a prolific cause. The campers and sportsmen seem to be responsible next, while the remaining instances were due to various and sometimes unknown causes.

" In cases where the fire was discovered as soon as started, and a posse of men was promptly warned out, there was no difficulty in extinguishing it ; but where, by reason of delay, the fire got well started, it was impossible to put it out, and the efforts of the firewarden and his men were confined to checking any further spread, and to watching it until the rain could accomplish the work. The unfailing regularity with which the larger fires were followed by rain was a noticeable and interesting fact. The system and the work done under it, is however, far from perfect as yet, but each year has brought with it better methods and a more effective organization. The commission feels encouraged, and believes that the present system can be perfected and the laws enforced so that extensive forest fires in the State of New York will be a rare occurrence.

" In one case a serious fire was started by an incendiary. This occurred May 14th, near Indian Lake, Hamilton County, and the person started his fire in the windfall of dead timber and tree tops left by the cyclone of 1888. As this fire was clearly of incendiary origin, the forest commission offered a reward of $300 for information which would lead to the conviction of the person or persons who set fire to this slash. The printed hand-bills containing the announcement of this reward were conspicuously posted and freely distributed throughout the

country; but the officers of the commission were unable to obtain any clue what
ever to the perpetrator of the crime.

"Another incendiary fire was started in Franklin County, near Loon Lake,
May 2nd, a full report of which, made by firewarden Chase, will be found on a
subsequent page.

"The miscreants who start these fires have every opportunity for accomplishing their work without fear of detection. They enter the forest alone and
unobserved, start a fire, and then, aided by their knowledge of the wilderness,
emerge at some point many miles distant. These incendiary fires are the
cowardly means resorted to in revenge for fancied wrongs; and it is an unpleasant
fact to contemplate that in certain localities the forest is at the mercy of any
lawless man, who, brooding over imaginary grievances, or inflamed by ignorant
passion or drink, seeks to gratify a revengeful spirit by this dire resort. One
need not travel far in the Adirondacks before the ear will catch the discontented
muttering and significant threats which leave no doubts as to the danger from
this source.

"From the table given it appears that of eighty-nine fires reported, the
causes were:—

From clearing land	22
From railroad locomotives	10
From railroad track hands	3
From hunters and fishermen	9
From mischievous boys	5
From incendiaries	4
From camp fires	3
From mosquito smudge	1
From tobacco smoking	3
From sparks from saw-mill	1
From causes not stated	14
From causes unknown	14
Total	89 "

SANITARY BENEFITS OF THE ADIRONDACKS.

The advantages of the Adirondack Park as a health resort are becoming
recognized. On this point the Commissioners in their report for 1891 have the
following:—

"Information regarding the sanitary benefits of the great forest is so largely
sought after that, at the risk of repetition, we quote from one of our previous
reports:—

"The sanitary value of our forests cannot be over-estimated. In addition to
their furnishing a summer resort for the over-crowded population of our towns
and cities, a place where rest, recuperation and vigour may be gained by our
highly nervous and over-worked people, the healthful and purifying influence of
coniferous forests has been thoroughly established.

"The testimony, based on personal, careful and scientific investigation of such
men as Dr. E. L. Trudeau, of Saranac, cannot be set aside. Himself an invalid
restored to health by forest life, he has devoted himself to the question of
environment, in its relation to tuberculosis, and by experiment on living
animals, has demonstrated the value of the terebinthine forests of the Adirondack
region as a factor in warding off pulmonary diseases. He says:—

Twenty-five per cent. of the patients sent to the Adirondacks suffering from incipient consumption come back cured, a proportion only surpassed by the State of Colorado. As a sanitarium for the State and City of New York alone, the value of this region is inestimable, and many professional men will be at a loss where to send their suffering patients who are unable to pay the expenses of a trip to Colorado or California, unless some steps be immediately taken to save to the State this heritage that should be preserved for the people.

"Dr. Alfred L. Loomis, of New York, (an eminent authority) has also given valuable scientific testimony to the value of evergreen forests as a therapeutic agent in lung affections. He writes:—

'Having long since been convinced by my observations that evergreen forests have a powerful purifying effect upon the surrounding atmosphere, and that it is rendered antiseptic by the chemical combinations which are constantly going on in them, I invite attention to some conditions which may explain their therapeutic power. Such ambiguous terms as 'balsamic influence,' 'health-giving emanations' and 'aromatized atmosphere,' must be regarded as empty phrases, and meaningless as scientific explanations. The clinical evidence, however, of the beneficial effects of pine forests, on phthisical subjects is unquestionable. The changes attributable to the persistent inhalation of air impregnated with the emanations of evergreen forests are such as to indicate that the atmosphere is not only aseptic, but antiseptic, made antiseptic by some element which is not alone fatal to germ life, but at the same time is stimulant and tonic to normal physiological processes within the lungs. We are led to the conclusion that this antiseptic element of evergreen forests, an element which is not found elsewhere is the product of the atmospheric oxidization of turpentine. It is evident that the local and constitutional effects of turpentine are those of a powerful germicide, as well as a stimulant. Its presence in the atmosphere of the pine forests cannot be questioned. Again, ozone is said to be present in excess in the air of evergreen forests, and the beneficial effects of such an air have been ascribed to this substance alone. But it seems evident that there is a close relation between an excess of ozone in the atmosphere and turpentine exalation.

"'Recent developments in the treatment of phthisis by gaseous injections, if they are found beneficial, are apparently due to the arrest of septic poisoning, and not to the destruction of the tubercle bacilli. It is my belief that the atmosphere of evergreen forests acts in a similar manner, and facts seem to prove that the antiseptic agent which so successfully arrests putrefactive processes, and septic poisoning, is the peroxide of hydrogen, formed by the atmospheric exudation of turpentine vapours. It is stated that wherever the pine, with its constant exhalation of turpentine vapour and its never failing foliage can be distributed in a proper proportion to the population, the atmosphere can be kept not only aseptic but antiseptic by nature's own processes, independent of other influences, than a certain amount of sunshine and moisture. It is not possible for every one to take his weak lungs to an antiseptic air, but it is possible to render the air of most localities antiseptic. I would therefore, impress on the public the importance of preserving our evergreen forests, and of cultivating about our homes evergreen trees.'"

THE CUTTING OF TIMBER FOR PULP WOOD.

The conditions which obtain in the area covered by the Adirondack Park of the State of New York, in so far as the forest itself is concerned, are analogous to those in the wooded parts of Ontario, and the following extracts from the report of the New York Forest Commission for 1891, relating to the wood-pulp industry, the tendency to a natural regeneration of the forest under favorable circumstances, etc., are interesting in view of what is going on in our own Province : -

"The manufacture of paper from wood is a comparatively new industry in this country. Its rapid development and the consequent increase in the consumption of valuable forest product demands the attention of everyone interested in American forestry. The introduction of wood pulp was regarded with satisfaction by students of the forestry question, because they saw in its use a market for certain small-sized timber, the sale of which is necessary to an economic forestry management. The successful pecuniary results obtained in the management of European forests are due largely to the fact that there is a market for everything that is left after cutting the large-sized timber ; and so the advent of the wood-pulp industry encouraged our forestry people to believe that operations in afforestation could now be carried on, as the sale of the thinnings would cover the expense.

"But the consumption of timber by the pulp mills has increased so rapidly as to endanger, instead of promote, the welfare of our forests. In the last eight years the amount of timber used for this purpose has increased 500 per cent. In the year just passed, 1891, the timber cut for wood pulp in the Great Forest of Northern New York, was equal to one-third the amount cut by the lumber men.

"It is not the increased consumption of this forest product that is so noticeable, but the fact that the entire amount consumed is taken from young trees Only a small amount of pulp timber can be gathered from the limbs and tops left by lumbering operations. Spruce and balsams furnish the main supply, and owing to their excurrent growth, only the tree trunks of these varieties are available.

"The pulp mills on the eastern side of the Great Forest use timber whose diameter runs from fourteen down to six inches. On the west side, the mills on the Black River use wood with a diameter as low as three inches. It will thus be seen that the introduction of wood-pulp, while it might be a valuable factor in economic forestry under proper management and restrictions, now indicates a speedy extinction of the conifers

"The mills on the Upper Hudson use poplar to an extent of twenty-five per cent., and spruce for the balance ; but the proportion of poplar used is growing less each year. The mills on the Black River use spruce, balsam, poplar and some small second-growth pine. Hemlock is used to some extent, when mixed with other kinds of wood. In making chemical fibre, however, the sulphite mill can use one-third hemlock. Tamarac is also used in small quantities, but it is a dark-coloured wood, and makes a dark, although strong paper. No cedar is used, nor any hardwood. On the Hudson the pulp timber is cut in the same length as logs, and is floated down the streams with the log drives. It is cut thirteen feet long, and is sent to the mill with the bark on. The most of the pulp timber for the Black River mills comes from St. Lawrence and Lewis counties, where it is cut into four foot lengths, measured, and sold by the cord, and shipped then over the Carthage and Adirondack Railroad. A large proportion of the pulp timber cut in Lewis and St. Lawrence Counties is peeled before it is taken from the forest, thereby obviating the use of barking machines at the mills. This supply of peeled timber is cut during the bark season, which lasts from May twentieth to August fifteenth, before or after which time the bark will not peel.

"In estimates of a general character, one cord of timber is said to make one ton of brown pulp, dry weight; but the actual results indicate that a cord of wood will produce only 1,800 pounds. In the chemical process, two cords of wood are consumed in making a ton of dry pulp, or chemical fibre, as it is called.

"Wood pulp, or cellulose, when first manufactured in this country, was used for paper only, and to a comparatively small extent. But the industry has developed with surprising rapidity, and now almost the entire bulk of newspaper stock is made from wood. Other uses for it have been discovered, and these new adaptations are multiplying each year. Under the name of indurated fibre, it is used to a large extent in making tubs, pails, barrels, kitchen-ware, coffins, carriage bodies, furniture, and building material. In this State, there are pulp mills at Oswego and Lockport which manufacture various wares of indurated fibre, but these mills do not obtain their timber supply from the Adirondack forest. Wood pulp is also used to some extent in the manufacture of gun-powder.

"Prof. B. E. Fernow, of the Forestry Bureau, at Washington, says in his last annual report:—'While the use of timber has been superseded in ship building, the latest torpedo ram of the Austrian navy received a protective armour of cellulose, and our own new vessels are to be similarly provided. While this armour is to render the effect of shots less disastrous by stopping up leaks, on the other hand, bullets for rifle use are made from paper pulp. Of food products, sugar (glucose), and alcohol can be derived from it, and materials resembling leather, cloth, and silk, have been successfully manufactured from it. An entire hotel has been lately built in Hamburg, Germany, of material of which pulp forms the basis, and it also forms the basis of a superior lime mortar, fire and water proof for covering and finishing walls.'

"The State of New York leads all other States in the manufacture of wood pulp, having seventy-five mills engaged in the industry, out of the 237 mills in the United States. Wisconsin comes next, with twenty-six mills; then comes Maine, with twenty-four; and then New Hampshire and Vermont, with eighteen each. Canada has also a very large production of wood pulp from its thirty-three mills, besides supplying large quantities of timber to mills situated in the United States.

"Of the seventy-five mills in the State of New York, sixty-four mills draw their entire supply from the Great Forest of Northern New York, or what is known as the Adirondack woods." (pp. 221-227).

SPONTANEOUS RENEWAL OF THE FOREST.

Where the efforts of nature are not thwarted by the ravages of fire, the tendency of the forest is to re-assert itself. The commissioners note that this is particularly the case where the previous cuttings have been conducted systematically, as for instance in the operations of charcoal-burners.

"After crossing the divide, the road runs for a few miles through State land some of which was burned over about twenty-five years ago, but which is now covered with a new growth of small trees, indicating that if this land is protected from further damage by fire, it will in a short time completely reforest itself." (p. 113).

"From the Summit to Cedar River the appearance of the lands along the road is disappointing. There is too much open country, and too ittle of the forest scenery which one expects to see. The open country is due to unsuccessful efforts at farming, and the dwarfed condition of the trees to the disastrous fires which in some places have occurred repeatedly; but from Cedar River to Blue Mountain Lake the road runs for ten miles through an unbroken forest which, to an unpractised eye, shows no diminution of its primeval beauty. Though the lumberman cut off, years ago, the merchantable spruce and pine, they took so few trees to the acre that little trace remains of their operations, especially as the smaller evergreens that were left are fast taking the place of those which were cut." (p. 115).

"Part of the forest along this road was cut clean about twenty-five years ago by the charcoal burners. and persons interested in forestry matters will note with pleasure as they ride by, that the land has completely reforested itself, there being little in the present growth which would indicate to a casual observer that it differed from the original forest on the surrounding lands. There are large areas in this country which have been cleared by charcoal burners, but which are rapidly recovering their growth, their present condition affording an encouraging outlook for the future welfare of these forests." (p. 162).

"Beyond and west of this place are some abandoned charcoal kilns, which are responsible for the peculiar condition of the tree growth on either side for quite a distance. The forest was cut here by the charcoal burners, and every tree, large and small, was removed; but the land is now covered with a promising second growth.

"Throughout the entire region the lands which were cleared for charcoal reforest themselves quicker, and with a much more valuable growth, than those which have been denuded by fire. Fortunately, the cutting for charcoal resembles somewhat the coppice system, which is one of the recognized methods of forest management; and so, most of the stumps left by the charcoal axemen have sprouted persistently, and yielded a second growth exhibiting most of the original varieties, so far as the deciduous trees are concerned. But where the forest has been destroyed by successive burnings the soil and seeds are too badly scorched to reproduce the former trees, and so the land reforests itself with an inferior crop of small poplars and bird cherries. In driving through Essex County a good opportunity is offered for studying some of these phases of natural reforestation." (pp. 169-170).

"Before reaching Aiden Lair the road enters township 26, and for six miles passes through large tracts of State land, the most of this township having been acquired through defaulted taxes in 1877. Some of this land has been denuded by fire, but it was not burned so badly but that it is now reforesting itself. Of these burned tracts there is one, in particular, which offers an interesting study in reforestation, owing to a peculiarity in the process. The thick growth of small poplar and cherry which sprung up immediately after the fire is rapidly dying off and disappearing; but it in turn is being succeeded by a promising vigorous growth of spruce and balsam.

"In this vicinity there is another piece of second growth which is composed largely of white pine. The trees are strong and thrifty, and in a few years will be large enough for manufacturing purposes. This second growth white pine is inferior to the original. The trees are smaller, very knotty, and yield but little clear stuff. Still the knots are small, red and sound, and the lumber meets with a ready sale. The time is near when the propagation of this variety of pine must enter largely into our forest management.

"The tract of second growth white pine, just referred to, lies along the road that runs from Minerva, through the Hoffman Township, within and near the south eastern boundary of the park. The land was once cleared and used for farming purposes · but it was abandoned and it is now overgrown with a thrifty crop of conifers. Had these lands been denuded by fire, instead of farming, the resulting crop of trees would have been of a different kind. Poplars and pin cherries would probably have appeared in that case. The fire burns into the ground, and destroys every hidden seed. Other seeds, distributed by well known agencies are subsequently deposited on its arid surface, of which the poplar and bird cherry are the only ones that will germinate in the then unfruitful soil." (pp. 197-8)

THE YELLOWSTONE NATIONAL PARK.

The superintendent of the far-famed Yellowstone National Park, in his report for the year 1890, makes the following remarks :—

"That the dedication in 1872 of the Yellowstone National Park as a heritage wonder for the enjoyment of the people was a wis and timely act few will now question.

"The Snake River Fork of the Columbia, and Green River Fork of the Colorado of the Gulf of California (Pacific waters), and nearly all the other great rivers of that portion of the continent, including the Jefferson, Madison, and Gallatin Forks, and the Yellowstone, Big Horn and other branches of the Missouri-Mississippi Atlantic waters, to a great extent radiate from hot springs or spouting geysers within or adjacent to the great National Park, situated mainly in North-western Wyoming territory and also embracing portions of Idaho and Montana. There can be no doubt that the modern sulphur basins, sauces hissing fumeroles, and spouting geysers are only dwindled remnants of the ancient volcanoes and vast and long continued eruptions of lava, which, in the region of the National Park, characterized the elevation of the Great Plains and Rocky Mountain Ranges from the oozy bed of a shallow ancient sea.

"It is also evident that at some subsequent but remote period of time many of these mountain slopes were at an elevation of from 6,000 to 10,000 feet, covered with dense forests of timber, in size fairly rivalling those upon the Pacific coast, and that by some oscillation in the elevation of these regions, by eruptions of hot ashes, mud, and slime, like those which covered Pompeii and Herculaneum, or other all-powerful and long recurring agencies, forests have been crushed or covered, often many hundred feet deep, by conglomerate breccias or other volcanic material.

"Here erosion of the elements, or the blast, or pick and shovel of the tourist, unearth this ancient timber, which is often petrified entire into a perfect tree or log of stone : other timbers, while retaining their form, into opal or chalcedony, with amethyst or other crystalized cavities, matchless in shape, color and beauty, which, for cabinet specimens are unequalled elsewhere in nature and unrivalled by art.

"Many hot springs and mineral streams now petrify timber or coat it with sparkling lime or silica, build geyser cones and many beautiful forms of crystallization, but they are all clearly distinct, and mainly much inferior to those of the closing eruptive period. This wonderful region is really less one large park than a group of smaller ones, partially or wholly isolated, upon both sides of the continental divide, much lower in the park than the nearly unbroken surrounding mountain ranges. Its average altitude probably exceeds that of Yellowstone Lake, which is some 8,000 feet or nearly a half mile higher than Mount Washington. Its few yawning, ever difficult, often impassable, canon-approaches along foaming torrents. The superstitious awe inspired by the hissing springs, sulphur basins, and spouting geysers ; and the infrequent visits of the surrounding Pagan Indians have combined to singularly delay the exploration of this truly mystic land.

"The animals of the region are the bison or mountain buffalo, elk, white-tailed deer, black-tailed deer, prong-horned antelope, big-horn sheep, bears, mountain lion or cougar, wolves, foxes, skunks, badger, rockdog, procupine, rabbits, rats, mice, burrowing moles, squirrels, chipmunks, beaver, otter, mink, muskrat, etc., birds, fishes, reptiles, and insects.

TIMBER OF THE PARK.

"Black or bastard fir is far the largest variety of timber now growing in the park, and usually found scattered through forests of smaller timber near the Mammoth Hot Springs, Tower Falls, Upper Yellowstone, and other elevated terraces. It is often found from three to five feet in diameter, and one hundred and fifty feet in height, and is not unlike the eastern hemlock in the irregular form of it branched top as well as the coarse grained, shaky, and inferior quality of its timber.

"Black spruce, growing on the moist, sheltered slopes of the mountains near the snow, though having a smaller trunk, is fully as tall as the black fir, and is a statelier tree and more valuable for timber or lumber.

"Red fir is the next in size (which nearly equals that of the Norway pine of Michigan), and the first in value of any tree in the park for hewn timber for building bridges, etc., for which purpose it is admirably adapted. It is abundant in all except the very elevated regions.

"White pine, rivalling in symmetrical beauty the white pine of the east, but much inferior in size, and somewhat in quality, is the prevailing timber of most of the elevated terrace groves, and occasionally of the narrow valleys and canon passes of the mountains. It grows very densely, often rendering travelling among it on horseback exceedingly difficult when standing and utterly impossible when burned and fallen, as it is over large areas of the park, proving one of the greatest impediments to exploring as well as to improvement by roads and bridle paths. It is the best material found in the park for lumber, shingles, small timber, rafters, fence poles, etc.

"Balsam fir, somewhat different from that of the Alleghanies, is abundant and very beautiful, singly or in dense groves or isolated clumps scattered over the grassy slopes, just below the mountain snow-fields.

"Cedar of a red or spotted variety, growing low, and very branched, but with timber valuable for fence posts, is abundant.

"Poplar or aspen is found in dense thickets among the sheltered foot-hills, dwarf-maple with leaves often scarlet with fungus, is sparingly found, and innumerable dense thickets of willow; the main value of all these last named varieties being for the food use of beaver or for bait."

ACT OF DEDICATION.

The Act by which the Yellowstone Park was dedicated or set apart is as follows:—

An act to set apart a certain tract of land lying near the head waters of the Yellowstone River as a public park.

Be it enacted by the Senate and House of Representatives of the United States of America in Congress assembled, that the tract of land in the Territories of Montana and Wyoming lying near the head waters of the Yellowstone River, and described as follows, to wit:—Commencing at the junction of Gardiner's River with the Yellowstone River and running east to the meridian passing ten miles to the eastward of the most eastern point of Yellowstone Lake, thence south along the said meridian to the parallel of latitude passing ten miles south of the most southern point of Yellowstone Lake; thence west along said parallel to the meridian passing fifteen miles west of the most western point of Madison

Lake; thence north along said meridian to the latitude of the junction of the Yellowstone and Gardiner's Rivers; thence east to the place of beginning, is hereby reserved and withdrawn from settlement, occupancy, or sale under the laws of the United States, and dedicated and set apart as a public park or pleasure ground for the benefit and enjoyment of the people; and all persons who shall locate, settle, upon, or occupy the same or any part thereof, except as hereinafter provided, shall be considered trespassers and removed therefrom.

Sec. 2. That said public park shall be under the exclusive control of the Secretary of the Interior, whose duty it shall be, as soon as practicable, to make and publish such rules and regulations as he may deem necessary or proper for the care and management of the same. Such regulations shall provide for the preservation from injury or spoliation of all timber, mineral deposits, natural curiosities, or wonders within said park, and their retention in their natural condition.

The Secretary may, in his discretion grant leases for building purposes, for terms not exceeding ten years, of small parcels of ground, at such places in said park as shall require the erection of buildings for the accommodation of visitors; all of the proceeds of said leases, and all other revenues that may be derived from any source connected with said park, to be expended under his direction in the management of the same, and the construction of roads and bridle-paths therein. He shall provide against the wanton destruction of the fish and game found within said park and against their capture or destruction for the purpose of merchandise or profit. He shall also cause all persons trespassing upon the same after the passage of this Act to be removed therefrom, and generally shall be authorized to take all such measures as shall be necessary or proper to fully carry out the objects and purposes of this Act.

Approved, March 1st, 1872.

NOTE.—The boundaries of the park had not then been surveyed, but they are mainly crests of snow-capped basaltic mountains encircling the wonderland of cataracts, canons, fire-hole basins, geysers, salses, fumeroles, etc., unique and matchless, with an entire area from fifty to seventy-five miles square.

RULES AND REGULATIONS.

The rules and regulations governing the conduct of visitors to the Yellowstone Park and the care of same generally, are the following:—

1st. All hunting, fishing, or trapping within the limits of the park, except for purposes of recreation, or to supply food for visitors or actual residents, is strictly prohibited; and no sales of fish or game taken within the park shall be made outside of its boundaries.

2nd. Persons residing within the park, or visiting it for any purpose whatever, are required under severe penalties to extinguish all fires, which it may be necessary to make, before leaving them. No fires must be made within the park except for necessary purposes.

3rd. No timber must be cut in the park without a written permit from the superintendent.

4th. Breaking the siliceous or calcareous borders or deposits surrounding or in the vicinity of the springs or geysers for any purpose, and all removal, carrying away, or sale of specimens found within the park, without the consent of the superintendent, is strictly prohibited.

5th. No person will be permitted to reside permanently within the limits of the park without permission from the Department of the Interior, and any person now living within the park shall vacate the premises occupied by him within thirty days after having been served with a written notice so to do by the superintendent or his deputy, said notice to be served upon him in person or left at his place of residence.

NOTE.—These rules and regulations are those adopted by the Hon. C. Delano, Secretary of the Interior, at the dedication of the park.

Additional rules subsequently issued are :

(1) The cutting or spoilation of timber within the park is strictly forbidden by law. Also the removing of mineral deposits, natural curiosities or wonders, or the displacement of the same from their natural condition.

(2) Permission to use the necessary timber for purposes of fuel and such temporary buildings as may be required for shelter and like uses, and for the collection of such specimens of natural curiosities as can be removed without injury to the natural features or beauty of the grounds, must be obtained from the superintendent, and must be subject at all times to his supervision and control.

(3) Fires shall only be kindled when actually necessary, and shall be immediately extinguished when no longer required. Under no circumstances must they be left burning when the place where they have been kindled shall be vacated by the party requiring their use.

(4) Hunting, trapping, and fishing, except for purposes of procuring food for visitors or actual residents are prohibited by law, and no sales of game or fish taken inside the park shall be made for purposes of profit within its boundaries or elsewhere.

(5) No person will be permitted to reside permanently within the park without permission from the Department of the Interior ; and any person residing therein except under lease, as provided in section 2,475 of the Revised Statutes, shall vacate the premises within thirty days after being notified in writing so to do by the person in charge ; notice to be served upon him in person or left at his place of residence.

(6) *The sale of intoxicating liquors is strictly prohibited.*

(7) All persons trespassing within the domain of said park, or violating any the foregoing rules, will be summarily removed therefrom by the superintendent and his authorized employees who are, by direction of the Secretary of the Interior, specially designated to carry into effect, all necessary regulations for the protection and preservation of the park, as required by the Statute ; which expressly provides that the same " shall be under the exclusive control of the Secretary of the Interior, whose duty it shall be to make and publish such rules and regulations as he shall deem necessary or proper," and who, " generally, shall be authorized to take all such measures as shall be necessary or proper to fully carry out the the objects and purposes of this Act."

Resistance to the authority of the superintendent or repetition of any offence against the foregoing regulations, shall subject the outfits of such offenders and all prohibited articles to seizure, at the discretion of the superintendent or his assistant in charge.

Approved,

S. J. KIRKWOOD,
Secretary.

P. W. NORRIS,
Superintendent.

AN ACT TO SET APART A CERTAIN TRACT OF LAND IN THE STATE OF CALIFORNIA AS A PUBLIC PARK

Whereas the rapid destruction of timber and ornamental trees in various parts of the United States, some of which trees are the wonders of the world on account of their size and the limited number growing, makes it a matter of importance that at least some of said forests should be preserved. Therefore, be it enacted by the Senate and House of Representatives of the United States of America in Congress assembled, that the tract of land in the State of California known and described as township number eighteen south, of range numbered thirty east; also township eighteen, south range thirty-one east, and sections thirty-one, thirty-two, thirty-three, and thirty-four, township seventeen, south range thirty east, all east of Mount Diablo meridian, is hereby reserved and withdrawn from settlement, occupancy, or sale under the laws of the United States and dedicated and set apart as a public park or pleasure ground, for the benefit and enjoyment of the people : and all persons who shall locate or settle upon or occupy the same or any part thereof, except as hereinafter provided, shall be considered trespassers and removed therefrom.

Section 2. That said public park shall be under the exclusive control of the Secretary of the Interior, whose duty it shall be, as soon as practicable, to make and publish such rules and regulations as he may deem necessary or proper for the care and management of the same. Such regulations shall provide for the preservation from injury of all timber, mineral deposits, natural curiosities, or wonders within said park, and their retention in their natural condition. The Secretary may, in his discretion, grant leases for building purposes for terms not exceeding ten years of small parcels of ground not exceeding five acres at such places in said park as shall require the erection of buildings for the accommodation of visitors ; all of the proceeds of said leases and other revenues that may be derived from any source connected with the said park to be expended under his direction in the management of the same and the construction of roads and paths therein. He shall provide against the wanton destruction of the fish and game found within said park, and against their capture or destruction for the purposes of merchandise or profit. He shall also cause all persons trespassing upon the same after the passage of this Act, to be removed therefrom and, generally shall be authorized to take all such measures as shall be necessary or proper to fully carry out the objects and purposes of this Act.

Passed the House of Representatives August 23rd 1890 : approved September 25th, 1890.

ANCIENT FOREST LAWS.

The evolution of law in its relation to forests, game, etc., from the earliest days to the present time, presents an interesting field of study, but one which cannot be elucidated here. The following notes have been compiled from Manwood's work and other sources, and may serve to throw some light on the reasons which at various times have moved men to restrain the cutting of timber, as well as on the means which they have from time to time adopted with the view of providing a sufficient store of what in all ages has been regarded as a prime necessity, and which was never more so than now :—

By a law of King Ina it was enacted, that if anyone set fire to a wood, he should be punished, besides paying a fine of three pounds (an immense sum in those days) ; and for those who clandestinely cut, of which the very sound of the axe was to be sufficient conviction, for every tree they were to be mulcted thirty shillings. For a tree so felled, under the shadow of which thirty hogs could stand the offender was to be mulcted three pounds.

It was a law at Frankfort that every young farmer must produce a certificate of his having planted a certain number of trees (probably in proportion to his circumstances in life) before he was allowed to marry. In the Duke of Luxemburg's dominions, no farmer was permitted to fell a tree, unless he could make it appear, that he had planted another. Louis the XIV. of France would permit no oak trees to be cut, to whomsoever they might belong, till his surveying officer had marked them out ; nor could they be felled beyond such a circuit as was sufficiently fenced in by him who bought them ; and then no cattle were allowed to be put in till the seedlings which sprung out of the ground, were perfectly out of danger.

In Saxon times, all beasts and birds that were wild by nature, were wholly the property of the king on whosever land or grounds they were found, whether any part of the realm, as well those that were out of the forests, chases, and warrens, as those that remained within any of them ; so that it was not lawful for any man to kill, take, or hunt within his own ground ; and if anyone did so, he was liable to be punished for the same. This law continued till Canute the Dane came to the English crown, who it appears appointed certain forests and chases and fixed their limits the first year of his reign. For the preservation of his own forests he made particular laws at Winchester, from which the following extracts are translated :—

(1) " Let there be then four men of the higher class who shall have the right, according to the custom which the English call *pegened*, followed in each Province of my kingdom, of distributing justice and of inflicting punishment, and of all matters concerning the forest, before all my people, whether English or Danes, throughout all the kingdoms of England ; which four we order to be called *primarii foresta*, chiefs (or earls) of the forest.

(2) Let there be under each of these, four of the middling class of men (which the English call *lespegend*, but the Danes *yoong men*, and which would now be called yeomen, or perhaps esquires) who shall undertake the care and custody as well of vert as of venison.

(3) In administering justice, these (*yoong men*) shall not interfere in the least ; and such middling persons, after having had the care of the wild animals, shall be held always as gentlemen, which the Danes call *ealdermen*.

(4) Again, under each of these, let there be two of the lower class of men which the English call *tinemen ;* (or in modern phrase, grooms): these shall take the night charge of vert and venison, and do the servile work.

(5) If anyone of this lower class shall be a slave, so soon as he is placed in our forest, let him be free, and we therefore discharge him from bondage.

(6, 7, 8) Relate to outfit.

(9) Relates to exemption from all summons and popular pleas (*hundred laghe*) hundred courts, and from all summons to any other court, except that of the forest.

(10) Let the causes of the middling and lower officers, and the correction of them, as well civil as criminal, be judged and decided by the provident reason and wisdom of the first class ; but the enormities of the first class, if any should happen, (lest any crime should go unpunished), we will punish ourselves in our royal anger.

(28) Let no one cut any of our wood, or underwood, without leave of the chiefs of the forest; which if anyone do, he shall be adjudged guilty of an infringement of the royal chase.

(29) But if anyone shall cut down an oak (*ilicen*) or any tree that furnishes food for the beasts of the forest, beside infringement of the royal chase, he shall pay to the king twenty shillings.

(30) I will that every free man shall have venison or vert at pleasure in his open grounds, (plana) on his own lands, but without chase (or the right of punishing intruders); and let all avoid mine (venison or vert), wherever I think proper to have it.

(31 to 34) concern dogs and mad dogs. The term forest law is to be understood as applying to large tracts of enclosed land, where deer used to be kept. As the deer in those forests, and consequently the right of hunting them, were deemed to belong to the Crown, the forest themselves were brought under the same class of laws without reference to the deer ; and they remained in this state, though the trees should fail the laws being executed by the crown with the right of forestage, and all the privilege of royal forests."

The laws of Canute were afterwards confirmed by divers succeeding kings, though in practice they generally appear to have been little if anything more than the will of the Crown, and were so administered until the barons and others encamped in hostile array on Runningmede from Monday the fifteenth to Friday the nineteenth of June, 1215 ; during which time they were actively employed in rough hewing the broad basis on which the bulwarks of our liberty are built, by forming the Magna Charta with King John.

The Magna Charta of 9th Henry III., Chap. 21, was the *Charta Foresta* of February 10, 1225.

Many liberties were then granted, and customs defined, but the restriction on cutting of wood appears to have been considerably felt.

The 13th of Edward the III., Chap. 1 and 2, gave considerable liberty for cutting and carrying wood, but it was to be done within view of the keepers of the forest.

In the 17th and 25th of Henry VIII., there are several acts respecting the forests, but they are principally modifications of former acts.

In the 35th of this rei n was passed an Act for the preservation of wood, but principally respecting coal and billet wood. In Chap. 17, an Act for the preservation of timber, we find : " The king, our sovereign, perceiving and right well knowing the great decay of timber and wood universally within the realm of England, and that, unless a speedy remedy in that behalf be provided, there is a great and manifest likelihood of scarcity and lack, as well for building houses and ships, as for firewood; it is enacted, that in copse of underwood felled at twenty-four years' growth there shall be left twelve standrells or store oaks, on each acre, or in default of oaks, so many elm, ash, or beech, etc., and that they be of such as are likely trees for timber, and such as have been left at former fellings, if there have been any left before; under pain of forfeiting 3s. 4d., for every such standard not left, one half to the crown, and the other to the party who may inform and may choose to sue for it in any court of record, which might be done as in an action for debt. When cut under fourteen years' growth, the ground shall be enclosed or protected for four years, by the proprietor or the lawful possessor of the wood under pain for not enclosing for every rood so left unenclosed 3s. 4d., for every month it may remain so unenclosed. No calves are to be put in for two years after felling, and no other cattle for four years. When cut from fourteen to twenty-four years of age to be six years enclosed under the same penalty after twenty-four years twelve trees to be left under penalty of 6s. 8d., each tree, the moiety to the Crown, and the informer may recover as before. The ground to be kept enclosed for seven years under the penalty of 3s. 4d. per rood per month as before." In the County of Cornwall, within two miles of the sea, trees might be felled when dead on the top.

No wood containing two or more acres, at the distance of two furlongs from the house of the owner was to be cut down under the pain of forfeiture of ten pounds for every acre of woodland so destroyed. Woods felled under fourteen years were afterwards not to have colts or calves put into them till eight years after cutting and enclosing. Most of these acts of Henry, etc., were only temporary till the 13th of Elizabeth, Chap. 25, when the time of protection was enlarged and the whole made permanent. By the 7th of Edward VI., Chap. 7, th. Act of the 35th of Henry VIII., Chap. 3, was confirmed, and a little modified.

It was then enacted, that every taleshide (bundle of cleft wood) be four feet long beside the carfe; and if named one, to be marked one, and to be sixteen inches circumference within a foot of the middle; if two marked two and twenty-three inches girt; if three, marked as such, and to be twenty-eight inches girt; if four, to girt thirty-three inches; if five, to girt thirty-eight inches; and so on in proportion. Billet wood was to be three feet four inches in length, the single one, to be seventeen inches and a half in girt, and every billet of one cast as they term the mark, to be ten inches about; and of two cast, to be fourteen inches girt, and to be marked within six inches of the middle, unless for the private use of the owner. Every bound fagot should be three feet long, and the band twenty-four inches in circumference beside the knot. This Act was principally for London, but the 43rd of Elizabeth, Chap. 14, rendered the statute more general; and ordered that the fagots should be every stick three feet in length except one to harden and wedge the binding of it. This was to prevent the abuse, then much practised, of filling the middle with short sticks. These Acts were confirmed by the 9th of Anne, Chap. 15, and the 10th of the same reign, Chap. 6, directs that the assize of billet shall not extend to beech, but that these shall not be sold in London or Westminster, unless the vendor make them of the same size as required by the Statutes for other wood. Chap. 17 of the 7th of Edward VI. is an Act for preventing unlawful hunting in parks, places, forests, etc.; and confirms the 38th of Henry VIII.

The 2nd and 3rd of Philip and Mary, Chap. 2, confirms that of Henry 7th and of the 20th of Henry VIII.; and in the 27th of Elizabeth, there is another Act to the same effect, nearly as that of Henry VIII., which was then made permanent; and to render it still more complete and effectual in promoting improvement, it further enacts, that timber of twenty-two years' growth shall be exempted from tithes. By the 1st of Elizabeth, timber shall not be felled for iron-workers of the breadth of one foot at the stub, and growing within fourteen miles of the sea or of the River Thames, Severn, Wye, Humber, Dee, Tyne, Tees, Trent, or any other navigable river or creek, under pain of forfeiture of forty shillings for every tree, one moiety to the Crown and the other to the informer, recoverable as before.

Second of Elizabeth, Chap., 19, is an Act for the preservation of timber in the wolds of Kent, Surrey and Sussex.

By the 43rd of Elizabeth, Chap., 7, it is enacted that if any idle person cut or spoil any wood or underwood, pales, or trees standing, and be convicted by the oath of one or more witnesses, if they cannot pay the satisfaction required, they shall be whipped. Receivers of wood so cut, knowing it to be so, to incur the same punishment.

The 2nd. of James I., Chap., 22, is an Act respecting bark, as it relates to tanners, curriers, shoemakers, and others concerned in leather. By Sec. 19, it is enacted, that no person shall contract for oak bark to sell again. By Sec. 20, that no person shall fell or cause to be felled any oak tree meet to be barked, where the bark is worth two shillings a cartload over and above the charges of barking and peeling, timber to be employed in building and repairing houses and mills excepted, but between the first day of April and the last day of June, upon pain of forfeiture of every such oak tree, or double the value thereof. And by Sec. 21, for the better preservation of timber, (which by the takers is spoiled through the desire of gain from the top and lop, or bark of timber trees,) it is therefore enacted, that no taker, purveyor, etc., or their deputies, shall fell for the use of the Crown any oak trees meet to be barked, but in the barking season except for the purposes before mentioned; or take or receive any profit, gain, or commodity, by any top, or lop, or bark, of any tree, to be taken or cut out of the barking season; and then only those for the king's house or ships, under pain of forfeiture to the party aggrieved (or on whose ground the tree may be cut) for every tree so felled forty shillings; and it shall be lawful for every party, of whom such tree shall be taken to retain all the bark, top and lop of the whole of such tree, notwithstanding any commission or other matter.

The 15th of Charles II., Chap. 2, is an Act to render the 43rd of Elizabeth more effective ;' and it enacts further punishment, on account that the destruction of wood tends to destroy the Commonwealth. It is therein declared that the officers of justice may apprehend even on suspicion of having carried, or in any way conveyed any burden or bundle of wood of any kind, underwood, poles, young trees, bark, or bast of any tree, gate, stile, post, rail or hedge-row wood, broom or firs, etc.

Chap. 3, of the 19th of Charles II., is an Act for the increase and preservation of timber in the forest of Dean. Eleven thousand acres are directed to be enclosed. Commissioners may sell decayed trees, to make good and maintain the said enclosures. When and how much shall be laid open, and by what authority as much shall be enclosed and has been opened, is declared. Wood fit for sale must be viewed and marked by the justices. Cutting wood contrary to this Act subjects the party offending to the penalties mentioned in former Acts. The enclosed land to be all re-afforested. All estates made out of it to any person whatever to be null and void. The king may retain game of deer, but not above eight hundred.

Proviso, for owners, tenants, and occupiers :—Former offences remitted ; pannage shall be re-enjoyed after Michaelmas, 1687 ; and when and in what manner all privileges to be enjoyed. Letters patent for certain woods and iron works saved. Coal mines and grindstone quarries may be leased.

The 9th. and 10th of William III., Chap. 36, is an Act for the preservation of wood in the New Forest, in the county of Southampton. Two hundred acres, part of this forest, to be enclosed for the growth of timber, after being set out by commissioners; two hundred acres more to be enclosed yearly for twenty years: and to remain in possession of the Crown for ever. Wood is not to be cut without sufficient authority. No coppice wood to be cut. Enclosures not to be ploughed or sown. The foresters to be fined, if they browse or lop any oak or beech tree in the forest. Charcoal not to be made within one thousand paces of the enclosure. Persons breaking down fences may be committed as rogues and vagabonds.

Ninth of Anne, Chap. 17, is for the preservation of white and other pine trees growing in Her Majesty's colonies of New Hampshire, Massachusetts Bay, and Province of Maine, Rhode Island, Providence plantation, and the new Narragansett country or king's province, and Connecticut in New England, New York, and New Jersey. No person within the said colonies shall presume to cut, sell or destroy, white or any other sort of pine tree fit for masts, not being the property of any private person, such trees being the growth of twenty-four inches and upwards at twelve inches from the ground, without the royal license for so doing, under the pain of forfeiting £100 for every such offence one moiety to the Crown and the other to the informer, who may recover the same in any court of record. The surveyor-general to mark the trees to be cut with the broad arrow; but no other person than he or his deputy to make any mark under the penalty of £5.

In the 12th of Anne, we find, an Act Chap. 9, for encouraging the importation of naval stores from America and Scotland for eleven years, and thence to the end of the next session.

Sec. 26 observes, "Whereas there are in several parts of North Britain, called Scotland, pine and fir trees fit for masts, and for making pitch, tar, resin, and other naval stores, but the land and wood which may yield such naval stores are mostly in parts mountainous and remote from navigable rivers, therefore for the encouragement of the proprietor of such lands and woods in making roads and passages in rivers in those northern parts useful and commodious to the public, as well as for conveying such naval stores to the seaports in North Britain, to be brought by sea to England; be it enacted, that there be given a premium for every ton of hemp, £6, of tar £4, of pitch £4, of resin £3, of masts 20s., to be paid by the officers of the navy on a certificate from the custom house officer where the stores are landed."

The first year of George I. presents us with an Act, Chap. 48, for the encouragement of planting and preserving woods. By it, maliciously setting fire to wood is made felony.

Sec. 17 of Chap. 2, 5th of George I., directs particular examination into the quality of Scotch tar.

The 6th of George I., Chap. 16, is another Act for the encouragement of planting and preserving wood. By it damage done to woods is made recoverable from the parish unless within a certain time it discovers and convicts the real offender.

Sec. 3 of Chap. 12 of the 8th of the same king directs, that the inspecting officer shall grant no certificate, unless the articles, of which tar is particularly mentioned, are of good quality. In it, many sorts of timber are enumerated as being imported from America; among them oak, wainscot, pine, etc.; and in consequence of these being imported from foreign countries, at very advanced prices, particularly in time of war, it is enacted that due encouragement be given to importation from the colonies. The law respecting the pine, is nearly the same as enacted by Anne, but the penalty is reduced.

The 6th of George III., Chap. 36, is an Act for the better preservation of timber and trees. It is enacted, that every person, not being the lawful owner, who shall lop or top, cut or spoil, split down, damage, or otherwise destroy any kind of wood, underwood, poles, stack of wood, green stubs, or young trees, or carry or convey away the same, or shall have in their custody any such and shall not be able to give a satisfactory account how they came by them, shall be convicted before a magistrate on the oath of one or more credible witnesses, and be fined for the first offence, any sum not exceeding 40s., with all costs; for the second, not exceeding £5; and for the third offence be deemed an incorrigible rogue; oak, beech, chestnut, walnut, ash, elm, cedar, fir, asp, lime, sycamore, and birch, to be considered as timber.

This Act was confirmed by Chap. 33 of the 13th of George III., which further enacted, that poplars, alder, larch, maple, and hornbeam, should be deemed timber trees.

48th George III., Chap. 72, was for the better preservation of wood in the forest of Dean, similar to that of the 19th of Charles II., Chap. 3, where 11,000 acres are directed to be kept enclosed in the forest, and this Act enjoins 6,000 acres to be kept enclosed in the new forest, to be called nurseries for wood and timber. When the wood in such enclosures is past danger from the browsing of deer, etc., they may be laid open and other quantities enclosed. Every person who shall unlawfully destroy, or take away, or break any timber shall forfeit for the first offence £10, for the second £20, but the third offence is felony, and incurs a punishment of transportation beyond seas for seven years.

In 50 George III., we have an Act to extend and amend that of the 39th and 40th of the present reign for the preservation of timber in the New Forest, and to ascertain its boundaries; and another, Chap. 218, for disforesting the forest of Bere in the County of Southampton. The waste land, it observes, had been of great value and utility from the timber and underwood thereon, which of late years, has been much injured, and in many parts totally destroyed. In Sec. 64, it is enacted that no sheep, lambs, etc., be kept for ten years in any of the enclosures of the forest of Bere, unless the owners protect their neighbors' fences from such sheep, etc.

In 52nd George III., an Act was passed for making perpetual that of the 12th for lowering the duty on bark, after it came to a certain price.

The 10th of Charles I., Sec. 2, Chap. 23, referred to Ireland. By this it is enacted, that for cutting, peeling, barking, or otherwise destroying trees, the offenders shall be punished; and if they be poor and unable to pay the fine, they shall be whipped. If the constable refuse to execute the order of the justice of the peace to whip the offender, he shall be imprisoned till he agrees to do so.

10th William III., Chap. 12, enacted, among other things, that every person having an estate of freehold of £10 a year, and every tenant for years having a lease of eleven of those years unexpired, and paying £10 a year, shall plant or cause to be planted, at seasonable times, yearly, and every year during the term of thirty-one years, ten plants of four years' growth or more, of oak, fir, elm, ash, walnut, poplar, abele, or elder, in some ditch or elsewhere on the said lands, and preserve them from destruction. Every person or society having iron works shall plant or cause to be planted in ground sufficiently well enclosed for this purpose, five hundred trees of the aforesaid sorts on some of their ground yearly, and every year during such time or term as they shall keep or have the said iron works. Any person having 100 or more acres of land (plantation measure) or other tenants in common, shall, over and above the ten trees, within seven years, enclose with a good sufficient fence of stone, wall, ditch, hedge, pale, or rail, one plantation acre thereof; and within seven years aforesaid plant at the least of the height of one foot above the ground when planted, and the age and at times before mentioned, for every ten feet square contained in such acre, in such method as they shall think fit, and keep the same enclosed and fenced from cattle for twenty years. No sheep or cattle shall be allowed to graze in these enclosures for twenty years under the penalty of 20s. for every such sheep, one moiety to the informer, and the other to the poor of the parish.

Then followed several other Acts, the most important of which is 23 and 24, Geo. III., Chap. 39, being an Act to amend the laws for the encouragement of planting timber trees.

It is worth while to notice the great encouragement which the legislatures of Scotland and England even in very ancient times held forth to the planting of trees. So far back as the year 1457, by a statute of James II, freeholders are enjoined to cause their tenants to plant woods, trees and hedges. This is followed by the Act of James IV., 1503, chap. 74; by the Act of James V., 1535, chap. 10; and by the Act of Charles II., 1661, chap. 41, which are all equally explicit.

Transcripts of these enactments are made and they are very curious in themselves, and were passed in time long before systematic forestry was thought of.

(1) Statute, James II, 14th parliament, 6th March, 1457: "Anent plantations of woodes and hedges, and sawing of broome; the lords thinks speedful, that the king charge all his freeholders, baith spiritual and temporal, that, in the making of their Whitsundayis set, they statute and ordine, that all their tennents plant woodes and trees, and make hedges, and saw broome, after the faculties of their mallinges, in places convenient therefor, under sik paine as law and unlaw of the barronne sall modifie."

(2) Statute, James IV, 6th parliament, 11th March, 1503, chap. 74, that hedges, parkes, and dowcottes and cunningares be made. "Item, it is statute and ordained, anent policie to be halden in the cuntrie, that everilk lord and laird, make them who have parkes with dears, stankes, cunningares, dowcottes, orchardes, hedges; and plant at the least, ane aicker of woode quhair there is na greate woodes nor forestes."

(3) Statutes, James V, 4th parliament, 7th June, 1535.

10. For planting of woodes, forrests, and orchardes. "Item, for policie to be had within the realme, in planting of woodes, making of hedges, orchardes, zairdes, and sawing of broome. It is statute and ordained, be the king's grace and his three Estaites of Parliament, that the actes maid thereupon of before by King James the First, and uthers our Soveraine Lorde's progenitoures, be observed, keiped, and put to sharpe execution in all poyntes, with this addition: That everie man, spiritual and temporal, within this realme, havand ane hundredth pounde land of new extent be zeir, and may expend sameikle, quhair there is na woodes nor forrestes, plant wood and forrest, and make hedges, and having for himself, extending to three aickers of land, and abone or under, as his heritage is mair or less in places maist convenient; and that they cause everie tennent of their landes, that has the same in tack or assedation, to plant upon their on-set, zeirly, for everie marke land, ane tree. Ilk laird of ane hundreth pound lande under the paine of ten pound, and lesse or mair, after the rate and quantity of their landes."

Forest fires in those days were of frequent occurrence, and involved, as now, very grave consequences. The Statutes of muirburn prescribed a period of the year when the burning of muirs was expressly prohibited under a penalty.

The first Scots Act for regulating muirburn is that of 1st James I, chap. 20, (Anno. 1424) In the following terms: "It is ordained, that na man mak muir-burning after the moneth of Marche quhir all cornes be schorne, under the paine of fortie shillings, to be raised to the lord of the land of the burner, etc.

Subsequent Acts were 10 James III, Chap. 75, Anno. 1477 ; 4th James IV, Chap. 48 ; 6th James IV. Chap. 71 ; 6th James VI, Chap. 84.

There are also several British statutes regulating the time for making muir-burn. The 6th George III proceeds thus: "And whereas the laws now in force in that part of Great Britain called Scotland for preventing muir-burn in forbidden time are found defective and insufficient, whereby not only the game, but *also many valuable woods and plantations have been destroyed*; for remedy whereof, be it enacted, by the authority aforesaid, that no person or persons shall make muir-burn, or set fire to any heath or muir in that part of Great Britain called Scotland, from the last day of March to the first day of November in any year, under the penalty of 40s. lawful money of Great Britain, for the first offence," etc. This Act was amended by 13 George III, Chap. 54, sec. 3.

FOREST LANDS OF THE UNITED STATES.

A bill to provide for the establishment, protection and administration of public forest reservations, and for other purposes, was introduced in the Senate of the United States in June, 1892. It is unnecessary to set out the bill here, as it did not actually become law, but the report thereon which was submitted by Mr. Paddock, from the Committee on Agriculture and Forestry, contains matter of much value and importance as embodying an authoritative expression of opinion on the condition of the forest lands of the United States. It is given in full :—

The Committee on Agriculture and Forestry, having had under consideration the bill (S. 3,235) to provide for the establishment, protection, and administration of public forest reservations and for other purposes, submit the following report :—

(*1*) *The United States Government retains somewhat less than 70,000,000 acres of public domain which is designated as timber or woodland, mostly situated on the slopes and crests of the western mountain ranges.*

So little regard to the character and condition of the public lands has been given that it is impossible without much labor to determine how much woodland is comprised in them. An estimate was made in 1883, which places the woodlands at 73,000,000 acres, of which of course an unknown quantity has since been disposed of. There are still some woodlands undisposed of in Minnesota, Wisconsin, probably a small amount in Michigan, Louisiana, Mississippi, Alabama, and perhaps Florida, but the bulk lies on the Rocky Mountains, Pacific Coast, and Sierra Mountain ranges, mostly of coniferous growth (pines, spruces, firs, cedars, and redwoods) and mostly in subarid regions.

(*2*) *This property is at present left without adequate administration, nor is there in existence any practicable system of management by which the timber on it can be utilized without detriment to the future condition of the forest growth.*

The public lands are all held for the purpose of disposal to private holders, hence no further administration or management of the same beyond that incident to their disposal has ever been attempted. In the case of timber lands, however, it was recognized to a small extent that there was some additional value to them that needed consideration and special legislative measures. These measures have, however, been rather detrimental than otherwise to the future of this property, besides discriminating unjustly and imposing conditions which cannot practically be enforced.

In California, Washington and Oregon the law permitted the purchase of 160-acre tracts each by private citizens *for their own use.* The object of this law, which was evidently to encourage small holdings of timber lands in connection with agricultural lands and insure consequent protection and management of the same, has never been attained. It is alleged that millions of acres have been taken up under this act without intention to hold them for the use of the entryman, and immediately transferred to lumber companies, often foreigners, and immense tracts are being thus held for the same wasteful lumbering operations that have exhausted the forests of the east.

In the Rocky Mountain States timber lands could not be sold, but the citizens were authorized " to fell and remove timber on the public domain for mining and domestic purposes from mineral lands." In addition, railroad companies are allowed to take timber for construction along their right of way. The impossibility of purchasing in a straightforward, honest way from the Government either timber or timber-bearing lands—

has compelled the citizens of these nine States and Territories to become trespassers and criminals on account of taking the timber necessary to enable them to exist.

Settlements upon timber lands in these States and Territories under the homestead and pre-emption laws are usually a mere pretense for getting the timber. Compliance with those laws in good faith where settlements are made on lands bearing timber of commercial value is well nigh impossible, as the lands in most cases possess no agricultural value, and hence a compliance with the law requiring cultivation is impracticable. As to cutting timber from mineral lands, perhaps not 1 acre in 5,000 in the States and Territories named is mineral, and perhaps not one in 5,000 of what may be mineral is known to be such.

By the provisions of the law approved March 3, 1891, the Secretary of the Interior is empowered to further regulate and restrict this cutting of timber for domestic and railroad use, but in the absence of officers to control and enforce these regulations and restrictions they are practically meaningless, especially since it is almost impossible to obtain convictions where all are equally violators by necessity, arising from absence of adequate and equitable legislation.

And even if it were possible to enforce the regulations, there could hardly be expected any method in the cutting performed by an unknown number of independent individuals, and such a system comes as near deserving the name of management as the pillaging of a city by a band of soldiers in war time deserves the name of municipal administration. To verify the general existence of these conditions the reports of the Secretaries of the Interior, the Commissioners of the Land Office for the last fifteen or twenty years, and the report of a special commission laid down in a volume called "The Public Domain," published in 1884 (House Ex. Doc. No. 47), may be consulted, or Bulletin II of the forestry division, Department of Agriculture, on the forest conditions of the Rocky Mountains.

(3) *In consequence of the absence of a well-developed system of administration, the value of this forest property is annually decimated by fire and by illegal and wasteful cutting.*

It is not necessary to argue this point, for it is a necessary corollary of the preceding.

The Senate Irrigation Committe, travelling two years ago in the western mountains, was for weeks precluded from any view by dense clouds of smoke from forest fires, and it is asserted that in that year more timber was burned than has been used legitimately since the settlement of that country.*

* The acres burned over and values destroyed during the census year 1880 were reported as follows :

States and Territories.	Acres burned over.	Value destroyed.	States and Territories.	Acres burned over.	Value destroyed.
California	356,895	$140,750	Utah	42,865	$1,042,800
Washington	37,910	713,200	Colorado	113,820	935,500
Oregon	132,320	593,850	Arizona	10,240	56,000
			New Mexico	64,034	142,075
Total Pacific slope	527,045	81,747,800			
			Total Rocky Mountains	432,464	86,780,371
Montana	88,020	$1,128,000			
Idaho	21,000	202,000	Grand total	959,509	88,528,171
Wyoming	83,780	3,255,000			
Nevada	8,710	19,000			

The worst damage of these fires is not so much to be sought in the destruction of the standing timber but in the destruction of the forest floor, by which the chance for germinating of seeds and natural reforestation is annihilated, and the water regulating capacity of the forest is destroyed.

As to the amount of depredations, the following table, prepared from reports of the Land Office, is instructive, not only in showing the enormous amounts thus lost to the public treasury, compared with which the cost of a well-organized administration would be a mere bagatelle, but also by corroborating the statement that the loss is rarely recovered in the courts.

It should also be borne in mind that the cases reported do not by any means cover all cases of trespass, presumably only a small part, since the number of agents to ferret out the cases are ridiculously out of proportion to the area to be covered.

DEPREDATIONS ON THE PUBLIC TIMBER DURING ELEVEN YEARS.

Year.	Estimated value of timber reported stolen.		Amounts actually recovered, partly by compromise.	Appropriations for protection service.	Agents employed (number calculated on the basis of twelve months each per annum).
	Market.	Stumpage.			
	$	$	$	$	
1881	891,888	225,472	41,680	40,000	17
1882	2,044,278	511,069	77,365	40,000	31
1883	8,144,658	1,709,824	27,741	75,000	25
1884	7,289,854	1,093,178	52,108	75,000	26
1885	2,862,530	489,255	49,451	75,000	23
1886	9,339,679	1,726,516	101,086	75,000	21
1887	6,146,935	1,138,320	128,642	75,000	26
1888	8,397,500	840,145	128,522	75,000	25
1889	3,603,534	1,182,987	185,002	75,000	23
1890	3,067,152	832,420	100,942	75,000	29
1891	2,347,473	349,441	116,704	100,000	33
Total	54,135,481	10,098,627	1,009,243	780,000	*25

(4) It is a well-know fact, demonstrated by European experience and practice, that by a proper system of cutting not only can a forest be reproduced without the necessity of expensive replanting and kept continuously productive, but its yield per acre and year, in quantity and quality, can by proper management be increased considerably beyond that of the virgin forest left without management.

The methods of management for natural reforestation, or "regeneration methods," are practiced, especially in France and Germany, in broad-leaved as well as coniferous forests. The cutting of the old timber is done with a view of giving chance for seeds of the desirable species to sprout and for the young growth to develop satisfactorily. These methods prevail especially in the mountain regions, where planting would be expensive and sometimes impracticable.

Since in the well-managed forests only such species as are valuable are allowed to grow, to the exclusion of the inferior kinds, which the forester treats as weeds, the composition of the forest is improved, the growth is kept at the

*Average.

most favorable density for development, not only more individual trees but these of more servicable shape are growing, so that at the harvest the percentage of waste and useless material is reduced, and it is for these reasons that the yield, not only in quantity but also in quality, is increased.

While in our virgin forests the percentage of useful saw material is estimated to rarely exceed 20 or 25 per cent, the percentage in the French Government forests is over 50, which in pine and spruce of 130 years of age in Germany may reach the high figure of 60 to even 70 per cent; that is to say, the management of the crop is such that the firewood, branches, and waste material are kept down to from 30 to at most 50 per cent, of the total crop of wood.

Most of the timber cut and sawed in the United States is from trees more than 200 years of age, while the rotation, i.e., the time during which the crop is allowed to grow in Germany, for most timber is not more than 100 years. Comparisons of absolute yield are therefore impossible to make.

But if we allow the high estimate of 10,000 feet board measure per acre to be an average for the United States, we learn from the large statistical material on hand for the German forest administrations that the yield of the German forests is at least three times as large and that produced in a shorter time. We leave out of consideration, of course, the yield of the Pacific Slope forests, which is beyond any average computation.

That it is judicious for the government to keep in view the question of timber supplies and to give, at least as far as it own holdings are concerned, timely attention to the future, if for nothing else than an example and object lesson, may be inferred from the following statement in regard to the outlook of available supplies and demand, which, while not claiming to present actual conditions, for which statistics are lacking, discusses possibilities or probabilities.

The chief of the forestry division, in an address before the real estate congress at Nashville, Tenn., in 1892, says :—

The area of timber land in the United States, although changing daily by clearing of new farms and by relapsing of old ones into woodlands, may roughly be placed at 500,000,000 acres. Even if we were to class as timber land all the land not occupied by farms or known to be without tree growth, this figure can not be increased more than 60 per cent.; that is, the utmost possibility of the area of natural woodlands in the United States must be within 800,000,000 acres. The former figure, however, comes probably much nearer the truth. How much of this area contains available merchantable timber it is impossible to tell, or even to guess at. We only know that supplies of certain kinds are wanting. For instance, the white pine of the North shows signs of exhaustion, the white ash has become scarce in many localities, the tulip poplar will not last long, and the black walnut has ceased to be abundant. All we can do is to estimate the range of possibilities.

With the utmost stretch of imagination as to the capacity of wood crops per acre, if we allow even the entire area of half a billion acres to be fully timbered, and keep in mind the enormous yield of the Pacific Coast forests, 1,250 or 1,500 billion cubic feet of wood is all that could be crowded upon that area. This figure would far exceed the most highly-colored advertisement of a dealer in timber land, except on the Pacific Coast; in fact he would be afraid to assert one-half as much, for it would make the average cut of timber per acre through the whole country 10,000 feet board measure.

The above figure in cubic feet represents wood of every description, allowing as high as 33⅓ per cent. for saw timber.

Since we consume between 20,000,000,000 and 25,000,000,000 cubic feet of wood of every description annually, fifty to sixty years would exhaust our supplies, even if they were as large as here assumed and if there were no additional growth to replace that cut and no additional increase of consumption. Regarding the latter it may be of interest to state that according to as careful an estimate as I have been able to make upon the basis of census figures and other means of information the increase in the rate of consumption of all kinds of forest products during three census years, expressed in money values, was from round $500,000,000 worth in 1860 to $700,-000,000 worth in 1870 and $900,000,000 in 1880, while for 1890 it may probably reach $1,200,-000,000, an increase of about 30 per cent. for every decade, or somewhat more than the increase of population, which may in part be explained by higher prices.

It will also aid us in our conception of the situation to know that the saw mill capacity of the country in 1887 was round 200,000,000 feet (board measure) daily, which again may be figured equivalent to a probable consumption of wood of all kinds to the amount of at least 20,000,000 cubic feet round.

It remains to be seen what the chances are of supplying ourselves from the natural reproduction of our present forest area.

I have shown elsewhere that, while under the careful management of the German forest administrations, the average yearly new growth is computed at 50 cubic feet per acre, or 2.3 cubic feet for every 160 cubic feet standing timber, we can here, where there is no management at all, where fire and cattle destroy not only young growth but also the fertility of the soil, in spite of the originally greater reproductive power, expect no such annual crop.

From my observations I would not admit that more than one-half such annual growth is realized on the average over the whole area of 500,000,000 acres, and the likelihood is that much less is reproduced per acre.

Hence, while 500,000,000 acres reserved as forest at the very best would satisfy our annual consumption of 25,000,000,000 cubic feet—we need some 5,000,000,000 feet to supply our annual conflagrations—we are presumably cutting into our capital at the rate of at least 50 per cent. of our annual consumption ; that is to say, only one-half the annual cut is represented in annual new growth. What do these figures mean with reference to the subject in question ? Simply this, that while as yet prices for timber lands, and still less the price of lumber, are by no means advancing in proportion to the constantly growing reduction of standing timber supplies, when the general truth of these figures is recognized, which cannot fail to occur soon, timber lands will appreciate rapidly in value, and lumbermen, especially in the South, will regret their folly of having marketed their best supplies at unprofitable and unsatisfactory margins.

Nevertheless, it may be possible by a common-sense management and more rational methods of utilizing the timber, having some regard to the young growth, inaugurated now, to avoid the necessity of replanting at great cost and to maintain the present forest resources of the United States in sufficient and ever increasing productiveness.

(5) *It is also established beyond controversy that the forest cover, and especially the forest floor of leaves, twigs, decaying vegetable matter, underbrush, and root system, influence the regularity of the waterflow in springs, brooks, and rivers, as well as the state of the ground water level, the presence or absence of an efficient forest cover determining the percentage of subterranean or superficial drainage. Whatever the theories or facts regarding the influence of forest areas upon meteorological phenomena and climatic conditions—and these are partly at least still in controversy—there exists but little doubt, if any, among students and observers in regard to the influence which a forest cover exerts over the water drainage and soil conditions.*

Since it is in part upon the assumption of the existence of such an influence that the government is called upon to look to the preservation of forest conditions, and since the ideas regarding such influence are still more or less confused, it may be proper to explain more at length the action of the forest in this direction.

So far as formation of springs is concerned, no doubt, geological conditions and structure are of primary importance. This does not, however, exclude that the vegetable cover of the soil has at least a secondary influence upon the feeding and regular flow of springs. Even if we exclude any action of the forest upon the increase of precipitation, such as is claimed and partly sustained by observation, there are various ways in which the supply of springs is influenced by forest cover.

The forest floor and the foliage breaking the force of the rain drops, prevent a compacting of the soil ; it remains porous and permits the water to percolate readily, changing a large amount of it from surface drainage into subterranean channels ; the root system, no doubt, works in the same direction. Forest floor and foliage also prevent rapid evaporation, and although the trees consume a large amount of water in their growth, evaporation is the worst dissipator of moisture, and the balance, between the consumption and the saving of evaporation by forest growth, is largely in favor of this kind of vegetation as compared with any other vegetable cover or with naked ground, provided the forest floor of decayed leaves, twigs, etc., is not destroyed. Furthermore, the melting of snows is retarded under the forest cover, and finally the mechanical retardation of the surface water flow promotes subterranean drainage, insuring to springs a greater supply for a longer time.

This observation, very generally made, used to be explained by popular writers as due to the sponge-like condition and action of the forest floor, being able to take up water and then gradually giving it up to the soil below. Fortunately, the forest floor is rarely like a sponge, for a sponge never gives up water below, but always by evaporation above after the supply has ceased. The simile was an unfortunate one.

The open runs, i. e., brooks, rivulets and rivers, receive their supply mainly from springs, but also from the surface waters which flow without definite channels down the slopes. The more the supply is derived from springs the more even is the water flow of the river; the greater the supply of the surface drainage the more dependent is the water flow on the changeful rains and on the melting of the snows, and the more changeful is the water flow. While, then, in the first place, the water flow in rivers is dependent upon the amount and frequency of rainfall and snow, the manner and time in which the water reaches the channels determines the greater or smaller extremes of water stages.

The retardation of the melting of the snow, which in a well-covered mountain district may be prolonged for two or three weeks under a forest cover, is of great significance in reducing the spring floods. The main influence, however, lies in the mechanical impediment which the forest floor opposes to the rapid surface drainage, promoting filtration to the soil and preventing the rapid filling of surface runs and lengthening the time during which the water is to run off. Observations in one of the reforested parts of the French Alps showed this retardation to be in the ratio of 5 to 3.

Thus, while in extreme cases, with excessive rainfalls or sudden rises of temperature in early spring, with steep declivities and impermeable rock formation, even a forest cover may have no practical effect in preventing a flood, it may be accepted as a generally true proposition that a forest cover has a tendency to lengthen the time to run off, and hence to reduce in amount and frequency flood conditions and to maintain the water flow more even with fewer excessively low and high stages.

Lastly, but of greater importance than has often been conceded to this influence, the forest cover prevents erosion of the soil and formation of the so-called detritus of rocks, gravels, and sands which, carried into the rivers, increase the danger from floods, impede navigation, and if deposited on fertile lands may, as in France, destroy the soil value of whole districts. Along the coast and in the sandy plains the protection of the loose soil and dunes against the disturbing action of the winds, and in the mountains which are liable to avalanches and snow-slides, as in Switzerland, the protective value of a forest is also well established. If there were any doubts regarding the influence of forest cover upon water and soil conditions before they have been entirely dispelled by the extensive reforestation work undertaken by the forest department of France.

There seventeen departments or counties had been impoverished and depopulated by the washing of the soil, torrential action of the rivers, and repeated floods, due to deforestation of the mountains, when the government adopted the policy of reclothing the denuded slopes with tree growth and sod. The population in these counties had diminished from 10 to 20 per cent. within less than 20 years, and fertile fields had been covered up for more than 100 miles from the source of the soil, with the debris brought from the mountains by the rushing torrents.

The French Government has expended for reforestation of these mountains, during the last thirty years, over $35,000,000 and expects to have to spend more than the same amount in addition before the damage is repaired. The result of this work, some of which is now long enough established to show effect, perfectly justifies the anticipations of its efficiency. In the "perimeters" which have been recuperated the waters are carried off more slowly and without damage. These works in their result must quiet all theoretical discussion of the efficiency of forest cover in this particular. They present ocular proof not only of the fact that deforestation invites floods, erosion, and untold damage, but that reforestation is the method of remedying the damage and proper attention in time to the forest cover the method of obviating it.

Recognizing the value, then, which a forest may have in preserving proper water conditions and soil conditions, and perhaps, too, in some degree in climatic conditions, the conception in Europe of "protective forests" as distinguished from the "economic forest," that is, a forest which has value only from a material point of view, a policy has grown up in the higher developed nations of placing the first class of forests, which have a significance as a natural condition rather than as a source of material supply for the whole community, under government control, direct or indirect.

(G) Aside, therefore, from the undesirability of destroying or unnecessarily impairing a valuable resource of material, which can be continuously reproduced on land otherwise useless, there is strong reason why, especially in regions dependent upon irrigation for their agricultural development, favorable forest conditions should be carefully maintained.

Modern experience and scientific research have confirmed the experience of antiquity, namely, that plant production is primarily dependent upon water and that the management of water supplies is much more essential to the farmer even in the humid regions, than management of mineral constituents of the soil, for the latter can be supplied with ease, but the former can be regulated and supplied properly only with difficulty. If, then, water management becomes more and more important in all sections of our country, it is particularly so in those regions where, from natural causes, the supply is scanty. No artificial reservoirs can supply the more easily and cheaply maintained natural reservoir of the forest floor.

In this connection it will be well to quote the following language from a memorial recently transmitted to the President of the United States by the Colorado State Forestry Association, to which the Secretary of State, State engineer, State treasurer, attorney-general, and other leading officers of the State, together with the chambers of commerce of Denver and Colorado Springs, and some 500 leading citizens of the State have appended their signatures, recommending the reservation of all the timber lands in their State.

To his Excellency the President of the United States:

Your memorialist, the Colorado State Forestry Association, respectfully represents that the agriculture of this State, now rapidly increasing in magnitude and importance, is almost entirely dependent upon systems of irrigation. At least $13,000,000 are invested here in reservoirs, canals, ditches, and other works for the storage and distribution of water. No less than 13,000 miles of irrigating canals and main ditches are in operation or in course of construction in the State.

The agricultural yield of Colorado (exclusive of live stock) for the year 1891 amounted to $53,900,000 ; the mineral output for the same period was $33,549,000—a large sum, but greatly inferior to the one first named.

It will thus be seen how vitally important to the growth and continued prosperity of this Commonwealth is an abundant supply of water for irrigation. In fact it may be said that henceforth the agricultural yield of the State will be limited only by its water resources.

The streams upon which the irrigation systems of Colorado depend are fed by the springs, rivulets, and melting snows of the mountains, which in turn are nourished and protected by the native forests. Where the forests have been destroyed and the mountain slopes laid bare most unfavorable conditions prevail. The springs and the rivulets have disappeared, the winter snows melt prematurely, and the flow of streams, hitherto equable and continuous, has become fitful and uncertain. Floods and drought alternating clearly indicate that the natural physical conditions of the region have been unduly disturbed. In winter and early spring, when heavy masses of snow have been accumulated on treeless precipitous slopes, snow and landslides frequently occur with disastrous results to life and property. Even thus early in the present season a considerable number of valuable lives have been sacrificed in this manner.

The main Rocky Mountain range extends throughout the State, from north to south, and is flanked on either side by numerous spurs and minor ranges. The average or mean elevation of Colorado, 7,000 feet above the sea level, is greater than that of any other portion of North America. The high and rugged interior region contains 140 peaks or more, exceeding 11,000 feet elevation, and comprises about one-fourth of the area of the State. Small portions of this region are used for agriculture and grazing, but in the main it is unsuited for such purposes. Its surface, below timber line, was originally quite generally covered with a coniferous forest growth, but has subsequently been marred and disfigured by fire and the axe. Vast areas have been thus desolated. Above timber line proper there are many gulches and sheltered places, in some of which exist a stunted growth of trees and shrubs, where the drifting snows find lodgment, melting only during the summer months.

At certain of these greater elevations are found morasses, alpine lakes, and during portions of the year, ice fields of limited extent. The region is mainly one of cold and humidity for long periods of each recurring season. This is one of the principal, if not the chief, of the distributing centres of the continental water system. It contains the sources of the North Platte (in part,) the South Platte, Arkansas, Rio Grande, Dolores, Gunnison, Grand, White, Yampa, and other powerful streams, the preservation of which is not only important to Colorado but to neighboring States and Territories. New Mexico would be uninhabited were it not for the life-giving waters of the Rio Grande, which flow from the snow mountains of Colorado.

In view of the above, and in consideration of many recognized evils which follow the reckless and inconsiderate denudation of timbered areas, we respectfully ask that you will, under the act of March 3, 1891, cause to have withdrawn from disposal and constituted a forest reserve, all public lands along the crests of the mountain ranges and spurs in this State, as above mentioned, and upon either side thereof for a distance of 6 miles, more or less, according to the width of the timber belts in different localities and as may be deemed advisable after due official examination of the same.

We beg to represent further, that in our opinion the rights of prospecting and mining and right of way for public roads within the territory in question should remain inviolate, and that the general government should inaugurate at the earliest practical period a careful and conservative administration of such public lands. We also believe that, under proper regulation, a prudent and economical use of the forest resources may be had without endangering the perpetuity of the forests. Forest conservation should promote, rather than retard, all legitimate industries.

In this connection it is also worth while to quote the language of the chief of the forestry division from the annual report of the Secretary of Agriculture for 1891 :—

Water Management the Problem of the Future.

Before even attempting the control of precipitation, our studies, in the opinion of the writer, should be directed to secure better management of the water supplies as they are precipitated and become available by natural causes. How poorly we understand the use of these supplies is evidenced yearly by destructive freshets and floods, with the accompanying washing of soil, followed by droughts, low waters, and deterioration of agricultural lands.

It may be thought heterodox, but it is nevertheless true, that the manner in which most of the water of the atmosphere becomes available for human use (namely, in the form of rain) is by no means the most satisfactory, not only on account of the irregularity in time and quantity, but also on account of its detrimental mechanical action in falling, for in the fall it compacts the ground, impeding percolation. A large amount of what would be carried off by underground drainage is thus changed into surface-drainage waters. At the same time, by this compacting of the soil, capillary action is increased and evaporation thereby accelerated. These surface waters also loosen rocks and soil, carrying these in their descent into the river courses and valleys, thus increasing dangers of high floods and destroying favorable cultural conditions.

Here it is that water management and, in connection with it or as part of it, forest management should be studied ; for *without forest management no rational water management is possible.*

(7) *Experience in the United States has shown that under private ownership, forest conditions are almost invariably destroyed or deteriorated, for the simple reason that the timber for present use is the only interest which private enterprize recognizes in the forest, not being concerned in the future or in the consequences of mismanagement to adjoiners, who have to suffer.*

It is therefore undesirable to transfer the ownership of the public timber lands to individual owners in the expectation of having them managed with a view to the broader interests of the community.

If there were need of other demonstration of this point beyond the history of the eastern forest lands, which have been for many years in the hands of private owners, we need only refer back to the working of the law in the Pacific Coast States, where such disposal to private holders has utterly failed in accomplishing its object. There is neither the interest nor even the knowledge to be found among the many to let us anticipate forest management by small holders. Besides foresty thrives best on large consolidated areas, from financial as well as technical considerations.

It will be necessary, in order to promote rational forest management, to do the same that all other nations have found necessary to do, namely, for the government to set the example and furnish the object lesson and opportunity for the others to follow.

The fact that a tree crop takes from fifty to one hundred years and more to grow to usefulness requires a patience and stability of ownership which our people have not yet attained, and hence the government must furnish the conservative elements where needed, as in our forest policy.

(8) *The cession of the public timber domain to the individual States with a view of having the States devise methods of conservative management, would fail in accomplishing the object for various reasons. Experience in the past with such cessions has not proved it practicable to place restrictions or conditions upon such cessions or to enforce them.*

Even if a cession, upon condition that the State provide efficient management, could be practically effected, lack of unity in the various systems and clashing of interests where watersheds are situated in more than one State, make retention of these lands in the general government desirable, or at least more promising of conservative results.

Other reasons of expediency make such a wholesale cession of timber lands impracticable. Among these may be mentioned the difficulty of segregating the timber lands from public lands of other description or transferring obligations of the general government toward railroad companies, resting upon such lands.

Nevertheless, co-operation with the State authorities in inaugurating a sound forest policy is most desirable, and should be made a prominent feature in whatever measures the general government may devise.

(9) *The present proposed legislation keeps in view the following principles :*
(a) That the retention of the public timber lands in the general government, and their administration as such, is the only proper policy for all wooded areas of the public domain which do not stock on agricultural land.

(*b*) *That only a fully developed and separate system of management and administration, carried on by competent men under expert advice, can accomplish the objects of a rational forest policy.*

(*c*) *That the object of the public forest reservations is twofold, namely, to maintain desirable forest conditions with regard to waterflow, and, at the same time, to furnish material to the communities in their neighborhood.*

(*d*) *That while the service of protection of watersheds would warrant an expenditure out of other funds for such service it should, nevertheless, pay for itself by the sale of surplus forest material.*

It is only necessary to add a few words of explanation on this latter point, says the chief of the forestry division in discussing the practicability of a government forest administration :—

To meet any objections on the score of expense, a rough estimate of this question may be made as follows :—

Allowing 50,000,000 acres of timber land reserved, I find that a tolerably efficient administration may be provided for a round $2,500,000, or 5 cents per acre. It would be satisfactory of course if only this expense be covered by the revenue. While the annual growth of wood per acre on the reserved area would exceed in value the assumed cost of administration the local market and consumption is restricted. But when we consider that the present saw-mill capacity of the region affected is over 3,000,000,000 feet B. M., and the resident population 3,000,000, requiring at least 50 cubic feet of wood material per capita, sufficient margin is assured even if only half of these amounts are furnished from the government reservations and the average charge for stumpage is taken at 10 cents.

And in another place (see Annual Report, 1886) :—

(7) The cost of the total service depends of course on the number of districts to be formed. Take Colorado alone, which we will assume contains about 5,000,000 acres of public domain. For this we may require three hundred rangers and ten inspectors, and the expense may be placed in round figures at $300,000. This amount could be saved by preventing only one-third of the forest fires, which seem to destroy over $900,000 worth of public property in that State yearly, and the 50,000,000 cubic feet or so of timber, which may be cut to satisfy the needs of the country for its development, would certainly, without hardship to any one, yield enough to help pay the expenses of less favorable localities and of the central bureau. The expense of the latter, with the necessary staff of clerks, etc., could certainly be kept within the sum of $50,000. Even if the whole forest area were as thoroughly organized as proposed for Colorado, the expense of the service would not be more than 30 per cent. of the income which might be derived from this domain, or, which could be saved, by preventing one-half of the fires that yearly destroy about an equal amount.

Referring to the operations of several European forest administrations we find that their expenditures represent from 37 to 58 per cent. of their gross income, or from $1.33 to $5 per acre, the net revenue being 96 cents to $4.40 per acre. These are results under conditions of very extensive management and under highest economic development. Taking Prussia alone, with a round 6,000,000 acres of forest and much poor and undeveloped country, the cut in 1890 amounted to round 333,000,000 cubic feet of wood, of which 215,000,000 feet went into cord-wood and 118,000,000 feet in saw-logs, or round 56 cubic feet of wood representing the annual growth per acre per year over the entire 6,000,000 acres, with a proportion of 45.6 per cent. in saw timber and wood for manufactures. The price received for this material in the woods, butt cut, was at the average rate of $10.63 per M feet, board measure, and $3.69 per cord, or both together about 5 cents per cubic foot of wood, the total income from wood being $16,225,000, of which 62 per cent. came from saw timber. Other revenues of the forest administration amounted to $17,632,810, or about $2.63 per acre, as against $10,888,893 in 1870.

The expenditures, amounting to $8,796,740, or, if special appropriations not recurring are deduced, to $8,582,268, represented 47.38 per cent. of the gross income. It may be of interest to indicate in what direction this large amount is expended :—

There are 122 officers in higher branches of administration, aggregating salaries to the amount of.....................................	$154,350
681 district officers or managers	588,276
3,755 underforesters or guards...................................	1,162,867
114 financial agents...	73,141
Other temporary employes and personal expenses	1,073,587
Total personalia	$3,052,221
Cost of harvesting wood crop (lumbering at a little less than 7 cents per cubic foot)..	$2,266,030
Buildings	599,834
Roads and water ways..	410,102
Surveys	110,226
Injurious insects	60,454
Culture	1,230,882
Sundries	280,073
Total salaries and administration	$8,009,822
Forestry schools and scientific research	* 48,130
Purchase of lands...	304,156
Sundries...	434,632
Grand total................	$8,796,740

Or $1.33 per acre, leaving a net revenue of $1.30 per acre, as against 97 cents in 1870, when the expenditure per acre was 34 per cent. less.

(10) The proposed legislation contemplates a segregation of the timber lands that are stocking on non-agricultural soil from the other public lands and the transfer of their administration from the Department of the Interior, where lands are held only for disposal, to the Department of Agriculture, which is designed to look after cultural matters and where a bureau in charge of forestry matters already exists.

To save expense in the beginning and to create as quickly as possible an efficient protective service, the army may well be employed for such duty. This service has been conferred upon the army in the Yellowstone and the California parks to the full satisfaction of both officers and men, with the anticipated results as far as the protection of the forest property is concerned.

Co-operation with State authorities, such as forest commissions or commissioners, is provided for with a view of enlisting the authorities of the States in the upholding of a rational forest policy.

Since these forest reservations are not to be in the nature of parks, they are to remain open to public use and entrance for all purposes, excepting so far as restrictions appear necessary in order to protect the property from damage and depredation. Prospecting and mining are to be permitted under proper regulations.

The main features of the legislation, however, are its provisions for the cutting of timber under a system of licenses and the creation of the necessary force of officers to attend to the business of a regular forest administration pro-

---

* We (*i.e.*, United States) appropriate for a similar purpose, namely, the forestry division in the Department of Agriculture, whose function it is to build up an interest in the subject and to supply information on forestry matters where none existed before, less than one-half of this amount.

perly. The attempts hitherto of regulating the cutting of timber have remained futile for the lack of an organized system, and of the necessary force to maintain a system.

The license system here provided recognizes the various demands of settlers, prospectors, miners, and lumbermen as legitimate, and necessary to be provided for differently according to the nature of their business and in an equitable manner.

When all needs of the population can be legitimately satisfied, with a sufficient force of officers to attend to the wants of the public in a business-like manner, there is no reason why the existing vandalism with which the public timber domain has been wasted should not cease, destructive fires be reduced to a minimum, a system of proper forest conservancy gradually be developed, and the American nation add to its civilization by a rational treatment of the forests of the public domain at least.

In conclusion the fact is recalled that, as long ago as 1879 the writer of this report took occasion to refer to this subject before the Senate in the following language (see Congressional Record, February 10, 1879) :—

There is another subject, Mr. President, not strictly agricultural, and yet so closely allied to that interest as to demand consideration always when agricultural questions are under discussion. I refer to the preservation of our forest lands from denudation. Those who have investigated and given much thought to the matter declare that the wholesale destruction of the forests of a country, without providing for a new growth, not only seriously affects the material interests but impairs the health and comfort of all the inhabitants thereof. Bitter experience long ago taught the people of the Old World that they could not with safety wage indiscriminate war against their trees. Nature is, indeed, a kind mother to those who exercise an intelligent regard for her habits and her laws, but she is at times terrible in her wrath against those who blindly defy her decrees. The laying waste of the forests of a country rudely disturbs that harmony between nature's forces which must be maintained if the earth is to be kept habitable for its teeming millions.

We have ourselves heretofore sadly neglected these considerations, but our government cannot and must not much longer refuse to give to them its most serious attention. If we may not with propriety restrain the individual from injuring his own property, we can and should at least furnish information and devise plans through intelligent legislation, which shall incite him to coöperate with his neighbor to protect their common interests. Most European Governments have elaborated methods whereby they exercise a supervisory control over the forests of their dominions, and one day the public welfare will demand that our government shall follow their example. The subject is a practical one ; it is not a dream of the theorist ; it concerns the pockets of the people and their welfare in many ways.

Considering the very great importance of this measure early passage of the bill is recommended.

INDEX.

www.ingramcontent.com/pod-product-compliance
Lightning Source LLC
Chambersburg PA
CBHW021514210326
41599CB00012B/1256